Soils of South Africa

Martin Fey

with contributions from
Jeff Hughes
Jan Lambrechts
Theo Dohse
Antoni Milewski
and Anthony Mills

Sponsored by the EJ Lombardi Trust

CAMBRIDGE UNIVERSITY PRESS

Cambridge, New York, Melbourne, Madrid, Cape Town, Singapore,
São Paulo, Delhi, Dubai, Tokyo

Cambridge University Press
The Water Club, Beach Road, Granger Bay, Cape Town 8005, South Africa

www.cambridge.org
Information on this title: www.cambridge.org/9781107000506

© Cambridge University Press 2010

This publication is in copyright. Subject to statutory exception
and to the provisions of relevant collective licensing agreements,
no reproduction of any part may take place without the written
permission of Cambridge University Press.

First published 2010

Printed by Tien Wah Press (PTE) Ltd, Singapore

Designer: Keith Melvin-Phillips
Proofreader: Karoline Hanks

ISBN 978-1-107-00050-6 Hardback

Cambridge University Press has no responsibility for the persistence or
accuracy of URLS for external or third-party internet websites referred
to in this publication, and does not guarantee that any content on such
websites is, or will remain, accurate or appropriate.

Contents

Preface and acknowledgements ... 5

Chapter 1 Introduction to the soil groups and the naming of forms and families ... 9
1.1 The key to soil groups ... 10
1.2 Description of soil groups ... 12
1.3 Descriptive nomenclature for soil forms and families ... 14

Chapter 2 The soil groups: distribution, properties, classification, genesis and use ... 17
2.1 Organic soils ... 19
2.2 Humic soils ... 25
2.3 Vertic soils ... 35
2.4 Melanic soils ... 45
2.5 Silicic soils ... 53
2.6 Calcic soils ... 63
2.7 Duplex soils ... 73
2.8 Podzolic soils ... 83
2.9 Plinthic soils ... 93
2.10 Oxidic soils ... 105
2.11 Gleyic soils ... 115
2.12 Cumulic soils ... 123
2.13 Lithic soils ... 135
2.14 Anthropic soils ... 143

Chapter 3 Animals in soil environments ... 149
3.1 Termites ... 150
3.2 Termite- and ant-eating animals (myrmecophages) ... 158
3.3 Earthworms ... 161
3.4 Golden moles ... 165
3.5 Rodents ... 166
3.6 Honey badger ... 168
3.7 Insects and spiders ... 169
3.8 Birds ... 171
3.9 Large herbivores ... 171
3.10 Elephants, trees and termites ... 174

Chapter 4 Profile descriptions and analytical data ... 177

References ... 260
Glossary ... 264
Appendix: Criteria for generating distribution maps for soil groups ... 282
Index ... 283

Preface and acknowledgements

Not since C.R. Van der Merwe's *Soil Groups and Sub-Groups of South Africa* in 1940 has there been a publication that integrates what is known about the distribution, genesis and management of South African soils. There have been some outstanding regional treatments, most notably *Soils of the Sugar Belt* (Beater, 1957; 1959; 1962) and *Soils of the Tugela Basin* (Van der Eyk, MacVicar and De Villiers, 1969), both of which were forerunners, via an earlier edition, of the current classification of South African soils (Soil Classification Working Group, 1991). In the latter publication, seventy-three soil forms are identified in terms of defined diagnostic horizons and materials within the soil profile and definitions are supported by explanations of the underlying concept so that users of the classification can infer genesis and behaviour of whole soils from an understanding of the origin, properties and behaviour of the individual horizons and materials that are present. This is sometimes difficult, however, even for those well versed in general soil science.

During the 1970s the Soil and Irrigation Research Institute of the Department of Agricultural Technical Services began the ambitious task of documenting the distribution of soils in South Africa. This was done by mapping land types (soil-landscape-climate associations) at a semi-reconnaissance scale of 1:250 000. Only now is the project in a reasonable state of completion. The maps are accompanied by memoirs that summarise soil distribution patterns in relation to terrain morphological units and provide profile descriptions and laboratory data for modal soil profiles in most land types. The maps and memoirs, which are available in digital format, are an excellent database for South African soils. No guidance is given, however, on either genesis or land use.

The main objective of this book is therefore to provide the detailed account of South African soils which is still lacking – one that embraces the main topics of distribution, properties, classification, genesis and environmental significance. In order to achieve this objective, a basis for grouping soil forms at a higher level was devised so that generalisations about them could more readily be made. This grouping led to the development of a nomenclature that is conceptually informative at all levels of the classification, adhering as far as possible to international pedological tradition while at the same time remaining faithful to local classification.

The first chapter of the book introduces the fourteen soil groups and the new, alternative nomenclature for soil forms and families. The second chapter then provides a systematic treatment of each soil group – beginning with a distribution map, describing essential properties (morphological, physical and chemical), summarising classification and its relation to international systems, discussing genesis, and concluding with land use. The third chapter, on the relationship between soils and animals, accommodates a modern trend towards ecological application of soil science and celebrates the concept of soil as a biomantle. It follows logically from the treatment of anthropic soils in the previous chapter, especially with regard to rehabilitation of disturbed land. This may inspire future contributions on similar topics such as soil-vegetation and soil-geomorphic relationships. An illustrative selection of soil profile descriptions and analytical data is provided in the fourth chapter. The descriptions and laboratory data in this chapter are from land type memoirs but the interpretations have been added. There is also a comprehensive glossary which, by including definitions of diagnostic horizons and materials, makes it possible to use the book without having to make frequent reference to the current classification system. A short appendix summarises the criteria that were used for generating the soil distribution and abundance maps. There were a few gaps created by the protocol of having to commence chapters and major sections on a right-hand page. These have been filled with text boxes of supplementary information or, in one or two places, a few lines of verse: my own self-indulgence.

It is anticipated that this book will be useful and interesting to a wide readership. My own students at university are the ones I most had in mind when preparing it but there have been fellow soil scientists, here and abroad, as well as colleagues in other disciplines – ecology, especially, but also hydrology, geology, geomorphology and agronomy – whose interest I have also thought to stimulate in the way the book is written and in the subject matter covered. There should also be much that will interest farmers, foresters, environmental engineers and consultants.

There have been numerous factors that have made publication possible including institutional support, contributing authorship, design and production, critical reviews, personal support and financial assistance. Stellenbosch University has functioned as the host institution through which the project has been funded and where most of the work has been carried out. The Agricultural Research Council's Institute for Soil, Climate and Water in Pretoria has been the leading gatherer and is the chief custodian of information on South African soils. The collaboration of ISCW staff and permission to make use of information from the land type database have been of crucial value.

In acknowledging contributions from co-authors, I thank Jeff Hughes in particular for his preparation of a large part of the first draft of Chapter 2. Although much of that has now been reworked and added to, his efforts during a month when we sat together writing were instrumental in breaking through that first barrier of having on paper something resembling the envisaged product. Jeff's critical evaluation of my groupings of soil forms was also fruitful. Jan Lambrechts, besides also offering valuable criticism, provided a wealth of information including photographs, text and the correlation with international classifications. Chapter 3 was initially prepared on invitation by Antoni Milewski and Anthony Mills and I then added to the material and rearranged it to ensure a good fit. I thank them for responding so enthusiastically and with such valuable ecological insights. The tables of data and profile descriptions in Chapter 4 were retyped from files out of land type memoir files provided by Garry Paterson of the ISCW. My interpretive comments were inspired by that delightful Australian book (Pam Hazelton's) on 'what those numbers mean'. They are somewhat off-the-cuff but hopefully useful even if not as accurate as they could be.

There were two main stages in the production of this book. At the beginning of 2006 a first draft of Chapters 1 and 2 was placed on the web site of the Soil Science Society of South Africa to facilitate comment and contributions from other pedologists. A revised version was then tabled for discussion at the Soil Classification Working Group meeting in Bloemfontein in 2007. Fellow members of this Group, past and present, are thanked collectively for their motions of support and encouragement. In particular Dave Turner, Garry Paterson, Keith Snyman, Theo Dohse, Piet le Roux, Cornie van Huyssteen, Martiens du Plessis, Jan Schoeman, Eben Verster and Chris MacVicar have offered valuable comments. Three significant improvements to the first draft were effected. The first of these was the development of soil maps in which Theo Dohse and Marjan van der Walt from the ISCW played an essential role. The second improvement was the chapter on animals in soil environments. The third was the compilation of Chapter 4 which, as indicated above, was made possible using land type data from the ISCW as well as some information on podzols provided by Jan Lambrechts. The data were transcribed by a group of soil science students led by Anneline Burger, while Ikenna Mbakwe provided additional help in standardising the layout of profile descriptions and analytical data. He and Julia Harper also assisted with the compilation of references and general proof reading. We are grateful to a range of people for providing photographs. Their names, and indication of copyright when required, appear adjacent to each illustration. Several satellite images appear in the book and I am very grateful to Vincent Garros of CNES/Spot Images and Amy Opperman of DigitalGlobe for providing copyright permission. Acquisition of the images was facilitated by Adriaan van Niekerk at Stellenbosch University and Corné Eloff at the CSIR, Pretoria.

The design and production of the book were carried out by Keith Melvin-Phillips with immense skill, attention to detail and professionalism – at an affordable price. The decision to use his services after discovering his work through the vegetation book of Mucina and Rutherford (2006) has been fully vindicated and it has been a pleasure working with him. A late decision was made to hand over publication of the book to Cambridge University Press and Tharlikha Krupandan, Ashley Parsraman and Simon Bekker in the Cape Town branch of CUP have done an excellent job of finalising the project.

My brothers Chris and Keith have shown tremendous enthusiasm for my project and my mother, Beth, has been the most interested of all, being so expectant and confident of a successful conclusion that it could not have been otherwise. She got me started on this a long time ago. Above all I would like to thank my wife, Eileen, for sustained and loving support over a long period.

A National Research Foundation (NRF) grant entitled Soils in an Ecosystem Context (FA2005040700027) provided partial support for some of the activities surrounding the production of this book at Stellenbosch University. The main financial support for this book has come from the E J Lombardi Trust. Created through the will of a prominent agricultural entrepreneur in the Western Cape, the Lombardi Trust has funded agricultural research at Stellenbosch University for a number of years and in 2004 the Trust approved a grant that was specifically aimed at the production of a book on the soils of South Africa. The confidence, patience and enthusiastic interest of the E J Lombardi trustees, Messrs David Macdonald and John Cragg and (the now late) Mrs Lombardi, have been enormously appreciated.

Martin Fey
Perth Australia
February 2010

Martin Fey is Professorial Fellow in the School of Earth and Environment at the University of Western Australia, and Professor Extraordinary in the Department of Soil Science at Stellenbosch University.

His co-authors who contributed to Chapter 2 are:
Jeff Hughes, Professor of Soil Science at the University of KwaZulu-Natal; Jan Lambrechts, Associate Professor Extraordinary in the Department of Soil Science at Stellenbosch University; and Theo Dohse, soil scientist at the Institute for Soil Climate and Water, Agricultural Research Council, Pretoria

Co-authors of Chapter 3 are:
Antoni Milewski, ecologist affiliated to the Percy FitzPatrick Institute of African Ornithology, University of Cape Town; and Anthony Mills, consultant and research ecologist in the Department of Soil Science, Stellenbosch University.

SOILS
OF
SOUTH AFRICA

Chapter 1

Introduction to the soil groups and the naming of forms and families

Martin Fey

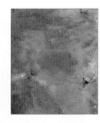

Soil diversity in the Free State near Bloemfontein
(© CNES 2008, Distribution Spot Image S.A., France,
all rights reserved) 29°13'56.86'S 25°03'29.79'E

Most soil classification systems are simple to begin with and then become more complex and elaborate as differences that at first were considered inconsequential assume heightened significance with the expansion of knowledge. The classification of South African soils (Soil Classification Working Group, 1991) is already quite elaborate, and the seventy three soil forms are too numerous for purposes of broad generalisation. Grouping into fewer classes is essential.

International classification systems such as Soil Taxonomy (Soil Survey Staff, 1999) and the World Reference Base (IUSS Working Group WRB, 2006) have a small number of soil orders or reference soil groups at their highest level of differentiation. Both employ a key that permits unambiguous identification, by a process of elimination, when passing through a fixed sequence of definitions. A similar scheme is presented in this chapter for organising the 73 soil forms of South Africa into fourteen groups. The grouping makes use of nomenclature that is internationally familiar, following pedological tradition, while at the same time being based on key diagnostic horizons as defined in the South African system.

Having introduced a conceptual system of nomenclature for soil groups it was decided to extend this to cover forms and families. This was done by making use of qualifier terms, similar in meaning to many that are used in the WRB and Soil Taxonomy but defined according to formal diagnostic criteria in the South African system. This conceptual nomenclature should (i) provide more intelligible legends for soil maps, (ii) conceivably facilitate future revision of the classification system by encouraging the use of WRB-style qualifiers and (iii) universalise the local soil classification system (thus promoting its wider interpretation in neighbouring African regions, for example) while preserving its essential structure and consequently the usefulness of existing soil maps. The fourteen soil groups have provided the basis for a systematic account of South African soils in terms of their properties, geography, genesis and use, which is presented in Chapter 2. In the present chapter, Section 1.1 presents the basis for arranging soil forms into soil groups, Section 1.2 provides a brief description of each group and Section 1.3 introduces and explains the alternative nomenclature at form and family level, the details of which appear in relevant sections of Chapter 2 and in a consolidated form in Appendices 1 and 2.

1.1 THE KEY TO SOIL GROUPS

The 73 soil forms can be organised into 14 groups based on either a distinctive topsoil (1 to 4) or, if the topsoil is orthic, a distinctive subsoil horizon or material (5 to 14). The group to which a soil belongs is arrived at by means of a key that works by elimination, proceeding through the groups until the defining characteristic is encountered (Table 1.1). As indicated earlier, this key is similar in concept to that used for grouping soils at the highest level in the WRB (IUSS Working Group WRB, 2006) and USDA (Soil Survey Staff, 1999) systems of classification.

It would be helpful at this point to explain how the criteria and their ranking were chosen for the key in Table 1.1. All soils have elements of all soil forming processes expressed in their make-up. Differing degrees of expression of these processes give rise to different soils (Simonson, 1959). There is no single, best method for constructing a key for identification: the precedence given to one criterion over another depends on the perspective of the classifier. As far as possible, however, the criteria for grouping soil forms are given an order of priority that is consistent with that currently used in the key to soil forms of the South African taxonomic system. Priority is given to topsoils (groups 1 to 4) and, within the remaining groups, to mature expression of diagnostic properties (hence cumulic, lithic and anthropic groups come last) and especially to the presence of horizons that reflect a marked pedogenic enrichment of secondary substances. Subsoil characteristics for groups 5 to 11 focus on the expression of a secondary accumulation of silica (5), carbonates (6), clay (7), metal-humate complexes (8), iron and manganese through redox redistribution (9), sesquioxides as a residue of weathering (10), and gleyed material as a consequence of reduction due to marked wetness (11). The absence of any of the horizons diagnostic for groups 1 to 11 leaves cumulic soils (12) in which youthfulness is evident in unconsolidated material, lithic soils (13) showing youthfulness in the form of saprolite and anthropic soils (14) in which recently disturbed material is the defining characteristic.

Table 1.1 Grouping of soil forms as an eliminative key* to identification based on the presence of specific diagnostic horizons or materials. The first four groups are distinguished by topsoil horizons and the remainder according to subsurface horizons and features.

Soil group		Concept	Identification	Soil forms
1	Organic	Wetland peat	Organic O	Champagne
2	Humic	Humus enrichment; free drainage; low base status	Humic A	Kranskop Magwa Inanda Lusiki Sweetwater Nomanci
3	Vertic	Swelling, cracking clay	Vertic A	Rensburg Arcadia
4	Melanic	Black clay; high base status	Melanic A	Willowbrook Bonheim Steendal Immerpan Mayo Milkwood Inhoek
5	Silicic	Silica enrichment; arid	Dorbank	Garies Oudtshoorn Trawal Knersvlakte
6	Calcic	Carbonate or gypsum enrichment; arid	Soft or hardpan carbonate or gypsic horizon	Molopo Askham Kimberley Plooysburg Etosha Gamoep Addo Prieska Brandvlei Coega
7	Duplex	Marked clay enrichment (duplex)	Pedocutanic or prismacutanic B	Estcourt Klapmuts Sterkspruit Sepane Valsrivier Swartland
8	Podzolic	Metal humate enrichment	Podzol B	Tsitsikamma Lamotte Concordia Houwhoek Jonkersberg Witfontein Pinegrove Groenkop
9	Plinthic	Iron enrichment (absolute); mottling or cementation	Soft or hard plinthic B	Longlands Westleigh Avalon Lichtenburg Bainsvlei Wasbank Glencoe Dresden
10	Oxidic	Iron enrichment (residual); uniform colour	Red apedal, yellow-brown apedal or red structured B	Pinedene Griffin Clovelly Bloemdal Hutton Shortlands Constantia
11	Gleyic	Reduction (aquic subsoil or wetland)	G	Kroonstad Katspruit
12	Cumulic	Young soil in unconsolidated sediment (colluvial, alluvial, aeolian)	Neocutanic or neocarbonate B, regic sand, deep E or stratified alluvium	Tukulu Oakleaf Montagu Augrabies Namib Vilafontes Kinkelbos Fernwood Dundee
13	Lithic	Young soil on weathered rock	Lithocutanic B or hard rock	Glenrosa Mispah Cartref
14	Anthropic	Human disturbance	Disturbed material	Witbank

*Identify by elimination, following the sequence from 1 to 14. Consult the concept as a guide then check the definition of diagnostic horizons for identification (See Glossary pp. 264-281).

Groups 1 to 4 have a distinctive topsoil horizon.
Groups 5 to 11 have orthic topsoil and are distinguished by the type of enrichment that dominates their subsoil horizon.
Groups 12 to 14 have orthic topsoil but are typically younger by virtue of recent deposition, erosion or human disturbance; subsurface enrichment is therefore weak.
All soils contain some expression of all processes. In many groups eluviation has produced E and bleached A horizons. These horizons and other criteria help define forms and families.

With respect to each of the diagnostic horizons, the soil groups are not mutually exclusive. They never can be, irrespective of the order in which the criteria are considered. The use of qualifying terms at the form and family level (see Section 1.2) maintains the conceptual connection between the soils of all groups. Despite this interconnectedness it is possible to focus, as will be done in the next section, on the archetype for each group.

1.2 DESCRIPTION OF SOIL GROUPS

1. Organic soils (organic O)
Peaty soils are typically associated with wetlands and bogs where saturation with water inhibits the breakdown of organic residues, especially under cooler conditions. They are usually more fertile than soils of surrounding uplands. The O horizon gradually decomposes when drained and cultivated, and may even turn severely acidic if sulfides are present.
Soil form: Champagne

2. Humic soils (humic A)
Intensely weathered, low base status soils with oxidic mineralogy and exceptional accumulation of humus develop in relatively cool areas of high rainfall, free drainage and plateau topography. They have stable, micro-aggregate structure and low erodibility. Biomass potential is large but limited by nutrient deficiency and acidity. Chemical amendment is needed to make them highly productive. Natural veld on these soils is sour. Some variants have an exceptionally thick humic A horizon, thought to be related to the depth of pedoturbation by soil fauna. Aluminium forms a complex with the humus, making it less retentive towards nutrient cations but more resilient.
Soil forms: Kranskop, Magwa, Inanda, Lusiki, Sweetwater, Nomanci

3. Vertic soils (vertic A)
Swelling, cracking, high activity clay (smectitic) soils are found on basic parent materials in semi-arid to sub-humid climates, especially in lower lying landscape positions. They characteristically exibit slickensides, haploidisation and pedogenic aggrandisement, with a tendency for the solum to become inverted (gilgai micro-relief is an extreme expression). Typically dark coloured, but sometimes red or grey; base-rich and chemically fertile, they are physically challenging to flora, fauna, farmers and engineers but can be highly productive under careful husbandry, especially when irrigated. They support sweet natural veld and are strongly buffering towards water and chemical substances. Smectitic clay is favoured as bulli for cricket pitches and as an impermeable liner for dams and landfills.
Soil forms: Rensburg, Arcadia

4. Melanic soils (melanic A)
These dark coloured, strongly structured, high base status soils are similar to the dark vertic soils but are less expansible for textural or mineralogical reasons. They are also associated with a semi-arid to sub-humid climate and are fertile but require irrigation to be highly productive. Natural veld on these soils provides sweet grazing. Although the structure is strong it is less stable than in humic soils and the soils may be more erodible. Melanic soils are well buffered and can be cropped intensively without needing lime to counteract acidification.
Soil forms: Willowbrook, Bonheim, Steendal, Immerpan, Mayo, Milkwood, Inhoek

5. Silicic soils (dorbank B)
Under arid conditions silica mobilised during weathering in the upper part of the soil precipitates as a massive or laminar dorbank (duripan) horizon, the depth of which appears to be related to the texture of the overlying soil. Other evaporite minerals such as gypsum, calcite, sepiolite and halite are sometimes also in evidence and many dorbank horizons are calcareous. The distribution of silicic soils locally and globally suggests that, for enrichment with silica, peculiar circumstances of silica activation such as volcanic ash showers or regular atmospheric additions of sodium in rain, fog or dust are required in combination with hydrolysis and intense evaporation.
Soil forms: Garies, Oudtshoorn, Trawal, Knersvlakte

6. Calcic soils (soft or hardpan carbonate)
A widely observed morphological feature of soils in arid climates is the accumulation of secondary lime as a distinctive horizon consisting chiefly of calcite. In the calcic soils either hardpan carbonate

(calcrete) or a soft carbonate horizon or, more rarely, a gypsic horizon dominates the morphology of the subsoil. A subtle interplay of calcium supplied by weathering, water movement, evaporation, pH and carbon dioxide from biological respiration produces an array of features such as nodules, pipe stems, vesicular structure and laminations. As with silicic soils, other evaporite minerals such as sepiolite may be found in association.
Soil forms: Molopo, Askham, Kimberley, Plooysburg, Etosha, Gamoep, Addo, Prieska, Brandvlei, Coega

7. Duplex soils (pedo- or prismacutanic B)

One of the properties by which a B horizon can be recognised is the accumulation of clay by illuviation. This may occur by either vertical eluviation from an overlying A or E horizon or lateral eluviation from upslope. The textural contrast may have been accentuated or even initiated by a lithological discontinuity arising from colluvial or aeolian deposition. In wetter variants, ferrolysis (reductive dissolution of iron, oxidative consumption of humus and acid weathering of clay minerals) may have further enhanced the textural contrast. Marked enrichment with clay results in strong blocky or prismatic structure, cutanic character (clay skins), and, defining the duplex character, a clear to abrupt transition from the overlying horizon. The B horizon of duplex soils is commonly an impediment to root growth, water movement and deep cultivation. Clay dispersibility entails unfavourable structure and the surface soil is prone to crusting and generally highly erodible. Climate is usually sub-humid to semi-arid.
Soil forms: Estcourt, Klapmuts, Sterkspruit, Sepane, Valsrivier, Swartland

8. Podzolic soils (podzol B)

In quartzitic materials the concentration of water soluble humic substances from decomposing vegetation exceeds the capacity of clay to adsorb them and they migrate downwards, within the soil profile and through the landscape, complexing metals (chiefly Al and Fe) and precipitating once a solubility threshold of metal saturation is reached (a gradient in soil pH is probably also involved). The result is a spectacular chromatogram in which a darker or conspicuously chromatic subsoil horizon, sometimes indurated, forms beneath a bleached E or Ae horizon. High rainfall and certain types of vegetation (e.g. fynbos, evergreen forest) are more conducive to podzolisation. Nutrient deficiencies are common especially for shallow-rooted plants.
Soil forms: Tsitsikamma, Lamotte, Concordia, Houwhoek, Jonkersberg, Witfontein, Pinegrove, Groenkop

9. Plinthic soils (soft or hard plinthic B)

An absolute enrichment with iron oxides can occur in situations where intermittent wetness from a fluctuating water table gives rise to the reduction and mobilisation of iron and its migration and reprecipitation as mottles, nodules, concretions and vesicular hardpan (ferricrete). A warm, sub-humid to humid climate with a distinct dry season is commonly associated with plinthite formation. Strongest expression occurs in middle to lower slope positions in the landscape. Old erosion surfaces may be preserved by a capping of hard plinthite (also termed laterite). Manganese is associated with iron in some plinthic materials.
Soil forms: Longlands, Wasbank, Westleigh, Dresden, Avalon, Glencoe, Bainsvlei

10. Oxidic soils (red or yellow-brown apedal B or red structured B)

Oxides of iron accumulate through weathering and impart to many soils a colour which is essentially uniform, at least in the upper solum, owing to the fact that conditions are well drained and aerated. The red colour of hematite signifies conditions that are warmer, drier, more base-rich and less affected by organic matter than those associated with the yellow-brown colour of goethite. Hematite is the stronger of the two pigments and many red soils contain more goethite than hematite. Some soils have an upper yellow-brown and a lower red B horizon. In the former, organic matter and redox conditions have preferentially removed the hematite. Most humic soils have a highly weathered oxidic subsurface horizon whereas oxidic soils may be but are not necessarily highly weathered and are found over a wider spectrum of climatic conditions. A blocky structure in the red structured B is thought mainly to arise from a more pronounced seasonal desiccation than that prevailing in apedal counterparts

from which it can be texturally and mineralogically indistinguishable.
Soil forms: Pinedene, Griffin, Clovelly, Constantia, Bloemdal, Hutton, Shortlands

11. Gley soils (G)

Soils with an orthic A horizon but none of the subsurface horizons that distinguish groups 5 to 10 may nevertheless exhibit a degree of maturity in the form of colours associated with intense reduction. The G horizon reflects relatively persistent (stagnant) wetness typically associated with clayey textures and hydraulic confinement, such as in wetlands or at sites with a high potential for vertical and lateral accumulation of water. Although the archetype G has bluish-green or olive tints reflecting the presence of iron in a reduced state, grey colours are more typical and some mottling may be evident. Comparable soils having an organic, vertic or melanic surface horizon occur in groups 1, 3 and 4.
Soil forms: Kroonstad, Katspruit

12. Cumulic soils (neocutanic or neocarbonate B, alluvium, or regic sand)

Many soils are youthful as a result of having formed in recent, unconsolidated sediments such as colluvium, alluvium, or aeolian sand. In some cases there is clear evidence of incipient soil development in the form of colour variegation caused by cutanic character (textural differentiation, clay skins or lamellae), carbonate accumulation and/or faunal incorporation of darker surface soil. In other cases this may be barely noticeable.
Soil forms: Tukulu, Oakleaf, Montagu, Augrabies, Dundee, Namib, Vilafontes, Kinkelbos, Fernwood

13. Lithic soils (lithocutanic B or hard rock)

Other soils are youthful either because of limited rock weathering or on account of rejuvenation through natural erosion on steeper, convex slopes, ensuring intimate contact between a surface horizon that is maintained by biological activity and the underlying rock or saprolite. Even where the rock is weathered the subsoil has a predominantly geogenic character, although tonguing of soil and illuviated clay (cutans) into the saprolite may be evident. The saprolitic material may have incipient features such as gleying, calcareousness or softening due to weathering but these are insufficiently expressed to qualify for one of the other distinctive subsurface horizons. Penetration of roots and water is typically non-uniform and restricted to spaces between fragments of rock or saprolite.
Soil forms: Glenrosa, Mispah, Cartref

14. Anthropic soils (disturbed deposits)

Incipient soil formation can be detected in most disturbed materials associated with mining or waste disposal. This subgroup is also 'cumulic' in that the parent material represents an unconsolidated accumulation, which ranges from municipal garbage to various mine spoils and metallurgical tailings on slimes dams and could include land that has been sufficiently disturbed by civil engineering projects to have lost any recognisable solum that might have been present. Deep cultivation in preparing land for some types of agriculture could also be included.
Soil form: Witbank

1.3 DESCRIPTIVE NOMENCLATURE FOR SOIL FORMS AND FAMILIES

Soil families in the South African taxonomic system are currently assigned a geographic name as well as a code, which combines the form abbreviation with a number to facilitate data storage. The large number of families and the anonymity of a numerical code make either of these options unsatisfactory for some types of communication. The ideal is to have names which mean something descriptively and conceptually even though some effort is still required to become familiar with the terminology.

The nomenclature for soil forms and families takes the form of qualifiers such as those used in the WRB classification. These qualifiers can be strung together to provide a compound name for a soil form or family. The variety and in some cases uniqueness of the differentiating criteria has required some innovation in developing terminology. As far as possible the traditions of soil classification have been adhered to in sourcing appropriate formative elements so that in most cases at least the approximate connotation is fairly clear. The actual nomenclature for forms and families is introduced for each soil group in Chapter 2. The Glossary presents the nomenclature for all forms and families in an extended table (begining on p. 273) for ease of reference and comparison. The following general points need to be kept in mind:

1. The nomenclature is not intended to replace, but rather to supplement, the existing names and abbreviations for forms and families, conveying information that is of value in constructing map legends and in communicating soils data internationally. All the criteria employed in grouping soil forms and describing soil families are based on current definitions of horizons and other diagnostic features of soils (see Glossary, pp. 264-281). Descriptor terminology follows pedological tradition and has been borrowed, where possible, from the WRB system (IUSS Working Group WRB, 2006) or the USDA Soil Taxonomy (Soil Survey Staff, 2003), with similar meaning but with altered definition. In some cases the terminology had to be invented.

2. All *italicised* descriptors refer to features of *surface horizons*; all others refer to features of subsurface horizons. Names and abbreviations in bold face refer to current forms and numbers refer to families. They are included to facilitate cross-referencing between descriptive terms and the existing names of forms and families.

3. Textural class of the topsoil should be employed when referring to a specific soil profile, pedon or mapped soil series; otherwise the rule should be to use the term 'soils' when referring to a group, form or family of soils. (Example: soils of the Swartland form are described as pedocutanic-lithic, duplex soils; whereas a particular pedon (e.g. in the family Sw 2212) with a sandy clay loam topsoil texture might be described as an *achromic*, rhodic, micropedal, calcic, pedocutanic-lithic, duplex, sandy clay loam. Preferably, it could be described as:

> **Duplex, sandy clay loam**: pedocutanic-lithic; *achromic*, rhodic, micropedal, calcic.

What this format achieves is the following: The first bold word identifies which of the fourteen soil groups the soil belongs; the remainder of the bold name provides us with the information we most want about our *tactile* relationship with the soil (i.e. topsoil texture); thereafter, the qualifying terms define, when in italics, further differentiae applying to the A horizon and, when not in italics, further differentiae applying to the presence, relative arrangement and properties of one or more subsurface horizons or materials. The terminology may be context specific, and to know exactly what it means requires, as in any event it should, that definitions be carefully consulted. With familiarity of use, however, the terms are likely to serve as a better *aide memoire* – on map legends and in research publications, for example – than would either a place name or an alpha-numeric code, even though when used collectively to define a particular soil they may appear to undermine the imperative of conciseness.

To summarise what has been presented in this chapter: we needed soil groups as a basis for generalisation; these groups were organised into an eliminative key for identification; they were named conceptually; and this naming process required a new terminology so that forms within groups, and families within forms, could be named consistently at the group level. In the next chapter each of the fourteen groups is presented in the same sequence as it appears in the key, and the properties, classification, geographic distribution, genesis and land use of soils within each group are discussed.

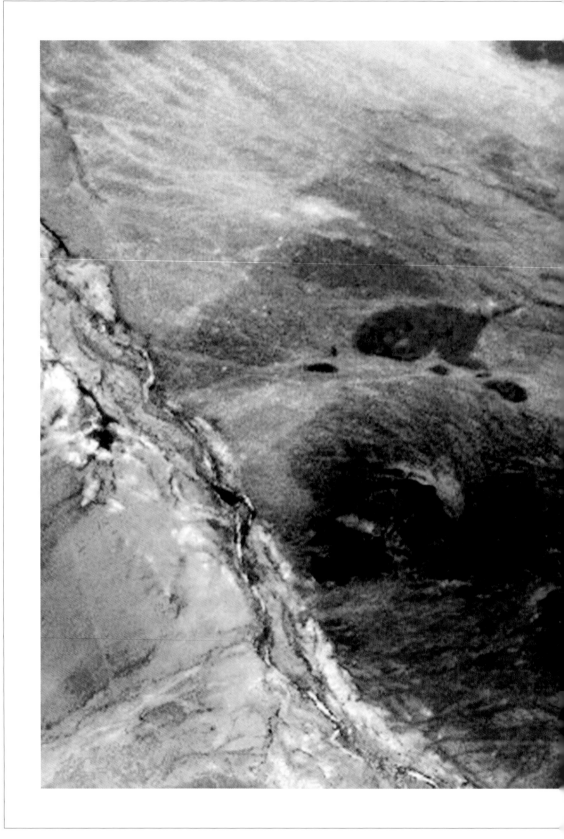

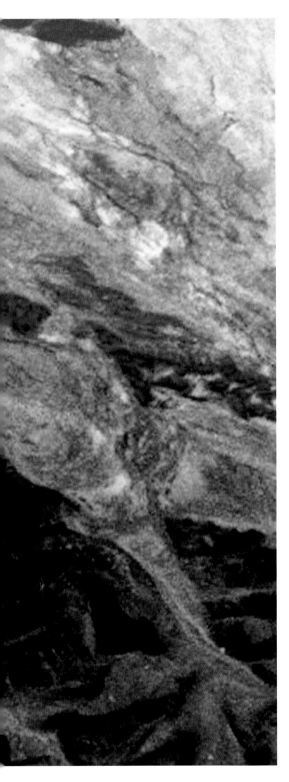

SOILS
OF
SOUTH AFRICA

Chapter 2

The soil groups: distribution, properties, classification, genesis and use

Martin Fey, Jeff Hughes,
Jan Lambrechts and Theo Dohse

Oblique aerial snapshot of the Karoo from a passenger plane. The mosaic of colours reflects the complexity of soils that exists even in arid environments. Although differences between soil types are sometimes dramatic and spur us to order them into different groups or classes, soils with different properties grade into one another to form a continuum in the landscape.

THE STRUCTURE OF THIS CHAPTER

This chapter contains fourteen sections each of which deals with one of the soil groups. In each case the group is first introduced and its mapped distribution is presented. This is followed by a discussion of properties, classification, genesis and use. The following table summarises the fourteen groups and the criteria for placing a soil into one of them by a process of elimination.

\multicolumn{4}{c}{Key to the soil groups}			
Principle	Soil group	Concept	Key horizon for identification
Soils with special topsoil characteristics	1 Organic	Wetland peat	Organic O
	2 Humic	Humus enrichment; free drainage; low base status	Humic A
	3 Vertic	Swelling, cracking clay	Vertic A
	4 Melanic	Black clay; high base status	Melanic A
Soils with an orthic A horizon and special subsoil characteristics relating to pedogenic accumulation	5 Silicic	Silica or sepiolite cementation; arid	Dorbank or sepiocrete
	6 Calcic	Carbonate or gypsum enrichment; arid	Soft or hardpan carbonate or gypsic horizon
	7 Duplex	Marked, clear clay enrichment	Pedocutanic or prismacutanic B
	8 Podzolic	Metal humate enrichment	Podzol B
	9 Plinthic	Iron enrichment (absolute); mottling or cementation	Soft or hard plinthic B
	10 Oxidic	Iron enrichment (residual); uniform colour	Red apedal, yellow-brown apedal or red structured B
	11 Gleyic	Reduction (aquic subsoil or wetland)	G
Young soils (inceptic)	12 Cumulic	Young soil in unconsolidated sediment (colluvial, alluvial, aeolian)	Neocutanic or neocarbonate B, regic sand, thick E or stratified alluvium
	13 Lithic	Young soil on weathered rock	Lithocutanic B or hard rock
	14 Anthropic	Human disturbance	Disturbed material

2.1 ORGANIC SOILS

Introduction

The term 'organic' has become popular for describing – inaccurately – a certain type of agricultural produce. Happily the traditional meaning of the term in soil classification is perfectly clear, referring to soils in which the natural accumulation of organic matter has been exceptional. As in other soil classifications organic soils are placed in a class of their own in order to distinguish them from so-called mineral soils. Unlike that of humic topsoils (Section 2.2) the accumulation of organic matter is related primarily to wetness and secondarily to low temperatures, both of which reduce the activity of decomposing micro-organisms.

The group is represented only by the Champagne form, originally defined by Van der Eyk et al. (1969) as a soil with a diagnostic O horizon in which the majority of plant forms are not macroscopically identifiable. The O horizon was subdivided into being 'friable' (Champagne series) or 'plastic' (Ivanhoe series) and was considered to overlie either a 'firm gley' or 'regic sand'. Soils of the Champagne form were only found in bottomland sites in the 'cool montane regions' of the Tugela Basin survey. Unlike all other soil forms, these soils occupy disparate areas and never constitute a major soil spatially within the landscape (Figure 2.1.1). Their importance does not lie therefore in the total area covered but rather in the particular ecological niche that they occupy. Their common association with wetlands gives them special environmental significance.

Properties

General

Organic soils are those that have a diagnostic organic O horizon. Such an horizon should not be confused with the identically named horizon in other classifications where it represents the above soil surface accumulation of organic material, usually under unsaturated conditions, and where it has

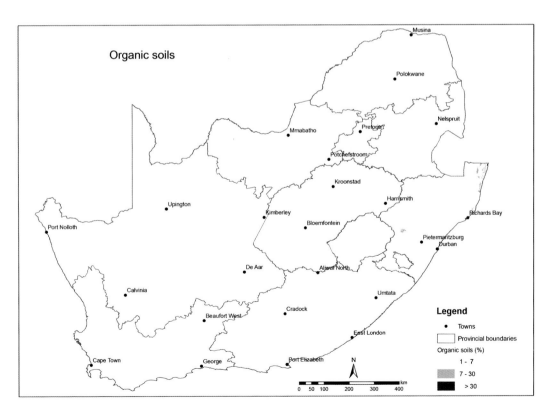

Figure 2.1.1 Organic soils in South Africa (abundance classes refer to estimated percentages within land types)

been subdivided into the Oi, Oe and Oa layers (e.g. Soil Survey Staff, 1999), and is underlain by a mineral A horizon. Indeed, the Soil Classification Working Group (1991) stresses that it excludes 'for example layers of pine needles in pine plantations'. These soils thus show a marked accumulation of organic material as a result of long-term wetness and/or cold temperatures. Organic and mineral soils are separated on the basis that the organic O must have at least 10% organic carbon throughout a vertical distance of 200 mm. Many such soils have much greater organic carbon contents and at the extreme would be almost totally organic with only minor inorganic material, and would be approximately equivalent to peat soils elsewhere. Crucially these soils are considered saturated with water for long periods in their natural state although many have been drained for agricultural use (see below).

Morphological

Because of the large amount of undecomposed organic material the soils are commonly dark brown or black (Figure 2.1.2). The fairly continuous wetness results in signs of gleying in the subsoil, which, when present is typically a transported material of either colluvial or alluvial origin (some are underlain by bedrock). In extreme cases the subsurface horizon shows bright blue or green hues reflecting the permanently reduced nature of the iron hydroxides. (A detailed discussion of gleying is presented in Section 2.11).

Chemical and physical

The properties of organic soils are dependent on the source of the organic material, its state of decomposition and their landscape position. They can thus be either strongly acid or have a near neutral pH and can contain variable amounts of inorganic material. These facts were recognised in the earlier classification (MacVicar et al., 1977) where the soil series were distinguished on the basis of a pH > or < 4.0 and a clay content of > or < 20%. The organic material may be finely divided (*sapric*) or more fibrous (*fibric*). Because the material is predominantly undecomposed they have high C:N ratios and low cation exchange capacities, though the latter will be affected by the nature of the inorganic material.

Fibric organic soils are loosely packed in their natural state with very low bulk densities (0.05 to 0.15 Mg m^{-3}) and extremely rapid saturated hydrau-

Figure 2.1.2 Organic soils (sapric, cumulic) with a characteristic high water table: a. in a swampy environment on the edge of a natural wetland in the Karkloof, KwaZulu-Natal; and b. in a *dambo* (seasonally wet grassland) near Likasi in Shaba Province, Democratic Republic of Congo; c. Sapric lithic organic soil.

lic conductivities (Ks >1.6 m d^{-1}) (FAO, 2001). Sapric materials, being more decomposed are somewhat more compact with finer pores, slightly higher bulk density and lower hydraulic conductivity. Due to their loose structure and high water content, organic soils have very poor load bearing capacity and trafficability. In their undisturbed state they have low erosion risk but this increases dramatically once drained unless careful management is applied. Part of this increased risk is as a result of the shrinkage that highly organic soils undergo on drying. Scotney (1970) found soils of the Ivanhoe series shrank by up to 48% after drying from a saturated state.

Classification

Organic soils are represented by the Champagne form which is subdivided into four soil families (Table 2.1.1) distinguished on the basis of, firstly, the predominant type of organic material (fibrous or humified) and, secondly, the type of underlying material (bedrock or saprolite, or unconsolidated material). A recent proposal still to be considered is the splitting of the Champagne form on the basis of lithic or gleyic character. International correlation is shown in Table 2.1.2.

Genesis

Organic soils are characteristic of 'wetlands' of various types such as marshes, bogs, mires, swamps and vleis; the last term will be used here. The organic material accumulates under conditions where plant material is produced by an adapted vegetation but decomposition is prevented or retarded by one or more of (a) permanent wetness; (b) low temperatures; (c) extreme acidity or lack of nutrients and (d) high concentrations of electrolytes or organic toxins (FAO, 2001), all of which retard microbial breakdown of the organic matter. However, not all such environments will produce the diagnostic organic O horizon. These soils thus occupy low points in the topography, at least locally, and could be expected to form widely. In fact such soils appear to be rare in South Africa since their occurrence is a result of a combination of topography and climate that ideally needs to be cool or cold throughout the year. Where summers are too hot, organic soils cannot form as the material breaks down rapidly in the warm temperatures. Thus in the western and southern parts of the Western Cape province organic soils are rare even though topography and winter temperatures are favourable (Maalgaten River, George, Kanetvlei, De Doorns). A similar situation exists on the Highveld and mineral hydromorphic soils dominate in such situations.

Organic soils are most commonly found at higher altitudes in KwaZulu-Natal, northern parts of the Eastern Cape and in the northern extension of the Drakensberg Mountain range in Eastern Mpumalanga. Within these areas they occur from the Low Berg upwards, anywhere where topogra-

Table 2.1.1 The form and families of organic soils (Nomenclature is shown alongside form names and abbreviations and family codes as defined by the Soil Classification Working Group, 1991)

Organic (Champagne Ch)

1100 *fibric*, lithic	1200 *fibric*, cumulic
2100 *sapric*, lithic	2200 *sapric*, cumulic

Table 2.1.2 Rough placement of organic forms in international classification systems (IUSS Working Group WRB, 2006; Soil Survey Staff, 2003)

WRB (Italicised qualifiers are those about which there is less certainty in correlation)

WRB Reference Group	Possible prefix qualifiers	Possible suffix qualifiers
Histosols (>20% organic carbon)	Fibric, Hemic, Sapric, Rheic, Leptic	*Thionic*, Dystric, Eutric, Drainic
Gleysols (10–20% organic carbon)	Histic, Haplic	Humic, Dystric, Eutric, Drainic

USDA Soil Taxonomy

Histosols

phy and bedrock are suitable, and are analogous to 'topogenous peat' (FAO, 2001). They are more common where rock types are less permeable so allowing water to persist. In addition, organic soils are patchily distributed at lower altitudes where conditions are favourable. However, organic soils are not spatially extensive even in vleis since they only form in the region between that occupied by permanent free-standing water and the mineral soils upslope and thus they never occupy more than small, localised areas. An exception to this is on the highest parts of the Drakensberg Mountains along the border of, and extending into, Lesotho where extensive areas of organic soils are found that are more closely akin to 'blanket peat', formed as they are under very high rainfall/low evapotranspiration conditions.

Organic soils in the George area (Figure 2.1.3) developed as a result of warping of the coastal platform. Upstream of the warp axis, about 200 m above sea level, the discharge rate of the rivers decreased, valleys were filled with light textured fluvial deposits, and hygrophilous vegetation flourished to such an extent that thick organic layers developed (mostly fibric). In the 1970s the fibric organic layers were mined and sold as peat to nurseries. Although dark, organic rich, acid, sandy soils are found in many low lying areas in the winter rainfall region of the Western Cape Province, genuine organic soils (>10 % C) are rare. This is probably due to the low annual production of organic residues in these nutrient poor environments coupled with high summer temperatures.

Use

The properties of the organic material (such as state of decomposition, texture and consistency) and the location of the area determine the use possibilities and management requirements for organic soils. As stated earlier, organic soils occupy a particular ecological niche and one that has become increasingly important as environmental awareness has increased. Vleis have come to be recognised as vital parts of the hydrological cycle and are now protected (Environment Conservation Act, 1989 (Act 73 of 1989); National Water Act (Act 36 of 1998); National Environment Management Act, 1998 (Act 107 of 1998). Their importance lies in the fact that they moderate stream flows, act as natural filters for sediment and pollutants and form a unique natural habitat for a range of flora and fauna. Thus organic soils play a vital part in natural ecosystems. It follows from this that the optimum land use for these soils is to be conserved as natural wetlands under natural vegetation that usually consists of reeds and sedges (e.g. *Phragmites* and *Cyperus* spp.).

Some vleis have been successfully farmed without causing undue ecological damage (Figure 2.1.4) but many have not. Their topographic position often makes them desirable agricultural or pastoral soils and many have been artificially drained. As a result of drainage one of the main driving forces for maintaining organic soils is removed and the inevitable breakdown of the organic matter follows. This in turn removes the buffering capacity of the soils to moderate water flow and in many cases severe erosion follows. In the past organic soils were drained in the Langkloof for the production of deciduous fruit under flood irrigation. Drainage combined with the application of lime, phosphate and potassium

Figure 2.1.3 Organic soil upstream of the coastal platform warp axis on the old flood plain of Maalgaten River, George. These soils have been mined as horticultural peat.

Figure 2.1.4 Organic soil under natural wetland in the Karkloof, KwaZulu-Natal, partially planted to poplar trees (background) and, on wetland margins, sown to soybeans (foreground). Such sites can add meaning (and rare authenticity) to claims of crops being grown organically.

accelerated the decomposition of organic material to such an extent that the soil surface was lowered as much as 300 mm. Tree roots were exposed and irrigation canals collapsed.

Another consequence of drainage and oxidation may be severe acidification, especially in peats of coastal areas where sulfides have accumulated during peat formation. Oxidation of sulfide minerals produces sulfuric acid and pH values of 1 or less have been measured (Figure 2.1.5). Unlike the peat bogs of the northern hemisphere, Champagne soils generally do not occupy extensive land areas and so are not specifically targeted for 'reclamation'. Rather they are 'caught in the net' when entire wetlands are drained. Van der Eyk et al. (1969) comment that 'greater than all of these (other research needs) is the need for research into vlei hydrology and vlei management'. A great deal of such research remains to be carried out.

Figure 2.1.5 Efflorescence of acid sulfate minerals in a drained peat at Suurbron, near Humansdorp, Eastern Cape. A pH of 1 was measured in a soil sample from the drain. Oxidation of pyrite is exothermic and a second visit to this site 30 years after the photo was taken revealed that much of the peat had been destroyed by (probably spontaneous) combustion. The remains were still smouldering.

PEATS OF THE SUB-ANTARCTIC TUNDRA ZONE: MARION ISLAND

Figure 2.1.6 Marion Island: Location (inset) and altitudinal zonation of vegetation (from Mucina and Rutherford, 2006)

Marion is a volcanic island, territorially South African but situated half-way between Africa and Antarctica. It is wet (about 2 000 mm annual precipitation) and cold (hyper-oceanic); an ideal environment for the formation of organic soils. Altitude reaches 1 230 m above sea level and there is consequently distinct zonation in terms of vegetation and weathering (Figure 2.1.6). The soils of Marion have recently been well documented in a thesis by N.R. Lubbe (University of Pretoria, 2009). One example is shown in Figure 2.1.7.

In terms of classification most of the soils of Marion Island fall into the organic group. International correlation would probably place some of them into Histic Andosols (IUSS Working Group WRB, 2006) because of their volcanic parent material.

Figure 2.1.7 Example of an organic soil on Marion Island, showing residual layering from volcanic ash deposition.

2.2 HUMIC SOILS

Introduction

The special significance of the humic horizon lies in its marked accumulation of humus, often to considerable depths but in a mineral as opposed to an organic surface horizon, and without the influence of prolonged wetness (cf. organic soils – Section 2.1). The solum is essentially well drained. The situations in which this occurs in South Africa are restricted to regions (Figure 2.2.1) in which high rainfall and cool temperatures are combined with gentle to moderate relief. Van der Eyk et al. (1969) reserved the term for soils, typically in an advanced stage of weathering, on elevated plateaus in the coast hinterland and midland mist belt regions of the Tugela Basin. Their definition specified a minimum thickness of 450 mm, containing an average carbon concentration of at least 2%, and a relatively low base status (as distinct from the melanic horizon; see Section 2.4). The Zulu word 'Dudus' for humic topsoil means fluffy or dusty and aptly describes this characteristic feature in dry weather, especially after cultivation.

Because humic soils are so commonly associated with a high degree of weathering and because such soils in the high rainfall uplands of South Africa are invariably characterised by a marked accumulation of humus, albeit not always to the great depths often found in the coast hinterlands of KwaZulu-Natal, the depth criterion was brought in (Soil Classification Working Group, 1991) as a criterion for defining soil families within the humic soil group, thus allowing some families that might have been included in the oxidic soil group to be included instead under humic soils. There will be many soils which contain insufficient organic matter to qualify as humic but which have a low base status and sesquioxidic mineralogy. These soils mostly belong in the oxidic group (Section 2.10) and should be referred to for additional discussion on the properties and genesis of soils with red and yellow-brown apedal horizons.

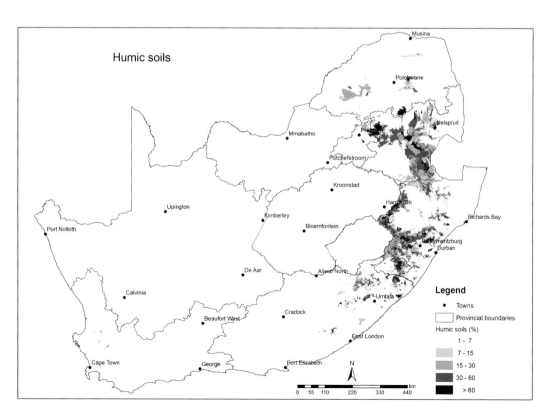

Figure 2.2.1 Humic soils in South Africa (abundance classes refer to estimated percentages within land types)

Properties

General

Humic soils are defined as all those having a diagnostic humic A horizon. This horizon is currently defined on the basis of carbon content (more than 1.8% OC), base status (less than 4 cmol$_c$ of exchangeable base cations per kg clay for every one percent organic carbon) and subsoil drainage. This may be followed by a yellow brown apedal B horizon and/or a red apedal B, or a pedocutanic, neocutanic or lithocutanic B. Descriptions and analytical data for a selection of representative humic soils are provided in Appendix 2.2.

Morphological

The humic A horizon is characteristically thick (Figure 2.2.2, a) but it can be thinner if the humus content is sufficiently high to justify inclusion on grounds of total profile organic matter content (Figure 2.2.2, b).

Subsurface horizons may vary although they are characteristically sesquioxide-rich with red (hematite) or yellow-brown (goethite) colours dominating as a result of advanced weathering and free drainage. A yellow-brown B1 horizon often overlies a red B2 horizon (Figure 2.2.3), the significance of which is discussed in Section 2.2.4. Such profiles are thought to represent a wetter variant of the Inanda form in Figure 2.2.2. Neocutanic, lithocutanic and pedocutanic horizons may occur in rarer situations where colour variegation due to clay illuviation and faunal mixing are marked and structure is moderately strong, often on the margins of the older surfaces where they are normally situated. Most commonly the B horizon has relatively uniform colour and the porous, friable, massive structure which is characteristic of apedal B horizons.

Chemical and physical

Low cation exchange capacity, a clay mineral composition dominated by kaolinite, aluminous chlorite, gibbsite, and iron oxides (mainly goethite and/or hematite, but sometimes including maghemite, either finely disseminated or in the form of small, dark nodules), low pH, a high phosphate fixing capacity and pH buffering capacity are some of the characteristic chemical properties of these soils. Shortages of some trace elements, especially zinc, are common. Physically they are well drained, have good water retention characteristics, are easily tilled and have a low tendency to become compacted. Although sometimes clayey in texture they do not show a marked tendency to shrink and swell with changes in water content. The topsoil has a peculiarly low bulk density as a result of not only the high organic matter content but also the micro-aggregating effect of iron and aluminium oxides.

Classification

Soil forms include Kranskop, Magwa, Inanda, Lusiki, Sweetwater and Nomanci. Some of these are illustrated in Figures 2.2.2 to 2.2.4. The first three forms are typical humic sola, in which there are deeply developed, strongly weathered, uniformly coloured, red and/or yellow-brown B horizons. The second three have characteristics that would qualify them as duplex, sedimentary or erosional were it not for the humic A horizon. Typically they would be associated with soils of the first three forms but in landscape positions

Figure 2.2.2 Two profiles of the rhodic form of humic soils (Inanda form) representing: a. *thick* families, from arkosic sandstone in the Wartburg district) and b. *thin* families (from dolerite in the Underberg district of KwaZulu-Natal) with different depths of humus enrichment.

Figure 2.2.3 Two profiles of the xanthirhodic form of humic soils (Kranskop form): a. representing *thick* families, near Bruyn's Hill, derived from arkosic sandstone; b. representing *thin* families, derived from basalt (relict amygdales are evident in the pink saprolite beneath the red B2 horizon) in the Kamberg Nature Reserve near Nottingham Road, KwaZulu-Natal.

Figure 2.2.4 Profiles of: a. xanthic (Magwa), b. lithic (Nomanci) and c. pedocutanic (Lusiki) forms of humic soils (a. and b. near Wartburg, KwaZulu-Natal; c. near Humansdorp, Eastern Cape).

that are somewhat less stable (midslopes and concavities) so that the solum is more youthful, with more evidence of an active, earlier weathering stage and of faunal- and argilli-pedoturbation than in the first three forms.

The names of forms and their constituent families are shown in Table 2.2.1. A recent proposal yet to be approved is to create new forms to accommodate the horizon sequences humic/hard rock (Ingeli), humic/neocutanic/soft plinthic (Umvoti) and humic/neocutanic/unconsolidated material with signs of wetness (Mooi River). Possible nomenclature for these three forms would be lithic, neoplinthic and neohydromorphic, respectively. Approximate correlation of the forms with WRB and USDA classes is presented in Table 2.2.2.

Table 2.2.1. Forms of humic soils and their constituent families (Nomenclature is shown alongside form names and abbreviations and family codes as defined by the Soil Classification Working Group, 1991)

Xanthirhodic (Kranskop Kp)

1100 *thin*, haplic	1200 *thin*, luvic
2100 *thick*, haplic	2200 *thick*, luvic

Xanthic (Magwa Ma)

1100 *thin*, haplic	1200 *thin*, luvic
2100 *thick*, haplic	2200 *thick*, luvic

Rhodic (Inanda Ia)

1100 *thin*, haplic	1200 *thin*, luvic
2100 *thick*, haplic	2200 *thick*, luvic

Pedocutanic (Lusiki Lu)

1110 *thin*, arhodic, micropedal	1120 *thin*, arhodic, macropedal
1210 *thin*, rhodic, micropedal	1220 *thin*, rhodic, macropedal
2110 *thick*, arhodic, micropedal	2120 *thick*, arhodic, macropedal
2210 *thick*, rhodic, micropedal	2220 *thick*, rhodic, macropedal

Neocutanic (Sweetwater Sr)

1110 *thin*, arhodic, haplic	1120 *thin*, arhodic, luvic
1210 *thin*, rhodic, haplic	1220 *thin*, rhodic, luvic
2110 *thick*, arhodic, haplic	2120 *thick*, arhodic, luvic
2210 *thick*, rhodic, haplic	2220 *thick*, rhodic, luvic

Lithocutanic (Nomanci No)

1100 *thin*, orthosaprolitic	1200 *thin*, hypersaprolitic
2100 *thick*, orthosaprolitic	2200 *thick*, hypersaprolitic

Table 2.2.2 Approximate placement of humic forms in international classification systems (IUSS Working Group WRB, 2006; Soil Survey Staff, 2003)

WRB (Italicised qualifiers are those about which there is less certainty in correlation)

Soil form	WRB Reference Soil Group	Possible prefix qualifiers	Possible suffix qualifiers
Xanthirhodic (Kranskop)	Ferralsols	Posic, Geric, Vetic, Acric, Umbric	Humic, Alumic, Dystric, Xanthirhodic
	Acrisols (those with a luvic B horizon)	Vetic, Umbric, *Fractiplinthic*, *Petroplinthic*, Haplic	Alumic, Humic, Hyperdystric, Xanthirhodic
Xanthic (Magwa)	Ferralsols	Posic, Geric, Vetic, Acric, Umbric	Humic, Alumic, Xanthic, Dystric
	Acrisols (those with a luvic B horizon)	Vetic, Umbric, *Fractiplinthic*, *Petroplinthic*, Haplic)	Alumic, Humic, Hyperdystric, Xanthic

(continued)

Rhodic (Inanda)	Ferralsols	Posic, Geric, Vetic, Acric, Umbric	Humic, Alumic, Rhodic, Dystric
	Acrisols (those with a luvic B horizon)	Vetic, Umbric, *Fractiplinthic, Petroplinthic*, Haplic	Alumic, Humic, Hyperdystric, Rhodic
Pedocutanic (Lusiki)	Acrisols	Cutanic, Umbric	Humic, Rhodic, Chromic
	Luvisols (those with CEC >24 cmol$_c$/kg ?)	Cutanic, Haplic	Humic, Rhodic, Chromic
	Lixisols (those with CEC >24 cmol$_c$/kg ?)	Cutanic, Haplic	Humic, Rhodic, Chromic
Neocutanic (Sweetwater)	Acrisols	*Cutanic*, Umbric, Haplic	Humic, Rhodic, Chromic
	Cambisols	Ferralic, Haplic	Humic, Dystric, Rhodic, Chromic
Lithocutanic (Nomanci)	Acrisols (those with a soft lithocutanic B)	Cutanic, Leptic, Umbric, Haplic	Skeletic
	Umbrisols	Leptic, Haplic	Humic

USDA Soil Taxonomy

Xanthirhodic (Kranskop): Oxisols (Humic Haplustox; Humic Hapludox); Ultisols (Typic Haplustults)
Xanthic (Magwa): Oxisols (Humic Xanthic Haplustox; Humic Xanthic Hapludox); Ultisols (Typic Haplustults)
Rhodic (Inanda): Oxisols (Humic Xanthic Haplustox; Humic Rhodic Hapludox); Ultisols (Typic Rhodustults)
Pedocutanic (Lusiki): Ultisols (Typic Haplustults); Alfisols (Haplustalfs – Ultic; Typic)
Neocutanic (Sweetwater): Inceptisols (Dystrustepts – Oxic; Humic; Typic; Haplustepts – Oxic; Dytsric; Typic)
Lithocutanic (Nomanci): Inceptisols (Dystrustepts – Lithic; Oxic; Humic; Typic; Haplustepts – Lithic; Oxic; Dystric; Typic)

Genesis

Humic soils are typically associated with old land surfaces (Figure 2.2.5) in the humid, eastern seaboard region of South Africa, especially in KwaZulu-Natal, the Pondoland coast and along the eastern escarpment region of Mpumalanga. They are deeply weathered, enriched with oxides of iron and aluminium (occasionally including nodules of gibbsite), and have formed as a result of prolonged weathering on stable land surfaces under conditions of good drainage in a cool, subtropical environment. Pristine vegetation of either forest or grassland has been suggested, although the history of fire from which formerly more extensive indigenous forests have retreated, and the tendency for humus accumulation to a significant depth to be more characteristic of grassland environments, have resulted in this issue still being debated by pedologists. Certainly there appears to be abundant evidence of soil fauna having been at least partly responsible for the unusually thick (500 mm or more) variants of the humic A horizon, with morphological evidence of faunal mixing being abundant in the variegated colours at the transition with the underlying B horizon. There is some evidence suggesting that thinner humic horizons occur at cooler, higher elevations further inland than on the coast hinterland plateaus where thick humic horizons are more frequently found (near Inanda, Mapumulo, Magwa and Babanango, for example). Depth of faunal activity (see krotovinas in Figure 2.2.3a for example) and root development, related to subsoil temperatures, may possibly be responsible for this difference.

The depth and intensity of weathering associ-

Figure 2.2.5 Humic soil landscapes: a. at the top of the Karkloof escarpment near Howick; b. Kingscliffe near Wartburg, KwaZulu-Natal. In both cases the plateaus are remnants of the early tertiary or African erosion surface.

ated with humic soils is considerable, with red apedal subsoils often extending to at least two or three metres before kaolinised saprolite is encountered (Figure 2.2.6a). This unusually deep solum has been interpreted as indicating the exceptional age of the land surfaces on which the humic soils are encountered. In the same landscape one finds occasional ferruginous laterite outcrops (Figure 2.2.6b) that are interpreted by Maud (1965), King (1972) and others as evidence for prolonged Tertiary weathering. Maud considers the deep red sola of the humic soils to have been derived by weathering of this ferriferous duricrust. Fey (1985), on the other hand, viewed the exceptional depth of the humic soils as being the product of colluvial aggradation on gentle slopes coupled with the protective effect of a dense vegetation canopy and the exceptionally stable structure of these sesquioxidic soils,

Figure 2.2.6 Deep, intense weathering, beneath a humic soil mantle, evident in: a. kaolinitic saprolite of arkosic Natal Group sandstone giving a pallid zone that often underlies; b. indurated secondary laterite or ferriferous duricrust. Both sites are on the road to Montebello Mission near Bruyn's Hill in the Wartburg district of KwaZulu-Natal.

minimising erosion. In a typical landscape of humic soils, such as in the Bruyn's Hill district near Pietermaritzburg, thick sola blanket most of the landscape except in extreme crest positions (e.g. those on which trigonometric survey beacons are located), where the solum is quite thin, consisting of a humic A horizon directly overlying and tonguing into weathered sandstone saprolite (Nomanci form). This suggests that little if any of the original land surface remains and that the surface is modern and equilibrated with the current climate. The question of how much of the evident weathering represents pre-weathering followed by colluvial transport and aggradation, as opposed to weathering *in situ*, requires further study. Uniformity of colour in the red and yellow-brown subsurface horizons is probably accounted for by pedoturbation.

The normally apedal nature of subsoil horizons in this group can be attributed to the clay mineral composition, with sesquioxides and kaolinite combining to form exceptionally strong microaggregates which stabilise the pore space and inhibit expansion and contraction as a result of wetting and drying. In addition, however, apedality (or at least a lack of macro-pedality) can be attributed to the moisture regime that remains remarkably uniform with little or no desiccation in the subsoil. This can be attributed to cool temperatures in the dry season (typically coinciding with winter), a dormant, non-transpiring grassland vegetation and a rainfall distribution which is characterised by as much as a third of the rainfall occurring during the drier half of the year, supplemented by frequent fog. That contraction and cracking are possible is clearly evident in areas where evergreen forest plantations with a high perennial water demand (pine, eucalyptus and wattle) have been established. Here, the subsoil often shows marked (and probably irreversible) shrinkage as a result of intense drying. This physical degradation still needs to be systematically characterised and quantified and its impact on forest productivity needs to be determined.

Yellow-brown and red colours in apedal B horizons assume special significance where there is a transition from the former to the latter with depth in the profile. This yellowing (Figure 2.2.7; see also Figure 2.2.3) is considered to be a product of organically-fuelled, redox stripping of hematite from a material (now remaining as the B2) which was formerly red and contained both hematite and a more stable, more Al-substituted goethite. Essentially the yellow B1 horizon represents a zone of compromise between supply of oxygen and availability of organic substrate for bacterial reduction of (mainly) hematite. The yellow colours imparted by this unmasking of the more stable goethite are seen as being diagnostic of hydromorphic conditions despite the absence of gley colours or mottling (Fey, 1982; Fey and Manson, 2004). In this sense some yellow-brown B horizons may have affinities with some E horizons and could well exhibit similar processes such as ferrolysis (see Section 2.11). Alternatively, de Villiers and Verster (1965) suggested that the yellow-red horizon sequence has some properties characteristic of subdued podzolisation.

A peculiar variant of the usually iron- and

Figure 2.2.7 Xanthirhodic yellowing beneath a thin humic A horizon in deeply weathered red soil (Kranskop form) derived from dolerite under grassland at the foot of Bamboo Mountain, near Drakensberg Garden, KwaZulu-Natal.

aluminium-dominated humic soils occurs in the Graskop district of Mpumalanga, as a result of the weathering of dolomite. The subsoil, in its most intense manifestation of Mn oxide enrichment (up to 17% Mn), is essentially a pitch black, soft, porous manganese wad (Figure 2.2.8). Hawker and Thompson (1988) and Dowding (2004) have investigated the special mineralogical and chemical attributes of these unusual soils that have virtually no parallel except perhaps in Hawaii.

Another special feature of many of the humic soils not currently accommodated in the classification is the occurrence of gibbsite nodules. These are sometimes platy or tubular, indicating secondary enrichment along cleavage planes or root channels, and are often much paler in colour than the matrix (occasionally almost white). The concentration of these nodules may be as high as 30% of the soil volume, easily qualifying them as gibbsihumox in Soil Taxonomy (Soil Survey Staff, 2003) (Figure 2.2.9, a,b). On dolerite or hornfels (but not sandstone or shale) this Al enrichment or bauxitisation can also manifest itself in the form of porous, speckled yellow-brown, brittle, jointed, saprolitic bauxite (Figure 2.2.9, c,d), with gibbsite and goethite pseudomorphs after feldspar and ferromagnesian minerals, respectively, preserving the texture of the original rock which has become almost completely desilicated. Cores of unweathered rock are often evident within blocks of this jointed saprolite, and are separated from the surrounding weathered material by a sharp boundary of only a few millimetres. Near the soil surface and on fragments of this material in stone lines, a convoluted patina of Fe oxide separates the material from the earthy soil matrix. Material of this kind is ubiquitous on dolerite-capped plateaus throughout the humic soil terrain on the eastern seaboard, from the Eastern Cape to Mpumalanga and the Soutpansberg. Particularly widespread occurrences near Ngome and Weza in KwaZulu-Natal have been the subject of detailed exploration and bauxite deposits of several million tons exist which are too scattered and of too low grade, however, to be of economic interest. They serve as a reminder of the special circumstances of climate and landscape stability with which the humic soils are typically associated.

On the high rainfall (up to 1 000 mm per annum) dissected upland Witzenberg and Koue Bokkeveld plateaus near Ceres, humic soils (Kranskop and Sweetwater) are common along well drained concave terrain positions. Upslope the soils are oxidic, duplex or inceptic lithic, all on deeply pre-weathered, highly leached shale saprolite. The humic soils mostly developed in oxidic colluvium from upslope. The more moist concave sites generally have high organic carbon content and dense fynbos (mostly Proteacae) vegetation and the cool conditions have retarded organic carbon decomposition. On sites with a thin colluvial cover lithocutanic, humic soils (Nomanci form) occur.

Use

Sugar cane, maize, forestry and vegetable crops are the main uses for these soils. In the sugar industry, Beater (1957) refers to them as 'Mistbelt TMS' soils (currently in the sugar industry they typify the so-called Mistbelt System and are described as dark brown, fluffy, humic loams of the NGS Mistbelt). The requirements for their successful utilisation are well documented (South African Sugar Association Experiment Station, 1999). A liberal phosphate application is paramount. Some of them are so strongly leached and acidic that a lime requirement of as much as 10 tonnes ha^{-1} is not uncommon,

Figure 2.2.8 Humic soil derived from dolomite near Graskop, Mpumalanga, with a black B horizon caused by enrichment with manganese oxide

HUMIC SOILS

Figure 2.2.9 Bauxitisation in humic soils: nodules, pipestems and plates of gibbsite with the same colour as the kaolinitic subsoil matrix in which they occur are common on stable crests of the escarpment plateau such as those near: a. Howick; b. World's View, Pietermaritzburg; c. a core of unweathered dolerite at World's View encased in brown saprolitic bauxite; and d. a brittle, porous, desilicated residue with an intergrowth of gibbsite and goethite pseudomorphs after feldspar and ferromagnesian minerals.

especially when old forest plantations are converted to crop land or pasture. Low subsoil calcium status has led some workers to expect returns from applying more soluble amendments such as gypsum that allows Ca to leach more readily than limestone. Calcium silicate slag has produced good responses in sugarcane. Gubevu (1996) has studied the buffer capacity of these soils which is considerable, especially following the acidifying effects of certain cropping practices such as tea (e.g. at Ngome in northern KwaZulu-Natal) and forest plantations. Under natural grassland the erosion risk is low because of both soil structural stability and plant cover. Terracettes are a common feature on steep slopes (Figure 2.2.10) brought about by a combination of mass slumping, grazing by cattle and inhospitability of the infertile subsoil to re-establishment of a grass sward. The leached soils give rise to naturally sour veld, palatable after spring burning but of decreasing nutritive value and digestibility as the summer sward matures. Invasion of pioneer species (*Aristida junciformis* or 'Ngongoni' in the coast hinterland and *Eragrostis plana* or 'Mchiki' at higher elevations) that compete with more palatable species such as *Themeda triandra* 'Rooigras' under heavy grazing, make veld management difficult. Much pasture on these soils has thus been degraded.

In many parts of rural KwaZulu-Natal, small cropping enterprises flourish on these soils, mostly for home consumption rather than the market (Figure 2.2.11). Because of humus mineralisation, nitrogen requirements to obtain adequate yields are relatively small while the humus probably makes a difference to the effectiveness of phosphate which, though still needed, would otherwise be fixed in large quantities. These soils can be worked repeatedly with great ease, neither compacting excessively when dry nor puddling badly in wet weather; this despite a clay content which often exceeds 50%. Thus the water holding capacity is far superior to that of a sandy loam of similar consistence. Gardeners in such verdant towns as Kloof, Hilton and Eshowe in KwaZulu-Natal and Sabie and Tzaneen in Mpumalanga province will attest to the horticultural benefits of humic soils for growing everything from lilies, roses and vegetables to avocados and other fruit trees. These soils are a pleasure to work by hand, retaining water but also draining well, providing clean root crops and, despite their acidity and shortage of some elements such as phosphorus, possessing a useful organic reserve of mineralisable nutrients.

An important land use question concerns the effect of liming combined with P and K fertilisation on decomposition of organic material to a level below that which defines humic A horizons. Does regular tillage increase the rate of decomposition? Will the decrease in organic carbon only take in the plough layer or even deeper? A related problem encountered in the Witzenberg and Koue Bokkeveld regions of the Western Cape with planting of pome fruit on 'virgin' humic soils is the high release of N due to organic matter decomposition and the application of lime, P and K. Even without any N fertilisation the young trees are vegetatively too vigorous and do not make the transition to a bearing phase. Certain farmers either reserve these soils for stone fruit or make use of very low vigour pome fruit rootstocks. The potential N mineralisation in a 300 mm thick A with 2% C, C:N-ratio of 10, and 5% annual mineralisation could be up to 400 kg N annually. This is at least five times the amount of N applied annually on young pome fruit trees.

To summarise, the concept of the humic soil group is one of marked accumulation of humus and of a high degree of weathering, in well-drained topographic positions, in landscapes of high rainfall, cool temperatures and a dense cover of grassland or forest. Nutrient deficiencies and marked soil acidity are to be expected. Physical attributes and climatic circumstances are near to ideal for forestry and dryland agriculture provided plant nutrient status is attended to. These soils are environmentally robust in that they can be subjected to a good deal of physical and chemical abuse without markedly eroding or deteriorating.

Figure 2.2.10 Terracettes on humic soils outside Underberg, KwaZulu-Natal

Figure 2.2.11 Small home gardens of root crops (*madumbis*) and maize on humic soils in the Inanda district, with commercial plantings of wattle, gum and sugarcane in the background

2.3 VERTIC SOILS

Introduction

The vertic soil group consists of all soils with a vertic A horizon (L. *vertere*, to turn). These are roughly equivalent to the Vertisols of international classification (Soil Survey Staff, 2003; FAO, 2005). Regionally named equivalents include black cotton soils (USA), regur (India), and tirs (Morocco). Shrink-swell properties are sufficiently developed to give rise to characteristic features such as cracks, slickensides and self-mulching. Although agricultural and engineering uses of these soils are beset with difficulties they can be very productive, especially when irrigated. The productivity of natural ecosystems is also greatly affected by the prevalence of vertic soils. A tropical or subtropical climate with a marked dry season seems to be the main factor determining their global distribution. This is confirmed in Figure 2.3.1 which shows that land types with a substantial component of vertic soils are confined largely to the northern parts of South Africa. The dry season usually coincides with winter months in southern Africa but the outlier at Nieuwoudtville (west of Calvinia) demonstrates that vertic soil formation can also occur in a xeric (i.e. summer dry season) climate.

Some forms, especially in the melanic, duplex and hydromorphic soil groups, have families with vertic properties in subsurface horizons. These do not key out in the vertic group because their surface horizon is non-vertic. Much of the ensuing discussion on properties, genesis and land use is, however, also pertinent to those soils.

Properties
General

In terms of their properties, vertic soils are perhaps the most well known of the four special topsoils recognised in the current South African classification. The extremely strong structure, strong tendency to shrink and swell with changes in water content that produces characteristic pressure faces known as slickensides, and the capacity to self-mulch and to develop extensive and deep cracks, are all features

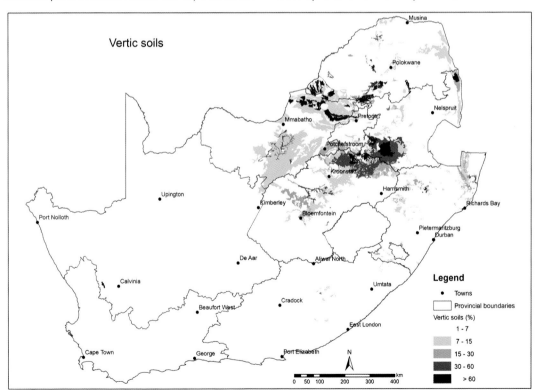

Figure 2.3.1 Vertic soils in South Africa (abundance classes refer to estimated percentages within land types)

that are peculiar (though not necessarily unique) to vertic soils.

In common with many of the soil horizon definitions, the vertic descriptors are under scrutiny, especially with regard to recognising a vertic B horizon. However, currently vertic soils must display a vertic A horizon that is defined on the basis of strongly developed structure and either visible slickensides (Figure 2.3.2) or a high plasticity index (PI) of 32 or 36, depending on the apparatus used to measure the liquid limit (see box). Although slickensides are a field-evident manifestation of the smectitic clays that give rise to the high PI they are not always present due to their being influenced by the depth of the vertic material. Shallow vertic topsoils may therefore not contain slickensides (Soil Classification Working Group, 1991).

Morphological

All vertic soils display a strong blocky structure with a tendency towards prismatic structure as a result of their wedge-shaped peds (so-called 'lenticular' or 'bicuneate' structure). Although many vertic materials are black (Figure 2.3.3), other colours are known especially red (Figure 2.3.4) and even yellow-brown and grey. Such colour differences are believed to be a function of drainage status as discussed in Section 2.3.4. Other visible features (see box) may include a self-mulching surface (Figure 2.3.10); a highly plastic and highly sticky consistence; vertical cracks that appear on drying and that run throughout the depth of the vertic material (and sometimes into the upper subsoil), opening to the surface when the material is fully dry (Figures 2.3.5; 2.3.6); slickensides (Figure 2.3.2); gilgai microtopography (see box); and accumulation of calcium carbonate, gypsum and even halite. The last feature is not defining but is quite common. Carbonate nodules are sometimes found scattered on the surface of those profiles known to invert.

Chemical and physical

The chemical and physical properties that characterise vertic soils are a function of their clay mineralogy. Vertic soils have a high clay content (typically >50%, although FAO, 2001, p. 81, mentions 'sandy

---

ATTERBERG LIMITS

These are tests that measure a series of material water contents and are standard tests in civil engineering laboratories. They form the basis for estimating the degree of movement in shrink/swell clay materials by characterizing material plasticity. The most commonly determined of the Atterberg limits, and the ones used in the South African soil classification, are those that determine the plastic limit (PL) and the liquid limit (LL).

Liquid limit

This is the water content at which the material is practically a liquid but at which it still retains some shearing strength. It is determined by mixing the soil with various amounts of water and placing the mixture in a special cup. A trapezoidal groove is cut into the material and the number of taps required to close the groove at its base measured for each of the soil:water ratios. A graph is constructed of the water content (determined on a sub-sample after over drying at 105 °C) vs. number of taps. The LL is the water content, obtained from the graph, at which 25 taps closes the groove. This is sometimes known as the upper plastic limit.

Plastic limit

This is the lowest amount of water required to make the material plastic. It is obtained by mixing the soil with decreasing amounts of water and determining the water content at which a 3 mm diameter thread just begins to crumble. This is sometimes known as the lower plastic limit.

Plasticity index (PI)

This is a measure of plastic behaviour and is determined as the difference between the LL and the PL.

Figure 2.3.2 Slickenside in a vertic soil (Rensburg form) derived from dolerite colluvium near Secunda, Mpumalanga Province

Figure 2.3.3 Black vertic soil derived from gabbro colluvium near Kokstad, KwaZulu-Natal

Vertisols') containing a relatively large proportion of fine clay (<0.1μm). Much of this clay is of the 2:1 type and typically smectitic. Some vertic soils, however, have considerable amounts of other clay minerals including interstratified clays (Bühmann et al., 1985; 1988). Vertic soils consequently have a high cation exchange capacity, a base saturation generally close to 100%, and pH which is mildly acid or moderately alkaline pH (6.0 to 8.5) unless sodium predominates on the exchange complex.

Amounts of organic carbon are variable but lower than might be suggested by the black colour of many vertic soils. This is due to the finely dispersed nature of the humic material that combines with calcium to form a black calcium-humate complex that thinly coats the mineral grains. Vertic soils may be amongst the blackest of all soils despite having <2% organic carbon in many cases. The high surface area, CEC and base status impart an exceptional capacity for plant nutrient retention, metal sorption and acid buffering to these soils.

The high smectite content of vertic soils results in a pronounced tendency to shrink and swell with changes in water content. In dry periods the shrinkage produces wide cracks that open to the surface, allowing an initially unlimited infiltration of water. As the soil swells and the cracks close on wetting, infiltration rate drops dramatically. Where the surface is self-mulching the contrast in infiltration between wet and dry sates is not so great because the cracks become clogged with fine peds. In both cases infiltration rate will become negligible once the surface seals. Most vertic soils have a high water holding capacity but the water is held at a high

Figure 2.3.4 a. red vertic soil (rhodic, calcic) derived from dolerite; b. same soil sloughing through shrinkage in aardvark burrow; c. surface exposure showing self-mulching below a dolerite outcrop, all in the wild flower reserve near Nieuwoudtville, Western Cape. This occurrence of red vertisols on dolerite in a winter rainfall climate is unique. An exceptionally high concentration (and variety) of geophytes (bulbous plants) occurs on these soils.

FEATURES PECULIAR TO VERTIC SOILS

1. Self-mulching

In the dry season vertic soils begin to desiccate and the extent of drying is greatest at the surface and decreases with depth. The drying initiates cracking and as drying intensifies with the length of the dry period so the cracking pattern becomes increasingly fine i.e. the cracks become narrower and more numerous. Towards the surface this results in the peds becoming smaller and smaller as drying proceeds, since the peds are defined by the cracking pattern. This results in the surface of some vertic soils developing a fine granular or crumb structure called a 'surface mulch' (see Section 2.3.5 on land use). This is in stark contrast to the medium/coarse angular/subangular blocky material that immediately underlies the mulch and the even coarser structure beneath that. Self-mulching is normally contrasted with crusting morphology (see below).

2. Crusting

Not all vertic soils develop a self-mulching surface; some develop a hard surface crust. The distinction between these different behaviours is often difficult as shown by Snyman et al. (1984) where a soil with a high modulus of rupture (MOR), thought to indicate a strong potential to crust, was actually self-mulching while a nearby soil with very low MOR exhibited crusting morphology. The soils were, however, mineralogically distinct with the self-mulching soil being dominated by well-crystalline smectite while the crusting soil contained predominantly a randomly interstratified smectite-chlorite-mica. In a later study, however, Bühmann (1995) showed that crusting and self-mulching soils shared a common mineralogy within broad ranges and no consistent differences could be established in layer charge characteristics of the smectite species between the two Vertisol types. This conclusion prompted the Soil Classification Working Group (1991) to remove the crusting criterion form the South African Soil Classification System. FAO (2001), however, comment that sandy Vertisols develop a surface crust. The cause of crusting behaviour in vertic soils therefore remains problematic, and is presumably related to factors that influence crust formation in other soil types such as texture and sodicity.

3. Cracking

As the process of drying out proceeds and the clays lose increasing amounts of interlayer water, the soil loses its plasticity and tension builds up until the tensile strength of the soil material is locally exceeded. The soil then cracks. It is the increasing number of such cracks at the surface that results in the self-mulching characteristic discussed above. Increasing drying allows these cracks to penetrate deeper and deeper into the vertic material. The finer material on the surface, when disturbed, falls into these cracks thus mixing the material by a process known as 'argillipedoturbation' and it is this repeated turning over of upper soil material and its placement into the lower parts

matric potential and the fraction of water that is plant available is consequently lower than in other soils. These soils have been considered to have only a low to moderate erosion hazard (Bühmann et al., 1999 (for self-mulching vertic soils); SASEX, 1999) in their undisturbed state. Once disturbed, however, and should rill or gully erosion be initiated, the self-mulching property can result in rapid collapse. On the other hand the shallow angle of repose of gully sides affords easier revegetation. Some erosion classifications tend to err on the safe side and assign a high erosion hazard (e.g. Scotney, 1970). Where there are erosion gullies the shallow angle of repose is quite valuable as a field indicator of vertic soils (e.g. in the eastern Free State near Vrede). The physical properties of a toposequence of Highveld vertic soils have been discussed in relation to classification by Snyman et al. (1984).

Classification

In earlier South African classification Van der Eyk et al. (1969) recognised two soils with vertic topsoils of the horizon that gives the vertic soils their name. When the vertic material rewets and tries to swell as the clays absorb water the swelling is prevented due to the space occupied by the surface material within the cracks. Continuing uptake of water generates pressures that are relieved by the soil masses sliding against each other producing slickensides (see below) and by upward pressure resulting in gilgai (see next box).

4. Slickensides

As the vertic soil mass undergoes swelling pressures caused by increasing water uptake, shearing forces occur and mass movement occurs along oblique planes at angles of 20 to 30 degrees to the horizontal to relieve the pressure (FAO, 2001). These shear planes are known as 'slickensides' – a geological term that describes similar features seen in rocks that underwent similar processes. As shown in Figure 2.3.2 the slickensides are polished pressure faces that are grooved in the direction of shear. Intersecting shear planes define wedge-shaped angular blocky peds. It is this type of structure that has come to be known as 'bicuneate' or 'lenticular' but which more generically is called 'vertic' structure. Slickensides are recognised worldwide as a characteristic that defines vertic soil material and in some cases can become major 'bowl-shaped' features (Figure 2.3.6).

Figure 2.3.5 Surface cracking in a vertic soil (near Kokstad, KwaZulu-Natal)

Figure 2.3.6 Bowl-shaped slickenside and vertical cracks in a vertic soil material (WRB: albi-vertic Planosol, eutric, bathygleyic, bathyhypocalcic) near Lake Albalossat, Kenya (depth to the base of the bowl is about 1 m)

namely the forms Rensburg (underlain by gleyed subsoil) and Arcadia (where the vertic A overlies a C or R horizon). Further subdivision was on the basis of calcareousness in the subsoil of the Rensburg form and, in the case of Arcadia, whether the vertic material showed self-mulching or crusting tendencies. MacVicar et al. (1977) maintained this distinction and additionally recognised that the colour of vertic topsoils may differ (Section 2.3.2). These basic distinctions remain in the current classification except that the crusting vs. self-mulching distinction in Arcadia has been omitted. Subdivision into families and corresponding nomenclature is shown in Table 2.3.1, and international correlation is given in Table 2.3.2.

Genesis

The prerequisite for vertic soil formation is a sufficiently sustained concentration of soluble silica, bases and alkalinity either to induce the formation of 2:1 layer silicates (the bisiallitisation of Millot, 1970) or to at least favour their persistence when

GILGAI TOPOGRAPHY

A unique feature of some areas of vertic soils is the development of a 'mound and basin' landscape called gilgai (an Australian Aboriginal word meaning 'waterhole' – a reference to the water-filled depressions during the wetter parts of the year). In South Africa the type area is probably the Springbok Flats in North-West Province though there are many other areas as shown in Figure 2.3.7. This topography is revealed especially well on aerial photos. Worldwide, three different topographic forms of gilgai are distinguished: normal or rounded, lattice and linear (FAO, 2001). The first is the best known, forming on generally level topography, with the latter two apparently being a function of increasing slope.

Formation

Although several hypotheses have been put forward to explain gilgai formation, all of them relate gilgai to shrink/swell clay minerals and it is considered to have its origins in the subsurface soil (FAO, 2001). For gilgai to form the soil must therefore have sufficient cohesion to transfer the pressures produced on swelling to the soil surface. Swelling pressures produced as the cracks close during the wet season are transferred in all directions outward from the base of the crack. However, the least restrained direction is upwards and so the slickensides produced are continuous from below the centre of the depression towards the (higher) centre of the mound i.e. the oblique shear planes show a preferential direction and subsoil material is pushed upwards along these sets of parallel slickensides, resulting in a mounded surface with interspersed hollows. Once initiated, the cracks open in the same place every dry season and thus the topography becomes increasingly magnified. While most South African gilgai consist of small mounds and hollows (e.g. Zuney Valley, Eastern Cape Province, Figure 2.3.7 a) such topography can become much more marked with mounds up to 2 m or more in height with intermound distances of over 100 m (e.g. Western Australian gilgai north of Perth). The example in Figure 2.3.7 (b and c) near Rustenburg is the best known occurrence of gilgai in South Africa (Verster, de Villiers and Scheepers, 1973). In the Zuney Valley both mounds and hollows are underlain by a layer of calcium carbonate nodules at a depth of about 1.2 m. Where the mounds become more developed and the vertical distance between the top of the mound and the base of the neighbouring depressions is greater then the soils begin to differ in their properties. In the area north of Perth, Western Australia for example, the soils on the mounds contain calcium carbonate while it is absent in the soils of the depressions. The electrical conductivity of the mound and hollow soils also differs.

Anthony Mills   Jan Lambrechts   © DigitalGlobe 2010 25°35'24.30'S 27°15'11.49'E

Figure 2.3.7 instances of gilgai: a. Gilgai on valley alluvium near Zuney, Eastern Cape; b. linear gilgai on gabbro north of Rustenburg on a gentle footslope; c. satellite image of the Rustenburg gilgai showing the linear or 'wavy' form below and normal or 'tank' gilgai (spotted pattern) above the stream. The darker zones, enhanced by a recent grass fire, are the hollows. Besides these examples there is some anecdotal evidence for gilgai at other South African locations, notably on black clays of the Pongola floodplain in northern KwaZulu-Natal.

Table 2.3.1 Forms of vertic soils and their constituent families (Nomenclature is shown alongside form names and abbreviations and family codes as defined by the Soil Classification Working Group (1991)

Hydromorphic (Rensburg Rg)

1000 acalcic	2000 calcic

Aeromorphic (Arcadia Ar)

1100 melanic, *acalcic*	1200 melanic, *calcic*
2100 rhodic, *acalcic*	2200 rhodic, *calcic*
3100 alterchromic, *acalcic*	3200 alterchromic, *calcic*

Table 2.3.2 Approximate placement of organic forms in international classification systems (IUSS Working Group WRB, 2006; Soil Survey Staff, 2003)

WRB

Soil form	WRB Reference Soil Group	Possible prefix qualifiers	Possible suffix qualifiers
Hydromorphic (Rensburg Rg)	Vertisols	Grumic, Mazic, Gleyic, Sodic, Calcic, Haplic	Calcaric, Eutric, Pellic
	Gleysols (soils with weakly developed cracks and wetness within 50 cm)	Calcic, Haplic	Calcaric, Eutric
Aeromorphic (Arcadia Ar)	Vertisols	Grumic, Mazic, Endoleptic, Calcic, Haplic	Calcaric, Eutric, Pellic, Chromic
	Phaeozems (soils with weakly developed cracks)	Leptic, Vertic, Calcic, Haplic	Calcaric

USDA Soil Taxonomy

Hydromorphic (Rensburg Rg): Vertisols (Hapluderts and Haplusterts [Aquic; Chromic; Typic]); Haploxererts [Aridic; Aquic; Chromic; Typic]); Mollisols (some Endoaquolls)
Aeromorphic (Arcadia Ar): Vertisols (Hapluderts and Haplusterts [Chromic; Leptic; Typic]; Haploxererts [Aridic; Chromic; Leptic; Typic])

formed diagenetically or inherited from certain parent materials (see box). Factors which encourage these circumstances include:
- base-rich parent material,
- a soil climate that can be typified as sub-humid to semi-arid (xeric or ustic in the terminology of Soil Survey Staff, 1999), and
- drainage impeded sufficiently to inhibit leaching and promote the accumulation of solutes, either topographically or geogenically.

The development of cracks, self-mulching and slickensides also presupposes a marked seasonality in soil water content.

Furthermore, a form of pedogenic feedback operates in the formation of vertic soils since smectite formation begets a more sluggish drainage regime. A pedogenic threshold (Chadwick and Chorover, 2001) is reached beyond which further smectite formation is assured. Jackson (1965) calls it a classic case of pedogenic aggrandisement.

Thus vertic soils are most likely to occur in regions with a seasonally contrasting climate, basic or ultrabasic igneous rocks, and in lower landscape positions where bases and silica can accumulate. Within South Africa these conditions are met predominantly in the provinces of Limpopo, Free State, Mpumalanga (Highveld areas), North-West and KwaZulu-Natal. Such areas (and indeed many similar regions in many parts of Africa and elsewhere) commonly display a characteristic red-black toposequence with red, relatively leached, freely draining soils on midslopes and crests and black, vertic soils in the bottomlands. On the Springbok Flats the mineralogical variation within an extensive complex of red and black vertic soils derived from basalt has received much attention (Oberholster 1969a,b; Taylor, 1972; Bühmann and Grubb, 1991). An interesting 'outlier' body of red

FORMATION PATHWAYS FOR SMECTITE CLAYS

1. Diagenesis (direct alteration) from mica

Loss of interlayer potassium from mica occurs during chemical weathering. The resultant vermiculite retains the 2:1 structure of the mica and the interlayer space is occupied by hydrated, exchangeable cations. The layer charge of the mineral may also decrease as a result of oxidation of structural Fe, for example. The lower charge determines the swelling potential and whether smectite rather than vermiculite is the end product of mica weathering. This process is considered to be exceedingly slow and takes place over centuries or even millennia (McBride, 1994).

2. Neoformation

Smectite can form quite rapidly under laboratory conditions from dilute alkaline solutions of sodium silicate and magnesium chloride (Borchardt, 1977; McBride, 1994). Wilson (1992) synthesised crystalline smectite in solutions of pH >8.0 that contained Ca^{2+} and Fe^{2+} and that were kept anoxic. These artificial conditions suggest the kind of natural conditions required for smectite to form in soil. The environment needed for montmorillonite (dominated by Ca or Mg) is thus alkaline as a result of restricted drainage and/or salt accumulation that allows the normally soluble bases (Ca, Mg, Na and K) to accumulate together with Si. These experiments also suggest that somewhat less alkaline conditions, perhaps coupled with a reducing environment, are necessary for Fe (nontronite) and Al rich (beidellite) dioctahedral smectites to form (McBride, 1994). However, given the almost ubiquitous occurrence of mica in soil parent materials it is not always certain that smectite found in vertic soils has indeed formed neogenetically (Borchardt, 1977). On the other hand, the formation of smectitic clays from basic igneous rocks that are poor in micaceous minerals confirms that neoformation can be the dominant process of smectite formation.

3. Inheritance

The trioctahedral smectites (e.g. saponite, sauconite and hectorite) appear to be inherited from the parent material though they are rare in soils (Borchardt, 1977). Allen and Fanning (1983) suggest that smectite in many Vertisols is 'mostly inherited'. A number of studies in South Africa have shown that soils may directly inherit smectite from parent rock. Many dolerite sills have been hydrothermally altered before being exposed to weathering and contain clay minerals including smectite (Bühmann & Bühmann, 1990). In some of South Africa's geological formations e.g. the Vryheid Formation, marine transgressions and regressions are preserved in the sedimentary record and this change in solution pH is accompanied by a change in clay mineral associations, with 2:1 silicates dominating the marine deposits (Bühmann, 1994). In Karoo sediments the thermal effects of burial diagenesis and contact metamorphism are preserved in various stages of smectite illitisation (Bühmann, 1991; 1992) resulting in varying degrees of interstratifications of illite and smectite, a common mineral in many vertic soils (Bühmann et al., 1988).

vertic soils derived from dolerite under winter rainfall conditions is found at Nieuwoudtville in the Western Cape (Figure 2.3.4).

Use

The extreme characteristics of vertic soils, especially with regard to their physical properties, make them one of the most problematic soil types from a management perspective. Their tendency to alternate from being either too dry and hard or excessively wet and sticky means that the 'workable' period is often very short, and in some years almost non-existent. Thus water management is of critical importance, the more so when climatic conditions dictate that these soils require irrigation to yield maximally. Difficulties in their agricultural management has meant that many areas of vertic soils are used for extensive grazing where the soils produce sweet veld that allows year-round grazing potential. One of the more successfully grown crops on a large scale is cotton in northern KwaZulu-Natal and Swaziland (similarly in the southern United States where the name 'black cotton soil' is commonly used); its success being ascribed to the structure of the cotton root system which can withstand the shrinking/swelling movements within the soil. It is estimated that 13% of the sugar growing soils of South Africa and Swaziland are black clays that include both vertic and melanic (Section 2.4) soils. Despite their problematic physical properties vertic soils are extremely fertile chemically and it is more a question of fertility maintenance than fertility creation (Section 2.10, Oxidic soils). Many are, however, deficient in nitrogen as a result of their low content of organic matter. For sustainable intensive agriculture, phosphorus is likely to be required (Figure 2.3.8). In the North-West Province a mosaic of vertic soils and red, apedal soils (Section 2.10) exists such that there are many places where the soils abut one another along a knife-edge. Farmers in the area therefore have a mixture of these totally contrasting soils to manage. They recognise the difficulties of managing the vertic soils but nevertheless prefer them since although perhaps in four years out of five they get no or minimal yield from the vertic soils, the one year that conditions are favourable more than makes up for the lean years. In common with all strongly structured soils, vertic soils are extremely poor soils for any sort of commercial forestry, partly because of their physical properties but also because they tend to reflect a seasonally variable climate which indicates that trees will suffer from water shortage for a substantial part of the year.

The shrink/swell characteristics of the clay minerals make engineering projects challenging on vertic soils. Foundations have to be built to be able to withstand the forces generated, otherwise buildings will develop substantial cracks and will need expensive underpinning or even demolition. A soil with a PI of only 22 and containing 25% clay will generate a force of about 98 kPa and can expand by 10%. Mielenz and King (1955) give values of swelling pressures as high as 10 kg cm^{-1} (about 980 kPa). Given that some highly expansive clays may expand by up to 50% in volume and that 3% is the cut-off for requiring specially designed foundations, the engineering problems become clearly acute. It is estimated in the United States that while one in ten people are affected by floods every year, one in five are affected by expansive soils (Jones

Figure 2.3.8 Small-scale maize cropping on a self-mulching vertic soil near Polokwane, Limpopo Province showing severe P deficiency (purpling of leaves).

Figure 2.3.9 Vertic soils are notorious for the engineering challenges they present. This vertic clay has developed in gabbro colluvium on a swampy footslope near Kokstad, KwaZulu-Natal.

and Holtz, 1973). In wet weather, vertic soil terrain is notoriously obstructive to traffic (both wheeled and pedestrian) because of both the stickiness and the plasticity of fully hydrated smectite (Figure 2.3.9).

Where substantial smectite formation has occurred as a result of weathering of base-rich igneous rocks, deep and extensive deposits of montmorillonite have formed. This material is often mined and used to construct cricket pitches for which it is known as 'bulli' e.g. near Eston in KwaZulu-Natal. Vertic soil material is also used for the core of earth dams where it is covered by other materials and therefore remains moist and in a hydrated state, thus preventing water movement through the dam. Similarly, vertic clays are valuable as clay liners in waste landfills and there is potential for exploiting South African vertic clays for a number of industrial and engineering applications that currently make use of imported material. One interesting application is the light claying of sandy soils that is widely practised in Western Australia (McKissock et al., 2002) and has been considered for improving infertile sands of Maputaland (Ceruti, 1999).

2.4 MELANIC SOILS

Introduction

We use the name melanic (black, dark) for soils that have three main features in their topsoil horizon: strong structure, high base status, and dark colours even when dry. Most vertic soils (section 2.3) have melanic properties but their vertic properties (high PI or slickensides) take precedence in classification. Whereas many melanic soils could be described as 'neovertic', others are far removed from vertic behaviour and have greater affinity with either organic or humic soils, the main difference being a lower carbon content or higher base status, respectively. The more generic term 'margalitic' might have been preferred to avoid confusion associated with international use of 'melanic' for certain Andisols rich in allophane (FAO, 2006). The local usage of melanic is well established, however, having first been coined with its current meaning by van der Eyk et al. (1969), while margalitic (as defined by Mohr and van Baren, 1954) includes dark vertic soils and its meaning is therefore too broad. In some but not all respects the melanic A resembles the mollic epipedon of Soil Taxonomy (Soil Survey Staff, 1999). As might be expected from the common affinity with vertic soils, the distribution of melanic soils in South Africa (Figure 2.4.1) shows a similar pattern to that of vertic soils (Figure 2.3.1) except for some notable areas occuring further south. Because much of the land type mapping was done prior to the use of Atterberg limits in defining the vertic and melanic horizons, many vertic soil bodies too shallow to exhibit diagnostic slickensides or cracking would have been mapped as melanic. This suggests that there could be even greater similarity in the distribution of the two soil groups than the land type data suggest.

Properties
General

The melanic A horizon may overlie a wide variety of subsurface horizons and the properties of melanic

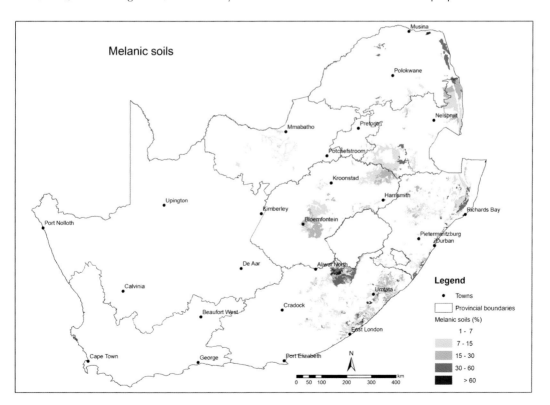

Figure 2.4.1 Melanic soils in South Africa (abundance classes refer to estimated percentages within land types)

soils are correspondingly diverse. Some examples are illustrated in Figure 2.4.2.

The melanic horizon itself is less expansible than the vertic A (in terms of plasticity index), is more base rich than the humic A (in terms of exchangeable basic cations per unit amounts of clay

Figure 2.4.2 Examples of melanic soils: a. over spheroidally weathered dolerite near Thornville; b. on weathered dolerite near Creighton; c. transition from a melanic A horizon via a prominent stoneline to a red pedocutanic B horizon near Stanger, KwaZulu-Natal; d. derived from basalt, Naude's Nek Pass, Eastern Cape.

and carbon present), has less organic carbon that that required for an organic O horizon, and is distinguished from the orthic A in retaining its dark colours in the dry state.

Morphological

Melanic soils are black or dark coloured (red colours with hues 5YR or redder are excluded) and typically have a strong, well-developed blocky structure (Fig. 2.4.3). Their dark colour has the same origin as that of dark vertic soils (section 2.3.2) and carbon content may similarly be much lower than the colour suggests. The structure is not necessarily strong blocky but is developed enough so that it is not both massive and hard when dry (Soil Classification Working Group, 1991).

Figure 2.4.3 Strong blocky structure and dark colour typical of the melanic horizon (scale indicated by match stick)

Chemical and physical

The minimum base status by which the melanic horizon is defined is specified relative to the content of both clay and organic carbon (4 cmol$_c$ basic cations per kg clay for each 1% carbon present). This is intended to distinguish degrees of leaching in a fashion similar to that of international classification, which employs base saturation of CEC determined at pH 7. Problems with both approaches may be encountered and alternatives have been explored (Fey and Donkin, 1994). Be that as it may, melanic soils are invariably base rich, with base saturation of CEC at field pH being close to 100%. This implies that their acidity status is low to neutral or slightly alkaline, and that the availability of plant nutrients, buffer capacity and metal attenuation capacity are all relatively high. Melanic soils are generally clayey or at least loamy in texture and retain water well. The clay fraction is sometimes smectitic but more often has mixed mineralogy with kaolinite sometimes being dominant (van der Merwe et al., 2002a); when smectite does dominate the clay fraction it tends to be of the high charge variety implying a smaller tendency to shrink and swell than is found in vertic horizons. Melanic soils tend therefore to have a higher infiltration rate than vertic soils, and their lower defined plasticity index (see box, section 2.3) reflects a lower capacity of the clay minerals to adsorb water in their interlayers; this implies that in melanic soils a higher proportion of water is available to plants than in vertic soils. Less expansive behaviour and slower surface sealing result in a lower erosion hazard compared to vertic soils (SASEX, 1999), although those that have a higher content of micaceous clay may be more susceptible to water erosion (Bühmann et al., 1998; 1999).

Classification

The seven soil forms listed in Table 2.4.1 are defined in terms of the horizon beneath the melanic A (G, pedocutanic B, soft carbonate, hardpan carbonate, lithocutanic B, hard rock and unspecified material, respectively). MacVicar et al. (1977) defined a plinthic form, Tambankulu (melanic A over soft plinthic B) which was subsequently discarded (Soil Classification Working Group, 1991). The unspecified material in the cumulic (Inhoek) form is usually colluvium or alluvium with minimal evidence of soil formation (although in some cases it might qualify as a neocutanic B). International correlation is shown in Table 2.4.2.

Genesis

Melanic soils correspond, broadly, to the margalitic soils of Mohr and van Baren (1954) or, more accurately, to a non-vertic subset of these soils. The paper

Table 2.4.1 Forms of melanic soils and their constituent families. (Nomenclature is shown alongside form names, abbreviations and family codes as defined by the Soil Classification Working Group, 1991).

Hydromorphic (Willowbrook Wo)

1000 acalcic	2000 calcic

Pedocutanic (Bonheim Bo)

1110 sombric, micropedal, acalcic	1120 sombric, micropedal, calcic
1210 sombric, macropedal, acalcic	1220 sombric, macropedal, calcic
2110 rhodic, micropedal, acalcic	2120 rhodic, micropedal, calcic
2210 rhodic, macropedal, acalcic	2220 rhodic, macropedal, calcic
3110 alterchromic, micropedal, acalcic	3120 alterchromic, micropedal, calcic
3210 alterchromic, macropedal, acalcic	3220 alterchromic, macropedal, calcic

Soft-calcic (Steendal Sn)

1000 acalcic	2000 calcic

Hard-calcic (Immerpan Im)

1000 acalcic	2000 calcic

Lithocutanic (Mayo My)

1100 orthosaprolitic, acalcic	1200 orthosaprolitic, calcic
2100 hypersaprolitic, acalcic	2200 hypersaprolitic, calcic

Lithic (Milkwood Mw)

1000 acalcic	2000 calcic

Cumulic (Inhoek Ik)

1100 aeromorphic, acalcic	1200 aeromorphic, calcic
2100 hydromorphic, acalcic	2200 hydromorphic, calcic

Table 2.4.2 Approximate placement of melanic forms in international classification systems (IUSS Working Group WRB, 2006; Soil Survey Staff, 2003)

WRB

Soil form	WRB Reference Soil Group	Possible prefix qualifiers	Possible suffix qualifiers
Hydromorphic (Willowbrook Wo)	Gleysols (G horizon < 50 cm)	Mollic, Calcic	Calcaric, Eutric
	Phaeozems (G horizon deeper than 50 cm)	Gleyic	Calcaric
	Umbrisols (G horizon deeper than 50 cm)	Endogleyic, Mollic, Haplic	

(continued)

Pedocutanic (Bonheim Bo)	Chernozems (very dark melanic)	Vertic, Calcic, Luvic, haplic	Clayic
	Kastanozems (paler coloured melanic)	Vertic, Calcic, Luvic, Haplic	Chromic
	Phaeozems	Vertic, Calcic, Luvic	Clayic, Chromic
	Luvisols	Cutanic, Vertic, Calcic, Haplic	Clayic, Rhodic, Chromic
Soft-calcic (Steendal Sn)	Chernozems	Calcic, Luvic, Haplic	Clayic
	Kastanozems	Calcic, Luvic, Haplic	Clayic
Hard-calcic (Immerpan Im)	Chernozems	Petrocalcic	
	Kastanozems	Petrocalcic	
	Phaeozems	Rendzic, Petrocalcic	Calcaric
	Calcisols	Petric, Hypercalcic, Hypocalcic, Haplic	
Lithocutanic (Mayo My)	Leptosols (shallower than 25 cm to hard rock)	Lithic, Mollic, Haplic	Calcaric, Eitic
	Phaeozems (deeper than 25 cm to hard rock)	Leptic, Calcic, Haplic	Calcaric
Lithic (Milkwood Mw)	Leptosols (shallower than 25 cm to hard rock)	Lithic, Mollic	Calcaric, Eutric
	Phaeozems (deeper than 25 cm to hard rock	Leptic, Calcic	Calcaric
Cumulic (Inhoek Ik)	Fluvisols (shallower than 25 cm to fluvic material)	Mollic	Calcaric, Eutric
	Phaeozems (deeper than 25 cm to fluvic material)	Gleyic, Haplic	Calcaric

USDA Soil Taxonomy

Hydromorphic (Willowbrook Wo): Mollisols (Calciaquolls; Endoaquolls); Alfisols (Endoaqualfs)
Pedocutanic (Bonheim Bo): Mollisols (Argiustolls; Argixerolls); Alfisols (Rhodustalfs; Haplustalfs; Rhodoxeralfs; Haploxeralfs)
Soft-calcic (Steendal Sn): Mollisols (Calcixerolls); Inceptisols (Calciustepts; Haplustepts; Calcixerepts; Haploxerepts)
Hard-calcic (Immerpan Im): Mollisols (Palexerolls); Inceptisols (Calciustepts; Haplustepts; Calcixerepts; Haploxerepts)
Lithocutanic (Mayo My): Haplustepts; Calcixerepts; Haploxerepts)
Lithic (Milkwood Mw): Mollisols (ustolls; xerolls)
Cumulic (Inhoek Ik): Mollisols (ustolls; xerolls); Inceptisols (Humaquepts; Haplustepts; Haploxerepts)

on formation of melanic soils by Van der Merwe et al. (2002b) presents a valuable statistical assessment of land type survey data in South Africa. Their conclusions are largely in agreement with the field experience of South African pedologists (MacVicar et al., 1977; Soil Classification Working Group, 1991) namely, that a semi-arid climate (mean annual rainfall of about 550–800 mm) is the most consistent factor related to distribution; that basic (especially mafic, igneous) rocks are the most suitable parent material (although not exclusively, provided the parent material has significant clay forming potential); and that, although there is no specific association with topographic position, there is a tendency, as is observed with vertic soils (section 2.3), for more frequent occurrence on sites of poorer drainage, especially bottomlands. Strong structure is promoted by both seasonal desiccation and clayey texture (especially when the clay fraction is at least partly smectitic). In some melanic soils the effects of grass roots and soil fauna, especially earthworms, in promoting the formation and stability of peds cannot be discounted. Base saturated humus, finely disseminated on the surface of clay particles, is the source of the dark pigmentation which remains dark when the soil dries out and is much darker than that of acid soils with higher carbon contents. This pigmenting effect remains something of an enigma.

The genesis of the several different forms of melanic soils is best considered by reference to those soil groups with orthic A horizons that are defined in terms of a particular subsurface horizon or material (calcic; duplex; gleyic; lithic; cumulic).

As indicated earlier, previous editions of the classification have accommodated additional soil forms with plinthic and oxidic (red structured) subsurface horizons beneath the melanic A. The former has been discarded although there may be some justification for reconsidering it, while the latter is now accommodated in red families of the duplex (Bonheim) form.

Gleyic (Willowbrook) form

This form consists of a melanic A underlain by a G horizon (see Section 2.11). As such it is typically a bottomland soil with its melanic features gained or maintained via inflow of bases from upslope. If the clay fraction is smectitic this is often a result of neoformation (Section 2.3.4). The G horizon is often a firm or very firm clay as a result of the formation of 2:1 clays *in situ* and/or the weathering of basic rocks. The family distinction depends on the presence or absence of calcareousness (nodular or disseminated; see Section 2.6) in the upper part of

Figure 2.4.4 Landscape of melanic soils: transition from a melanic to a red orthic surface horizon in cultivated maize land near Creighton

the G horizon. This form has pedogenic affinities with the gleyic (Rensburg) form of vertic soils with which it may be confused in the field.

Duplex (Bonheim) form

The Bonheim soil form is underlain by a pedocutanic horizon (see Section 2.7) that is distinguished on the basis of an increase in clay as a result primarily of illuviation and accumulation and visually expressed as cutans. These soils are often found in similar topographic positions as vertic soils but commonly are slightly higher upslope. They may therefore occur as members of red/black toposequences as described in Section 2.3 (Figure 2.4.4).

The families are distinguished on the basis of the B horizon in terms of colour, structure and calcareousness. Some of these distinctions are drawn on the basis of empirical observations in some regions concerning suitability for different kinds of land use and do not necessarily carry a genetic connotation. Nevertheless redder colours signify better drainage while calcareousness points to greater aridity, although topographic and parent material effects will also play a role. The B horizon of many Bonheim soils may have a plasticity index that would qualify it as vertic if it was a topsoil horizon (indeed many pedocutanic B horizons contain slickensides). Vertic properties of the B are currently being incorporated at family level in a revised edition of the classification.

Calcic (Steendal (soft carbonate) and Immerpan (hardpan carbonate)) forms

These soil forms represent the melanic group at the driest end of the climatic spectrum (see Section 2.6). In some cases the calcic condition may be associated with a limestone parent material and less so with aridity (Figure 2.4.5). In such cases it is useful to distinguish between a purely lithic form and one in which there has been a secondary enrichment of carbonate in a B horizon above the weathered limestone.

Figure 2.4.5 Soft-calcic, melanic soil derived from limestone (Steendal form; Rendzina by FAO classification) near Alexandria, Eastern Cape

Lithic (Mayo and Milkwood) forms

In these two soil forms there is either a gradual (via saprolite, with cutans and tongueing) or direct transition to underlying hard rock. Figure 2.4.2 (a,b,d) shows different variants of the Mayo form. Degree of weathering, and the texture and nature of jointing and cleavage in the parent rock, have a marked effect on the properties of the lithocutanic B horizon. These soil forms typically occupy convex crests and steep transportational midslopes where rates of colluvial deposition and horizonation do not exceed loss of material through erosion. Youthfulness (incipient soil formation) is the main indication, as discussed in Section 2.13.

Cumulic (Inhoek) form

Incipient soil formation is also indicated here but the parent material is unconsolidated colluvium or alluvium with little evidence of soil formation beneath the melanic A. It is possible, however, for the material beneath the A to have properties of a neocutanic B. Lower lying concave footslopes and alluvial terraces are the most likely places to find this soil form.

Plinthic (Tambankulu) form

As indicated earlier, this form was recognised and then dropped from local classification. Situations in which a plinthic horizon underlies a melanic A horizon are rare and this is probably because plinthite either signifies a relatively high degree of weathering (the lateritic variety) which is inconsistent with the occurrence of a melanic surface horizon or (more commonly) the presence of a fluctuating water table with lateral migration of iron in groundwater and its accumulation in lower lying parts of the landscape. The latter is more likely to occur in coarser textured, more permeable materials than those which commonly give rise to the formation of melanic soils.

Use

In general melanic soils are fertile but require irrigation to be highly productive. Natural veld on these soils provides sweet grazing and ecosystems dominated by melanic soils are highly productive. Melanic soils are not considered suitable for commercial forestry, mainly because of the semi-arid environment with which they are associated. They are well buffered and can be cropped intensively without needing lime to counteract acidification. In these respects they are similar to vertic soils but have the advantage of not having the same degree of shrink/swell and thus are much easier to manage. Not all the soils are of equal value agriculturally with some being restricted by shallow depth (Milkwood, Immerpan); wetness (Willowbrook); excessive alkalinity and even salinity (Steendal, Immerpan); and slope (Mayo, Milkwood). While none of these constraints need prevent the use of these soils for certain types of agriculture, they should be considered before decisions are taken as to land use viability. The remaining two soils, Bonheim and Inhoek, are likely to have fewer constraints with the former often being one of the most productive soils within its climatic area. If irrigation is possible, arable crops, pastures and horticultural crops can be cultivated. Their landscape position near or in valley bottoms often means that water for irrigation is easily accessible and their relatively freely draining subsoils (except in the 'wet' families of Inhoek form and some of the non-red families of Bonheim form) allow a certain amount of leeway in application of water. The strong structure of the melanic soils is able to withstand repeated cultivation though the low amount of organic matter may become problematic with continuous cultivation.

Although these soils lack the high PI and the extreme movement that characterises vertic soils they may still prove to be somewhat problematic. Materials with a PI >12 and a liquid limit >30 can create a shrink/swell hazard for engineering structures (Soil Classification Working Group, 1991). With melanic soils covering such a wide range of physical properties (e.g. PI is only defined as being <32 or 36) it is certain that many melanic soils will give potential engineering problems. There may thus be a case to be made for further subdividing these soils on the basis of their physical properties, in addition to the grade of structure currently recognised since this a somewhat subjective estimate of shrink/swell potential. The general rule in exploiting melanic soils should be that most of them deserve the epithet 'neovertic', and that due caution should be exercised in relation to their physical behaviour.

2.5 SILICIC SOILS

Introduction

The preceding four soil groups are defined in terms of a characteristic surface horizon. The remaining ten groups all have an orthic A horizon and are distinguished in terms of a key subsurface horizon. Because a geochemical rationale was followed in constructing the key to soil groups with a distinctive subsurface horizon, the silicic soil group precedes the others by virtue of an extreme retention of weathering products within the solum (silica, definitively, but also a variety of other evaporite minerals that would be absent from more leached environments).

Silicic soils have a subsurface horizon cemented by silica. In international classification this is commonly called a duripan (from L. *durus*, hard) or duric horizon (Soil Survey Staff, 1999, 2003; IUSS Working Group WRB, 2006). Locally it is known (Soil Classification Working Group, 1991) as dorbank (from Afrikaans *dor*, dry; *bank*, hard layer). The distribution of these soils (Figure 2.5.1) is associated exclusively with arid landscapes (Figure 2.5.2). The recent doctoral study by Francis (2007) revealed that some dorbank-like horizons, occurring in similar environments, are cemented mainly by the fibrous mineral sepiolite (possibly with accessory amorphous silica) and the name sepiolitic (or petrosepiolitic) was proposed for such horizons. The silicic soil group is therefore best defined as one having a silicate-cemented subsurface horizon. This section will nevertheless focus on the four currently recognised soil forms that have a dorbank horizon. Dorbank is distinct from silcrete, the latter generally being a paleo-feature which forms a resistant capping on very old (Tertiary), weathered remnants of dissected landscapes.

Properties
General

Dorbank is either hard or extremely hard and the silica cementation is such that it does not slake in either water or acid, and dissolution of the siliceous cement requires treatment with a hot, concentrated

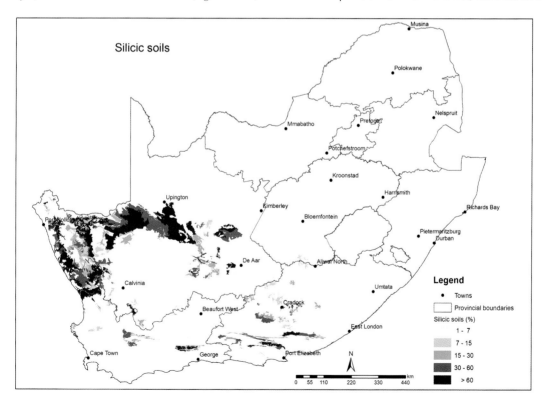

Figure 2.5.1 Silicic soils in South Africa (abundance classes refer to estimated percentages within land types)

Figure 2.5.2 Arid landscape with abundant silicic soils north of Vanrhynsdorp, Namaqualand

alkaline solution. A laminar or platy structure is common (Figure 2.5.3a) but it may also be massive (Figure 2.5.3b). Although it appears to be a paleofeature under certain conditions, dorbank is considered to be an active, currently forming horizon and is thus distinct from silcrete that is generally taken to be a relict, silica-cemented material that formed in an environment different from the present one. Silcrete often is exposed at the surface and caps local high points in the landscape (Figure 2.5.4) due to it being much harder and more resistant than the surrounding country strata. The environment in which dorbank is found is usually a level or gently sloping old alluvial plain or erosion terrace (Figure 2.5.5) and the parent material is typically colluvium or alluvium.

Morphological

Silicic soils are generally medium to coarse textured (sandy clay loam to sand) and are generally well to moderately drained – though this latter character depends on the coherence and depth of the underlying dorbank. Shallow silicic soils with a continuously indurated dorbank may, however, become saturated in low lying positions for a week or longer after heavy rainfall. IUSS Working Group WRB (2006) distinguish between a duric and a petroduric horizon with the former being somewhat less coherent, and less common, than the latter. Although all dorbank horizons are hardened and brittle, they may be either platy or massive (Figure 2.5.3a, b). Typically they have an abrupt upper boundary with the overlying material. The individual plates are between 5 and 15mm thick, while total

Figure 2.5.3 a. platy dorbank horizon near Oudtshoorn, Little Karoo; b. massive dorbank (below hammer) in a rhodic (Garies) form of silicic soil

Figure 2.5.4 Silcrete: a. capping above deeply weathered pallid zone near De Rust, Little Karoo; b. polygonal structure evident in outcrop near Heidelberg, Western Cape

thickness of dorbank may vary from a few centimetres to (rarely) a metre or more. Roots can grow between the plates or become concentrated on top of a coherent dorbank. Rodents are sometimes able to burrow through dorbank along cracks or fissures and their burrows subsequently become filled with material from above (krotovinas). Such pedoturbation improves infiltration of water and promotes root growth. The cementing silica is amorphous or microcrystalline and is often mixed with accessory cements including mainly calcium carbonate and possibly iron oxides (Figure 2.5.6). Iron oxides are responsible for the red colours often seen in dorbank.

Where the dorbank is less coherent, or in the earlier stages of its formation, the material consists of silica-indurated nodules ('durinodes' from Latin *durus*, hard, *nodus*, knot) that are subsequently cemented together by fresh accumulations of silica. This is analogous to the growth of septaria in some hardpan carbonate horizons (Section 2.6). These durinodes (Figure 2.5.7) often show a concentric growth pattern and thus presumably each 'ring' represents a growth period and as such they are analogous to the iron oxide concretions that constitute the morphology of some plinthic materials (Section 2.9). Slight differences in appearance occur between duripans formed in arid climates and those formed in more seasonally differentiated climates (Mediterranean) as a result of the slight expansion and contraction that occurs in the latter forming more prism-shaped blocks within the duripan (Soil Survey Staff, 1999).

In the Oudtshoorn and Trawal soil forms the soil families are distinguished on the basis of (a) bleaching of the A horizon (see box); (b) red or non-red colours in the B horizon; and (c) luvic or non-luvic B horizon (an indication of the amount of clay illuviation that has taken place and thus of effective rainfall; Section 2.7). The latter also distinguishes the two families of the Garies form. In the Knersvlakte

Figure 2.5.5 Exposure of dorbank in eroded footslope colluvium near Garies, Namaqualand

Figure 2.5.6 Close-up of platy (laminar) dorbank with some localised durinodes and pale coloured calcium carbonate lenses between plates

The physical properties of silicic soils are governed by the depth at which dorbank occurs in the profile, its thickness and the thickness of any overlying soil material. Dorbank obstructs the vertical flow of water through the soil so reducing the hydraulic conductivity. The often coarse texture of the overlying horizon(s) means that plant available water contents are also low. The bulk density of dorbank can vary from 1 200 to 2 000 kg m^{-3} (FAO, 2001). Dorbank is either structureless (massive) or more commonly platy (laminated) with hard to extremely hard consistence. Durinodes are indurated once fully formed but in the early stages of formation may only be weakly cemented (Soil Survey Staff, 2003). The erosion susceptibility of silicic soils is low to moderate since they are most common on gentle slopes and generally have sufficient vegetation cover

form, soil families are distinguished on the basis of presence or absence of free lime in the topsoil.

Chemical and physical

The pH (measured in water) of dorbank varies from about 5 to 10, but typical values are between 7.5 and 9.0 (FAO, 2001). Electrical conductivity is usually < 4 dS m^{-1} though higher values are not uncommon. Many silicic soils have high amounts of exchangeable sodium (though they may not be sodic as a result of similarly high amounts of calcium) and low organic carbon and extractable iron (IUSS Working Group WRB, 2006), despite the red colour of many of these soils. Base saturation is generally high as a result of low degree of leaching in these soils. The clay mineralogy of silicic soils in South Africa has been little studied but recently Francis (2007) found complex mixtures including sepiolite, palygorskite, smectite, kaolinite and perhaps mica and kerolite in some silicic soils on the Namaqualand coastal plain.

In the Vredendal-Vanrhynsdorp area dorbank without lime is less saline (average soil paste resistance 215±28 ohms) and contains less hot water-extractable boron (8±2 mg/kg) than dorbank with lime (65±9 ohms; 14±5 mg B/kg) (Provincial Government Western Cape, 2003). Extractable boron in dorbank is always far higher than the toxic limit (>1 mg/kg) for agricultural crops.

Figure 2.5.7 Red sand (aeolian) on coarse durinodes

BLEACHING, CRUSTS AND OTHER FEATURES OF ORTHIC A HORIZONS IN SILICIC SOILS AND ARID ENVIRONMENTS

Although not unique to silicic soils (see, for example, the calcic and duplex soils in Sections 2.6 and 2.7), certain features commonly associated with orthic topsoils in arid environments are of special relevance to understanding the genesis, properties and use of soils in this group. Bleaching of the orthic A is employed as a family criterion (chromic/achromic – Table 2.5.1) in a number of soil forms (Soil Classification Working Group, 1991). The bleached appearance is due to a high proportion of silicate mineral particles, especially quartz, lacking continuous coatings of humus or iron oxides. Removal of such coatings can occur through clay dispersion and eluviation (Section 2.7), podsolisation (Section 2.8) and/or ferrolysis (Section 2.9). If this occurs mainly beneath an orthic A then the result is a diagnostic E horizon. Sometimes bleaching in the orthic A is localised near the surface (Figure 2.5.8a), associated with a vesicular crust (Figure 2.5.8b), or may be discontinuous (Figure 2.5.8c) and equivalent to the sporadic bleach defined in Australian classification (Northcote, 1979). In Figure 2.5.8a the bleached layer is only a few millimetres thick and is capped by a desert pavement consisting of hard, residual fragments too coarse and stable to be removed by wind or water. The pavement performs an important function in reducing raindrop impact and thus dispersion and crust formation. Where crusts do form they sometimes contain a layer of bubbles or vesicles (Figure 2.5.8b) which originate from air entrapment beneath the sealed surface when the soil is saturated with water. In such cases they are referred to as vesicular crusts. In addition to mineral or depositional crusts, biological crusts consisting of lichens (Figure 2.5.8d) are also widely encountered on arid soils. Such localised surface modification of the topsoil has prompted Fey, Mills and Yaalon (2006) to call for greater research focus on this very thin surface layer (pedoderm) which, especially in arid environments, is often well differentiated from the remaining depth of topsoil.

Figure 2.5.8 Features of the soil surface in arid environments: a. thin bleached layer beneath desert pavement; b. vesicular crust with surface bleaching; c. sporadic bleach (discontinuous); d. lichen crust

(grass or short succulent shrubs). However, where erosion does occur, usually due to removal of the vegetation by overgrazing followed by intermittent but heavy rainfall, it rapidly becomes widespread with removal of all horizons above the resistant dorbank.

Classification

In the four silicic soil forms (Soil Classification Working Group, 1991) the dorbank occurs either beneath a B horizon (red apedal – Garies; neocarbonate – Trawal; or neocutanic – Oudtshoorn) or directly beneath an orthic A horizon (Knersvlakte). Subdivision of these forms into families and their nomenclature are given in Table 2.5.1 and international correlation in Table 2.5.2. The proposed new (petro)sepiolitic horizon (Francis, 2007) would entail additions to both local and international classification systems.

Genesis

Silicic soils are formed under somewhat incompletely understood conditions though clearly an environment that encourages both the eluviation of silica from the overlying surface horizons and its accumulation at some depth in the profile is a prerequisite. Since silica is only very slightly soluble at pH < 8.5, a high pH, and probably a long time span, would seem necessary. The length of time is also implied by the fact that these soils are found in areas with very low annual rainfall and thus the accumulation of the silica (and clay) must have taken many thousands (or even millions) of years. Most silicic soils in South Africa have formed in colluvial, alluvial, aeolian or marine terrace materials that have undergone long, uninterrupted periods of weathering under an arid climate. Under these climatic conditions silica mobilised during weathering in the upper part of the soil precipitates as dorbank, the depth of which appears to be related to the texture of the overlying soil; the lower the clay content the deeper the dorbank.

The platy structure of many dorbanks suggests that much of the accumulation takes place through successive laminations associated with downward moving wetting fronts which, as horizon development progresses, would undergo a forced degree of lateral movement across the upper boundary of the dorbank, allowing additional silica enrichment through seepage from upslope and not only from overlying horizons. A similar platy structure is sometimes evident in hardpan carbonate horizons (calcic soils – Section 2.6). The positive relationship between the depth of the dorbank and the permeability of the overlying soil suggests that the dorbank is not a relict feature but is forming under present-day climatic conditions. However, in most of the Garies form soils in the Vredendal-Vanrhynsdorp area the dorbank could well be a paleo-feature since the texture of the overlying red aeolian sand is markedly different from that of the dorbank and underlying soil material, indicating a lithological discontinuity. Other evaporite minerals such as gypsum, calcite, sepiolite and halite are sometimes also in evidence and many dorbank horizons are either calcareous or are closely associated with carbonate-rich horizons (Section 2.6). In South Africa silicic soils are mostly restricted to the western and southern parts of the Cape Province as well as the southern zone of the Karoo, where their generally maritime distribution suggests that, for enrichment with silica, regular atmospheric additions of sodium in rain, fog or dust are required in combination with hydrolysis and intense evaporation. The role of dust would assume greater importance further inland (e.g. in the region of Upington). An exception to the south-westerly distribution pattern has been observed near Steelpoort (Mpumalanga Province; personal communication of D P Turner) and closer examination of arid regions elsewhere in South Africa could well reveal further exceptions. Nevertheless there seems to be a strong correlation of silicic soil distribution with not only aridity but also a winter rainfall pattern (Figure 2.5.1) confirming the common association which is observed between Durisols and Mediterranean climates (IUSS Working Group WRB, 2006).

The biogenic aspect of dorbank formation warrants more attention. A particularly interesting association, in certain climatic zones, of dorbank lenses (often doughnut shaped) with heuweltjies (the broad, Mima-like mounds attributed to termite activity) has been attributed (Ellis, 2002) to hydrological intensification of the arid soil climate by mound topography. It could also reflect to some extent the biogenic accumulation of opaline silica from plants harvested by termites. This biogenic silica is more reactive than

Table 2.5.1 Forms of silicic soils and their constituent families (Nomenclature is shown alongside form names, abbreviations and family codes as defined by the Soil Classification Working Group, 1991). Terms in italics refer to properties of the orthic A.

Rhodic (Garies Gr)

1000 haplic	2000 luvic

Neocutanic (Oudtshoorn Ou)

1110 *chromic, arhodic, haplic*	1120 *chromic, arhodic, luvic*
1210 *chromic, rhodic, haplic*	1220 *chromic, rhodic, luvic*
2110 *achromic, arhodic, haplic*	2120 *achromic, arhodic, luvic*
2210 *achromic, rhodic, haplic*	2220 *achromic, rhodic, luvic*

Neocalcic (Trawal Tr)

1110 *chromic, arhodic, haplic*	1120 *chromic, arhodic, luvic*
1210 *chromic, rhodic, haplic*	1220 *chromic, rhodic, luvic*
2110 *achromic, arhodic, haplic*	2120 *achromic, arhodic, luvic*
2210 *achromic, rhodic, haplic*	2220 *achromic, rhodic, luvic*

Orthic (Knersvlakte Kn)

1000 *acalcic*	2000 *calcic*

Table 2.5.2 Rough placement of silicic forms in international classification systems (IUSS Working Group WRB, 2006; Soil Survey Staff, 2003)

WRB (IUSS Working Group WRB, 2006)

Soil form	WRB Reference Soil Group	Possible prefix qualifiers	Possible suffix qualifiers
Rhodic (Garies Gr)	Durisols	Petric, Fractipetric, Luvic, Haplic	Arenic, Chromic
Neocutanic (Oudtshoorn Ou)	Durisols	Petric, Fractipetric, Luvic, Haplic	Hyperochric, Chromic
Neocalcic (Trawal Tr)	Durisols	Petric, Fractipetric, Calcic, Luvic, Haplic	Chromic
Orthic (Knersvlakte Kn)	Durisols	Petric, Fractipetric, Calcic, Haplic	

USDA Soil Taxonomy

Rhodic (Garies Gr): Aridisols (Typic Argidurids); Alfisols (Durixeralfs); Inceptisols (Durixerepts)
Neocutanic (Oudtshoorn Ou): Typic Petrocambids; Cambidic Haplodurids
Neocalcic (Trawal Tr): Typic Haplodurids; Durixeralfs (Haplic; Typic)
Orthic (Knersvlakte Kn): Typic Haplodurids; Entic Durixerepts; Haplic Durixeralfs

quartz and other primary silicates and may well constitute an intermediate, labile pool from which silica is then remobilised for subsequent formation of dorbank. Thus phytocycling of silica (and sodium too) could play a more important role than direct pathways of enrichment (weathering, atmospheric addition) of raw material for dorbank.

The properties of the orthic A horizon vary widely but most silicic soils have a red or brown, calcareous or non-calcareous orthic A horizon which, as indicated in Section 2.5.2.2, may show signs of bleaching through ferrolysis.

Rhodic form

The red apedal B of the Garies form (Figure 2.5.3b) is usually eutrophic and is often formed in aeolian material that appears to be lithologically different to the dorbank and underlying material.

Neocutanic form

The occurrence of the neocutanic B horizon (Oudtshoorn form), which represents incipient pedogenesis, i.e. in young soils, is somewhat of a contradiction when underlain by dorbank which is formed only after long weathering. As such it is probable that the B horizon material is either unrelated to the underlying dorbank and represents a more recent addition to the profile or it is affected by persistent pedoturbation and/or additions of material that prevents the formation of more strongly defined horizons.

Neocarbonate form

The Trawal form is similar in its formation conditions to the neocutanic soil form except that the B horizon contains sufficient calcium carbonate to qualify as neocarbonate (Soil Classification Working Group, 1991). The youthfulness of this horizon (expressed by the 'neo' prefix) is again suggestive of similar genesis and relationship with the underlying dorbank. The increase in calcium carbonate in this soil compared to the Oudtshoorn form is either a function of initially higher calcium content in the parent material or greater addition of calcium, either from the atmosphere or via lateral leaching from higher in the landscape or from calcareous termite mounds (generally called 'heuweltjies') that are abundant along the west coast, the Breede River valley and the Klein Karoo.

Orthic form

The absence of a B horizon means that this soil (Knersvlakte form) is shallow (Figure 2.5.3a) and is found either in areas of gentle slope where removal of material from the surface by wind and/or water is common or on steeper sections of the landscape. In some variants of this soil form the topsoil has been completely removed by erosion and the dorbank is exposed at the surface. Wind and water erosion also lead naturally to the concentration of stones and gravel at the surface in the form of desert pavement, some spectacular examples of which may be found on the Knersvlakte and in the Tankwa Karoo (Figure 2.5.8a).

Use

Soil Survey Staff (2003) recognise Great Group categories that contain a duripan in six of the soil orders (Alfisols, Andisols, Inceptisols, Mollisols, Spodosols and Vertisols) and a Subgroup category within Aridisols. In addition these soils span climatic regimes from aridic to udic. The land use potential of such a wide range of soils is therefore a function of the presence of the duripan itself combined with the other salient properties of these soils. However, within South Africa the agricultural use of silicic soils is limited by a number of factors – not least of which is the climate (aridic or xeric at best) in which they are found. The limitations of the soils include shallow depth, especially Knersvlakte, excessively high pH in those with sodic (natric) properties and low water holding capacity due to the sandy texture. Depending on the depth and the thickness of the dorbank it is usually mechanically ripped to a depth of 1 m or more (Figure 2.5.9) and incorporated into the solum, to increase usable soil depth. If the dorbank is very hard and breaks into large fragments; the coarser fragments are usually removed from the upper soil layer. If irrigation is available these ripped soils may be successfully cropped, e.g. wine and table grapes along the lower Olifants River; otherwise the undisturbed silicic soils are used for extensive grazing. The dorbank material is often quarried (Figure 2.5.10) and used for road construction. Some ecological implications of silicic soils, and especially those related to restoration of land disturbed by open-cast mining on the west coast, have been discussed by Francis et al. (2007) who argued in particular that revegetation is likely to be hampered if the water retaining properties associated with subsurface cemented horizons such as dorbank are not taken into account.

Figure 2.5.9 Massive dorbank ripped for irrigation development near Klawer, Namaqualand

Figure 2.5.10 Quarry showing dorbank (red material) overlying white pallid zone material in weathered granite near Garies, Namaqualand

ADDENDA: SOME PECULIARITIES OF SILICEOUS DURICRUSTS

Shrinkage polygons

In discussing earlier the morphology of dorbank horizons, comparison was drawn between platy and massive types with little attention having been given to vertical cleavage that may occur in dorbank. In Figure 2.5.11 a polygonal structure is evident similar to that of the silcrete in Figure 2.5.4b. Polygonal cleavage is also widely found in columnar structured B horizons of duplex soils (Figure 2.7.3). A unified theory of how such jointing occurs in a variety of materials as a result of volume changes upon desiccation or solidification has been comprehensively presented by DeGraff and Aydin (1987). Suffice it to say that in soil and sedimentary environments shrinkage through desiccation is probably the common denominator. Because horizontal sections through dorbank horizons are seldom exposed it is not possible to say how commonly the polygonal cracking occurs especially in those dorbanks that also have platy structure.

Sepiocrete

As mentioned at the beginning of this section, a recent study by Francis (2007) has revealed a particular variety of dorbank in which sepiolite (fibrous magnesium silicate) plays a large role in cementing the dorbank. The material in Figure 2.5.12 is barely and only patchily effervescent when tested with acid, and does not respond unequivocally to the qualitative lab test for materials cemented by silica. Alternating, successive, acid and alkali treatments are typically needed to soften the cement. Electron micrographs show interlocking fibres of sepiolite. A methyl orange field test can be used for identifying sepiolitic materials (Francis, 2007). It has been proposed that such material be called sepiocrete when it occurs pedogenically, although it is essentially the same as the meerschaum which is mined for making ornaments.

Figure 2.5.11 Hexagonal cracks demarcating dorbank polygons viewed aerially on a road verge, near Oudtshoorn, Western Cape

Figure 2.5.12 Massive sepiocrete fragment in dorbank horizon near Vanrhynsdorp, Western Cape

2.6 CALCIC SOILS

Introduction

Calcium is another of the dissolved products of rock weathering that, like silica, will remain behind to form a cemented soil through evaporation of water in arid environments. Most commonly it is encountered as calcite, forming either a soft or hardpan carbonate horizon (Soil Classification Working Group, 1991). Less commonly it combines with sulfate to form gypsum. Occasionally magnesium carbonate (magnesite) will also form but rarely in more than accessory amounts because of the strong tendency of magnesium to be sequestered as secondary silicates (e.g. sepiolite, in silicic soils, or smectite, in vertic soils) when concentrated through evaporation.

The distribution of calcic soils is similar to that of silicic soils, only broader (Figure 2.6.1). This wider occurrence reflects the less intense aridity needed for Ca retention within the solum although such a statement is probably only true for the carbonate forms of calcic soils; gypsum accumulation requires more intense aridity similar to that associated with silicic soils. Gypsic (soft and hardpan) horizon definitions suitable for local use are still being considered and the ensuing section will focus on the more widely encountered carbonate forms of calcic soils. The gypsic variant is nevertheless well documented in international classification (IUSS Working Group WRB, 2006). Hardpan carbonate horizons are pedogenic variants of calcrete or secondary limestone. Limestone is all too familiar: as a raw material for cement and lime, in buildings and sculptures, and for making roads and ameliorating acid soils. Even poetic inspiration (W.H. Auden, *In Praise of Limestone*) has been drawn from its impact on landscape. As a pedogenic accumulation it is no less interesting.

Properties
General

The carbonate-rich horizons that characterise calcic soils are a result of the continuing accumulation of

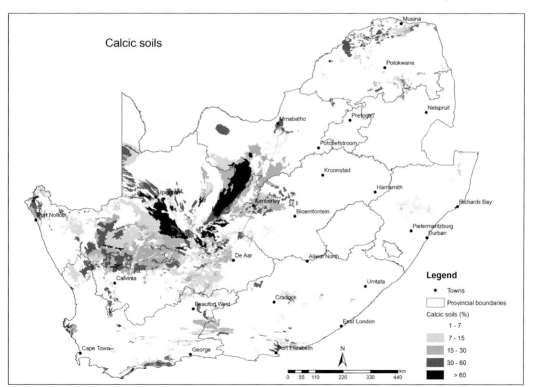

Figure 2.6.1 Calcic soils in South Africa (abundance classes refer to estimated percentages within land types)

calcium carbonate over a long period. Thus one can conceive of a progressive accumulation from neocarbonate through soft to hardpan carbonate over time. In both soft and hard calcic soils the overriding presence of calcium carbonates renders all other characteristics of secondary importance: the colour and morphology of the carbonate horizons result mainly from the constituent carbonates.

Morphological

Calcic soils often have only a thin, pale brown topsoil that has well-developed crumb or granular structure (Figure 2.6.2). Occasionally massive or platy structure is observed which may be enhanced by the presence of higher exchangeable magnesium content (FAO, 2001). Subsoil blocky structure is often more strongly defined in soils that have a clay increase with depth than in those where clay distribution is more uniform. Subsoil colour varies from brown (Figure 2.6.3a) to yellow to red (Figure 2.6.3b), mainly as a function of the origin of the parent material. Although the presence of carbonate is a criterion for the neocarbonate B horizon the level of carbonate accumulation is insufficient to dominate morphology. Some diagnostic B horizons (e.g. lithocutanic, pedocutanic) may also show evidence of carbonate accumulation in the form of whitish mottles (this is used as a family criterion), while other diagnostic B horizons have free carbonate absent by definition (e.g. red and yellow-brown apedal B, podzol B). Only in the soft carbonate soils is the presence or absence of 'signs of wetness' in the carbonate-rich horizon an important criterion for classification in South Africa. The presence of bleached grains in the A horizon is similarly a feature singled out in those soil forms with neocutanic and neocarbonate B horizons but not the other soil forms. Faunal activity is high in calcic soils and infilled animal burrows (krotovinas) and old root channels are common. The orthic A horizon may have properties such as vesicular or biological crusts, bleaching and desert pavement which are common in most arid environments and were

Figure 2.6.2 Hard- and soft-calcic forms of calcic soils: a. Coega form (Cg 1000) on the Bredasdorp coastal plain near Struisbaai and b. Brandvlei form (Br 1000) on old alluvial terrace at Sendelingsdrif near Rosh Pinah in southern Namibia

Figure 2.6.3 Soft-neocutanic and hard-rhodic forms of calcic soils: a. Etosha (Et 1111) in Namibia and b. Plooysburg (Py 1000) near Kimberley, Northern Cape

described under silicic soils in Section 2.5. Some special morphological features of carbonate horizons are described later under genesis.

Chemical and physical

Most calcic soils, as indicated by the pale surface colours, are low in organic matter (perhaps only 0.5 to 1%) as a result of generally sparse vegetation cover and rapid decomposition of organic matter in the arid environments where they occur. Not only the carbonate horizon but generally the whole solum is base-rich with little leaching of plant nutrients. The exchange complex is nearly always fully saturated with bases comprising mostly Ca and Mg. The pH is close to neutral in the topsoil and in the subsoil, with free carbonate, the pH typically approaches 8 to 8.5. Strong buffering and a high capacity for retention of metal cations have not only agricultural implications (deficiency of phosphate and certain trace elements) but also environmental ones (e.g. waste disposal).

Most calcic soils have a medium or fine texture and favourable water holding capacity (FAO, 2001). They are generally well drained with only temporary wetness during the occasional heavy rains that often characterise environments in which calcic soils are found. If the surface soil is silty, crusts may form that restrict or prevent rapid infiltration of rain or irrigation water. Hardpan forms of calcic soils not only have limited hydraulic conductivity but root penetration is also restricted.

The clay mineralogy of calcic soils reflects their formation in an environment of only slow or minimal chemical weathering. Thus 2:1 clay minerals are dominant and include mixed-layer clays, mica, and smectite, together with small amounts of kaolin and accessory minerals that, as in silicic soils, may include evaporite minerals such as sepiolite and halite. Calcite is the commonest Ca mineral in carbonate horizons and gypsum is the main Ca mineral in gypsic variants of calcic soils. In both cases, especially the latter, the particle size of these minerals is coarser than clay and in gypsic horizons the crystals may be macroscopic. A number of features, including micromorphology and mineralogy, of calcic materials have been described in soils of Namaqualand (Francis et al., 2007).

Classification

The calcic soil group has ten soil forms based on whether or not the carbonate horizon is soft or hard and on the nature of associated subsurface horizons (Table 2.6.1). International equivalents are summarised in Table 2.6.2. At least two more forms will be added to this group once definitions have been finalised for soft and hard gypsic horizons. The descriptive terminology for soil families in Table 2.6.1 corresponds to the diagnostic criteria that are currently employed (Soil Classification Working Group, 1991).

Genesis

Many calcic soils are considered extremely old and polygenetic i.e., their formation has taken place over many millennia and thus during geological periods with differing climates. Their development has been intermittent with presumably recurrent droughts slowing down soil forming processes such as chemical weathering and translocation of clay. However, over time calcium carbonate has been leached from the upper solum and has accumulated at depth. The dissolution and precipitation of calcite are governed by the partial pressure of carbon dioxide in the soil air and the concentration of dissolved ions. The dissolution of calcium carbonate is caused by its reaction with carbonic acid that forms by reaction of carbon dioxide with water as follows:

$$H_2O + CO_2 \leftrightarrows H_2CO_3$$

This then reacts with calcium carbonate to form soluble calcium bicarbonate viz:

$$CaCO_3 + H_2CO_3 \leftrightarrows Ca^{2+} + 2HCO_3$$

Precipitation of calcite takes place as a result of lowering of the pressure of CO_2 and/or by a reduction in the amount of soil water that causes an increase in the ionic concentration above that of the solubility product of calcium carbonate. A change in pH may also trigger the precipitation or dissolution of calcite. The partial pressure of CO_2 is highest in the surface horizon where respiration of soil organisms and plant roots raises the CO_2 content to between 10 and 100 times that in the above-ground atmosphere. Frequency of wetting is likewise highest in the surface horizon as a result of incoming rainfall. These factors result in a forcing of the above reactions to the right and the dissolution of calcite.

At depth in the profile, evaporative forces and a decrease in the CO_2 partial pressure, as a result of fewer soil organisms and roots, reverse the reaction and calcite reprecipitates. Such reprecipitation is not always evenly distributed through the soil matrix. Larger channels and pores, produced by roots and faunal activity such as earthworms, act as preferential pathways for air access into the soil. Therefore the partial pressure of CO_2 in these airways is much lower than in the bulk of the soil matrix. When bicarbonate-laden water reaches such an airway it loses CO_2 and calcite precipitates on the channel wall. These accumulations gradually restrict the flow of water causing more calcite to precipitate until the channel becomes completely infilled, forming characteristic pseudomorphs with the shape of the original channel (Figure 2.6.4). These calcium carbonate casts or pipestems (pedotubules) are the arid climate equivalent of those of similar shape formed by iron oxide and even gibbsite in more humid climates (see Figure 2.2.9). They are also commonly found in calcareous coastal sands around roots. Other forms of accumulation are soft or hard nodules, platy (laminar) or continuous layers (Figure 2.6.2), and calcite 'pendants' below stones (FAO, 2001). Some discrete carbonate concretions have concentric and radial crack (void) patterns and a central cavity that presumably formed through shrinkage at a stage when the concretions had only partially hardened (Figure 2.6.5).

Soft and certain hard calcic horizons are considered to be in phase with the present day environmental conditions and depending on the addition of new calcium from outside sources the accumulation of calcium carbonate in the subsoil is a continuous process. In many Askam and Plooysburg form soils with an aeolian sand overburden that is

Figure 2.6.4 Calcium carbonate infilling along a vertical pore, Bonamanzi Game Reserve, KwaZulu-Natal

lithologically different to the underlying hard calcic, the calcic horizon appears to be out of phase with present day conditions and is usually considered a paleo-feature.

In many forms of hardpan carbonate horizons it can be seen that the horizon has developed by the initial formation of small, hard concretions of calcium carbonate that in turn have been cemented together to create larger nodules. These can become cemented together to form large spheroidal concretions called septarian nodules. Such a growth pattern results in a 'boulder' form of hardpan carbonate. Some of the internodule cement may consist of silica that originates from the dissolution of feldspars and ferromagnesian minerals that are susceptible to the prevailing high temperatures and pH conditions (FAO, 2001).

Hard calcic horizons are commonly characterised by thin laminar capping where carbonates normally constitute half, or more, by weight, of the laminar horizon. Sand and gravel have been separated by the crystallisation of carbonates in parts of the laminar horizon and pushed aside at the surface of the capping. The smooth, normally undulating surface of the capping and decrease in carbonate concentration with depth is an indication of capillary rise of bicarbonate containing soil water and precipitation of carbonates at the upper boundary of the capping with the more porous overlying soil material. The resistant nature of such hardpans is illustrated in Figures 2.6.2a and 2.6.6.

The origin of the calcium in calcic soils is believed to be from both internal and external sources. Thus some will originate from the soil parent material that in many cases is calcareous (often colluvial or alluvial sediments or aeolian deposits from base-rich rocks) and some from outside sources such as lateral drainage and atmospheric dust. Birkeland (1984) considered that an external

Figure 2.6.5 Cross section through broken calcium carbonate nodule with radial shrinkage cracks (horizontal bar ~ 20 mm)

Table 2.6.1 Forms of calcic soils and their constituent families. Nomenclature is shown alongside form names, abbreviations and family codes as defined by the Soil Classification Working Group (1991). Terms in italics refer to the orthic A horizon.

Soft-xanthic (Molopo Mp)

1100 haplic, aeromorphic	2100 luvic, aeromorphic
1200 haplic, hydromorphic	2200 luvic, hydromorphic

Hard-xanthic (Askham Ak)

1000 haplic	2000 luvic

Soft-rhodic (Kimberley Ky)

1100 haplic, aeromorphic	2100 luvic, aeromorphic
1200 haplic, hydromorphic	2200 luvic, hydromorphic

Hard-rhodic (Plooysburg Py)

1000 haplic	2000 luvic

Soft-neocutanic (Etosha Et)

1111 *chromic*, arhodic, haplic, aeromorphic	2111 *achromic*, arhodic, haplic, aeromorphic
1112 *chromic*, arhodic, haplic, hydromorphic	2112 *achromic*, arhodic, haplic, hydromorphic
1121 *chromic*, arhodic, luvic, aeromorphic	2121 *achromic*, arhodic, luvic, aeromorphic
1122 *chromic*, arhodic, luvic, hydromorphic	2122 *achromic*, arhodic, luvic, hydromorphic
1211 *chromic*, rhodic, haplic, aeromorphic	2211 *achromic*, rhodic, haplic, aeromorphic
1212 *chromic*, rhodic, haplic, hydromorphic	2212 *achromic*, rhodic, haplic, hydromorphic
1221 *chromic*, rhodic, luvic, aeromorphic	2221 *achromic*, rhodic, luvic, aeromorphic
1222 *chromic*, rhodic, luvic, hydromorphic	2222 *achromic*, rhodic, luvic, hydromorphic

Hard-neocutanic (Gamoep Gm)

1110 *chromic*, arhodic, haplic	2110 *achromic*, arhodic, haplic
1120 *chromic*, arhodic, luvic	2120 *achromic*, arhodic, luvic
1210 *chromic*, rhodic, haplic	2210 *achromic*, rhodic, haplic
1220 *chromic*, rhodic, luvic	2220 *achromic*, rhodic, luvic

Soft-neocalcic (Addo Ad)

1111 *chromic*, arhodic, haplic, aeromorphic	2111 *achromic*, arhodic, haplic, aeromorphic
1112 *chromic*, arhodic, haplic, hydromorphic	2112 *achromic*, arhodic, haplic, hydromorphic
1121 *chromic*, arhodic, luvic, aeromorphic	2121 *achromic*, arhodic, luvic, aeromorphic
1122 *chromic*, arhodic, luvic, hydromorphic	2122 *achromic*, arhodic, luvic, hydromorphic
1211 *chromic*, rhodic, haplic, aeromorphic	2211 *achromic*, rhodic, haplic, aeromorphic
1212 *chromic*, rhodic, haplic, hydromorphic	2212 *achromic*, rhodic, haplic, hydromorphic
1221 *chromic*, rhodic, luvic, aeromorphic	2221 *achromic*, rhodic, luvic, aeromorphic
1222 *chromic*, rhodic, luvic, hydromorphic	2222 *achromic*, rhodic, luvic, hydromorphic

Hard-neocalcic (Prieska Pr)

1110 *chromic, arhodic, haplic*	1210 *chromic, rhodic, haplic*
2110 *achromic, arhodic, haplic*	2210 *achromic, rhodic, haplic*
1120 *chromic, arhodic, luvic*	1220 *chromic, rhodic, luvic*
2120 *achromic, arhodic, luvic*	2220 *achromic, rhodic, luvic*

Soft (Brandvlei Br)

| 1000 *aeromorphic* | 2000 *hydromorphic* |

Hard (Coega Cg)

| 1000 *acalcic* | 2000 *calcic* |

Table 2.6.2 Approximate placement of calcic forms in international classification systems* (IUSS Working Group WRB, 2006; Soil Survey Staff, 2003)

WRB

Soil form	WRB Reference Soil Group	Possible prefix qualifiers	Possible suffix qualifiers
Soft-xanthic (Molopo Mp)	Calcisols	Hypercalcic, Hypocalcic, Endogleyic, Luvic, Haplic	
	Luvisols	Cutanic, Calcic, Haplic	Chromic
	Lixisols	Cutanic, Calcic, Haplic	Chromic
Hard-xanthic (Askham Ak)	Calcisols	Petric, Hypercalcic, Luvic, Haplic	
Soft-rhodic (Kimberley Ky)	Calcisols	Hypercalcic, Hypocalcic, Endogleyic, Luvic, Haplic	Chromic
	Luvisols	Cutanic, Calcic, Haplic	Chromic, Rhodic
	Lixisols	Cutanic, Calcic, Haplic	Chromic, Rhodic
Hard-rhodic (Plooysburg Py)	Calcisols	Petric, Hypercalcic, Luvic, Haplic	Chromic
Soft-neocutanic (Etosha Et)	Calcisols	Hypercalcic, Hypocalcic, Endogleyic, Luvic, Haplic	Chromic
	Luvisols	Cutanic, Stagnic, Calcic, Haplic	Chromic, Rhodic
	Lixisols	Cutanic, Stagnic, Calcic, Haplic	Chromic, Rhodic
Hard-neocutanic (Gamoep Gm)	Calcisols	Petric, Hypercalcic, Luvic, Haplic	Chromic
Soft-neocalcic (Addo Ad)	Calcisols	Hypercalcic, Hypocalcic, Endogleyic, Luvic, Haplip	Chromic
Hard-neocalcic (Prieska Pr)	Calcisols	Petric, Hypercalcic, Hypocalcic, Luvic, Haplic	Chromic
Soft (Brandvlei Br)	Calcisols	Hypercalcic, Hypocalcic, Endogleyic	
Hard (Coega Cg)	Calcisols	Petric, Hypercalcic	

*Calcic soils correlate with the Gypsisol reference soil group if the accumulation is predominantly gypsum and not calcium carbonate.

USDA Soil Taxonomy

Soft-xanthic (Molopo Mp): Aridisols (Calciargids; Haplocalcids); Inceptisols (Calcixerepts – Aquic; Typic)
Hard-xanthic (Askham Ak): Aridisols (Petroargids; Petrocalcids); Inceptisols (Petrocalcic calcixerepts)
Soft-rhodic (Kimberley Ky): Aridisols (Calciargids; Haplocalcids); Inceptisols (Calcixerepts – Aquic; Typic)
Hard-rhodic (Plooysburg Py): Aridisols (Petroargids; Petrocalcids); Inceptisols (Petrocalcic calcixerepts)
Soft-neocutanic (Etosha Et): Aridisols (Calciargids; Aquicambids); Inceptisols (Calcixerepts – Aquic; Typic)
Hard-neocutanic (Gamoep Gm): Aridisols (Petroargids; Petrocalcids; Petrocambids); Inceptisols (Petrocalcic calcixerepts)
Soft-neocalcic (Addo Ad): Aridisols (Calciargids; Haplocalcids; Aquicambids; Haplocambids); Inceptisols (Calcixerepts – Aquic; Typic)
Hard-neocalcic (Prieska Pr): Aridisols (Petroargids; Petrocalcids; Petrocambids); Inceptisols (Petrocalcic calcixerepts)
Soft (Brandvlei Br): Aridisols (Haplocalcids); Inceptisols (Calcixerepts – Aquic; Typic)
Hard (Coega Cg): Aridisols (Petrocalcids); Inceptisols (Petrocalcic calcixerepts)

*These soils correlate to the Gypsids or Gypsic Haploxerepts if the accumulation is predominantly gypsum and not calcium carbonate.

source was essential in the light of the vast quantity of primary rock that would need to be dissolved to yield the volume of carbonates found in many calcic soils. Given the formation conditions needed for the development of these calcic soils, their distribution is strongly governed by an arid or semi-arid climate. They are found most abundantly in South Africa in the dry areas of the Cape, Limpopo and North West Provinces (Figure 2.6.1), but scattered examples may be found throughout the country (e.g. in KwaZulu-Natal near Mhlabatini) where local climate and/or topography is suitable.

Soft and hard calcic horizons are often associated with termite mounds. Besides Ca and CO_2 enrichment from termite activity, the added aridity and altered pattern of soil water accumulation arising from the mound structure are probably also involved, as was described for dorbank lenses in heuweltjies in Section 2.5.

In the Cape west coast region extending northwards into Namibia, and at other isolated locations (e.g. near Kimberley, and in the lower Thukela valley in KwaZulu-Natal) there are limited areas of soils that have the same features as calcic soils except that the accumulating salt is gypsum ($CaSO_4.2H_2O$). The source of sulfate may be either through oxidative weathering of pyrite which is a common mineral in many rock types or, in maritime locations, through inputs of marine aerosols which contain sulfate in comparable concentration to that of magnesium (see discussion of sepiolitic horizons in Section 2.5). Because gypsum is more soluble than calcite the aridity associated with the gypsic forms of calcic soils is likely to be more intense than that required for carbonate horizons. Many soils contain both gypsic and carbonate horizons. In some of these, associated with saline pans and shallow depressions with a permanent water table along the west coast (e.g. at Ysterfontein), the gypsic horizon lies above the carbonate horizon, reflecting upward accumulation in the phreatic zone of a saline water table. In others, where a water table is absent and downward leaching is the main mechanism of accumulation, the (less soluble) carbonate horizon lies above the gypsic horizon. Hardpan (petric or petrogypsic– see Table 2.6.2) forms of gypsic horizons often have interlocking, macroscopic crystals of gypsum (Figure 2.6.7).

Use

As in silicic soils (Section 2.5), calcic properties are recognised in a number of soil orders in Soil

Figure 2.6.6 Outcropping of a resistant hardpan carbonate horizon in a road cut (Hard-rhodic form, Et 1220) from calcareous Tertiary Enon parent material on the coastal plain south of Grahamstown

Taxonomy (Soil Survey Staff, 1999; 2003) including Aridisols, Inceptisols, Mollisols and Vertisols in all moisture regimes. In South Africa the land use of calcic soils is strongly limited by climate in addition to other factors such as shallow effective depth, high pH, high salinity, low plant available P and trace elements (especially Fe), toxic levels of extractable B and stoniness. Certain pome fruit varieties (e.g. peaches), that are very sensitive to Fe deficiency exhibit severe foliar chlorosis, especially on soft calcic soils.

Such properties generally restrict calcic soils to extensive grazing unless irrigation is available when crops may be grown. The soil forms most likely to be productive under such conditions are those with a B horizon overlying a soft carbonate horizon. (Hardpan carbonate is in most instances a barrier to water movement.) The value of such soils then becomes a function of the thickness of the B horizon which determines the effective depth of the soil. On older terraces along the Breede River, and in the Little Karoo with abundant termite mounds containing a hardpan carbonate horizon within the potential rooting depth of grapevines and fruit trees, the hardpan is ripped with a deep tine implement to increase effective depth and improve drainage. Similar ripping is employed for table grapes near Upington (Figure 2.6.8). Grapevines grafted onto a high pH-tolerant rootstock do particularly well on these ripped calcic soils. All calcic soils, however, are highly susceptible to water erosion and so man-

Figure 2.6.7 a. large gypsum crystals in massive petrogypsic horizon at lookout point on D1991, Namibia; b. fragment of petrogypsic horizon at Steenkampskraal near Garies, Namaqualand

Figure 2.6.8 Hardpan carbonate ripped in preparation for extending irrigated table grape vineyards (background) in the lower Orange River valley

agement for whatever use has to be of an extremely high level to prevent exposure of the soil surface, accelerated runoff and soil loss (Figure 2.6.9).

Carbonate horizons have environmental significance in terms of the vulnerability of groundwater to pollution. Contaminated recharge water is likely to be cleansed of metal cations as a result of their adsorption on carbonate mineral surfaces and their lower solubility at high pH.

Calcic soils are mined both as a stable road surfacing material and, in the case of gypsic forms, for building and agricultural use. Near Vanrhynsdorp in Namaqualand the gypsum is first beneficiated by washing and screening. This section would not be complete without referring the reader to the extensive work by Netterberg (1969, 1980, 1985, et al.) on South African calcretes, from both a geological and an engineering perspective.

Figure 2.6.9 Hardpan carbonate horizon exposed through erosion near Noup, Namibia

2.7 DUPLEX SOILS

Introduction
In the previous two groups of soils with an orthic A horizon we encountered marked silica and calcium enrichment in the subsoil as the defining characteristic; in this group the corresponding criterion is enrichment with clay.

Illuvial accumulation of clay is used universally to distinguish B horizons. It occurs to some extent in all soils (Soil Classification Working Group, 1991), constituting the basis for defining luvic families of some forms, being incipiently expressed in lithocutanic and neocutanic horizons, and reaching full expression (thus defining soils in the duplex group) in pedocutanic and prismacutanic horizons. Earlier classification included a gleycutanic horizon (MacVicar et al., 1977); this has subsequently fallen away with diagnostic priority being given to gley morphology. Terms such as cutanic, abruptic and argillic have been suggested as alternatives to duplex.

Duplex soils are most common in the sub-humid and drier parts of South Africa (Figure 2.7.1). In some landscapes they are the dominant group (Figure 2.7.2).

Properties
General
Duplex soils have in common the development of strong structure in the B horizon and a marked increase in clay compared to the overlying horizon from which it is separated by a clear or abrupt boundary. The B horizon is often sufficiently hard and dense to be an impediment to both root growth and water movement and these soils commonly exhibit a high susceptibility to erosion.

Morphological
The marked enrichment with clay in the subsoil results in strong blocky, prismatic or columnar structure and cutanic character (clay skins). The cutans give the peds shiny surfaces that reflect light and often have a different colour to that of the ped

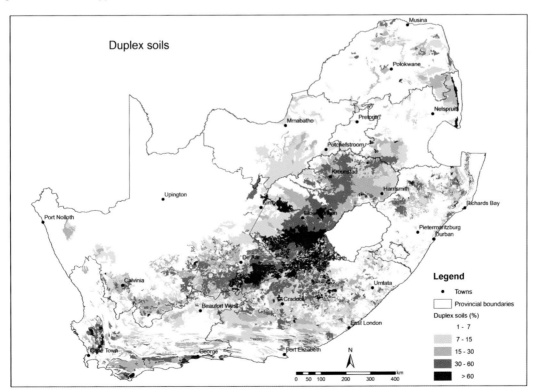

Figure 2.7.1 Duplex soils in South Africa (abundance classes refer to estimated percentages within land types)

Figure 2.7.2 Examples of landscapes dominated by duplex soils: a. the Swartland near Moorreesburg in the lower Berg River Basin, Western Cape; b. duplex soils eroded down to bedrock in tall grass veld near Hlobane, northern KwaZulu-Natal

interiors. Illuvial clay is recognised both in the field and in thin section by the occurrence of cutans (clay skins or argillans (from Fr. *argille* – clay) that line pores and coat peds). The orthic A horizon often has a weak structure and when it contains sufficient fine particles (especially silt and fine sand with some clay) it may become hard or very hard when dry—a feature known as 'hard-setting'. Examples of duplex soils are shown in Figures 2.7.3 and 2.7.4.

Chemical and physical

Few general statements are possible for the group as a whole except that the amount of organic matter is usually low and the textural horizon contrast ensures that permeability is often limited by that of the B horizon (although surface crusting may also impede infiltration), while CEC, pH and reserves of plant nutrients are typically greater in the B horizon than in the orthic A. Base status may vary considerably. High levels of exchangeable Na (and sometimes Mg) are common especially in the prismacutanic forms and in those families of pedocutanic forms that are macropedal (i.e. with coarser, more angular blocky structure). Calcareousness (calcic families) and oxidic character (rhodic families) may modify both physical and chemical characteristics of the B horizon. Salinity may be evident in the more arid duplex soils, especially within or immediately below the B horizon. Acidity and depletion of nutrients are more likely to prevail in achromic families with a bleached orthic A. Retention of anions such as phosphate is likely to be weak. Many duplex soils in South Africa have slickensides in the lower B horizon with a plasticity index that would qualify as vertic material (see Section 2.3; some examples are reported by Snyman et al., 1984).

Classification

Soil forms and families of the duplex group are listed in Table 2.7.1. International correlation is given in Table 2.7.2.

The features which the soil forms and families of the duplex group have in common as well as those which differentiate them warrant some discussion. Firstly it is useful to point out that the distinction between prismacutanic and pedocutanic horizons is conceptually similar to that of the natric and argillic subsurface horizons, respectively, in Soil Taxonomy, with natric being a special type of argillic that is sodic and usually has prismatic structure (Soil Survey Staff, 1999). Although the prismacutanic and pedocutanic horizons are defined differently to the natric and argillic horizons, respectively, the analogy between them is useful. Similarly the boundary between the duplex soil group on the one hand and the cumulic and lithic soil groups on the other is analogous in certain respects to that between alfisols and inceptisols in Soil Taxonomy, although obviously there are additional criteria which tend to blur the comparison. The distinction between prismatic and blocky structure is not always clear-cut (see, for example, Figure 2.7.4).

A second point relates to the form and family criteria (Table 2.7.1). The six forms consist of two which have an E horizon (eluvic) and four that do

Figure 2.7.3 Examples of duplex soils: a. Estcourt form (family Es 1100 – eluvic-prismacutanic, albic, asombric) near Humansdorp, Eastern Cape; b. prismacutanic B displaying exceptionally coarse columnar structure with bleached capping, near Mossel Bay, Western Cape; c. Swartland form (family Sw 1211 — pedocutanic-lithic, *chromic*, arhodic, micropedal, acalcic) overlying weathered Dwyka tillite near Stanger, KwaZulu-Natal; d. bleached orthic A variant of the Swartland form (family Sw 2111 – pedocutanic-lithic, achromic, arhodic, micropedal, acalcic) developed in quartzitic colluvium over weathered shale near Grahamstown, Eastern Cape

not. The two eluvic forms are each split into families with a yellowish (hypoxanthic) E and others with a grey (albic) E. The four non-eluvic forms are each split into families that have a bleached orthic A (*achromic*) and those that do not (*chromic*). Those families that have a bleached orthic A horizon are transitional to some extent in that their bleaching corresponds to eluvic character, but this is insufficient to have given rise to a subsurface E horizon (e.g. Figure 2.7.4).

The increasing recognition of vertic character in the B horizon of many duplex soils has given rise to proposals that the family criteria be revised or expanded to accommodate this important engineering property. One of the advantages of the functional soil nomenclature introduced in this book is that such revisions are relatively easily accommodated by simply making use of additional qualifier terms as suffixes, as is done in the WRB classification (IUSS Working Group WRB, 2006).

Genesis

Duplex character often manifests itself not only in a clear or abrupt increase in clay content with depth but also in contrasting mineralogical composition of the coarser fractions – suggesting a lithological dis-

Table 2.7.1 Forms of duplex soils and their constituent families (Nomenclature is shown alongside form names, abbreviations and family codes as defined by the Soil Classification Working Group, 1991). Terms in italics refer to the orthic A horizon.

Eluvic-prismacutanic (Estcourt Es)

1100 albic, asombric	2100 hypoxanthic, asombric
1200 albic, sombric	2200 hypoxanthic, sombric

Eluvic-pedocutanic (Klapmuts Km)

1110 albic, arhodic, micropedal	2110 hypoxanthic, arhodic, micropedal
1120 albic, arhodic, macropedal	2120 hypoxanthic, arhodic, macropedal
1210 albic, rhodic, micropedal	2210 hypoxanthic, rhodic, micropedal
1220 albic, rhodic, macropedal	2220 hypoxanthic, rhodic, macropedal

Prismacutanic (Sterkspruit Ss)

1100 *chromic*, arhodic	2100 *achromic*, arhodic
1200 *chromic*, rhodic	2200 *achromic*, rhodic

Pedocutanic-cumulic-hydromorphic (Sepane Se)

1110 *chromic*, micropedal, acalcic	2110 *achromic*, micropedal, acalcic
1120 *chromic*, micropedal, calcic	2120 *achromic*, micropedal, calcic
1210 *chromic*, macropedal, acalcic	2210 *achromic*, macropedal, acalcic
1220 *chromic*, macropedal, calcic	2220 *achromic*, macropedal, calcic

Pedocutanic-cumulic-aeromorphic (Valsrivier Va)

1111 *chromic*, arhodic, micropedal, acalcic	2111 *achromic*, arhodic, micropedal, acalcic
1112 *chromic*, arhodic, micropedal, calcic	2112 *achromic*, arhodic, micropedal, calcic
1121 *chromic*, arhodic, macropedal, acalcic	2121 *achromic*, arhodic, macropedal, acalcic
1122 *chromic*, arhodic, macropedal, calcic	2122 *achromic*, arhodic, macropedal, calcic
1211 *chromic*, rhodic, micropedal, acalcic	2211 *achromic*, rhodic, micropedal, acalcic
1212 *chromic*, rhodic, micropedal, calcic	2212 *achromic*, rhodic, micropedal, calcic
1221 *chromic*, rhodic, macropedal, acalcic	2221 *achromic*, rhodic, macropedal, acalcic
1222 *chromic*, rhodic, macropedal, calcic	2222 *achromic*, rhodic, macropedal, calcic

Pedocutanic-lithic (Swartland Sw)

1111 *chromic*, arhodic, micropedal, acalcic	2111 *achromic*, arhodic, micropedal, acalcic
1112 *chromic*, arhodic, micropedal, calcic	2112 *achromic*, arhodic, micropedal, calcic
1121 *chromic*, arhodic, macropedal, acalcic	2121 *achromic*, arhodic, macropedal, acalcic
1122 *chromic*, arhodic, macropedal, calcic	2122 *achromic*, arhodic, macropedal, calcic
1211 *chromic*, rhodic, micropedal, acalcic	2211 *achromic*, rhodic, micropedal, acalcic

(continued)

1212 *chromic*, rhodic, micropedal, calcic	2212 *achromic*, rhodic, micropedal, calcic
1221 *chromic*, rhodic, macropedal, acalcic	2221 *achromic*, rhodic, macropedal, acalcic
1222 *chromic*, rhodic, macropedal, calcic	2222 *achromic*, rhodic, macropedal, calcic

Table 2.7.2 Approximate placement of duplex forms in international classification systems (IUSS Working Group WRB, 2006; Soil Survey Staff, 2003)

WRB (Italicised qualifier terms reflect less certainty)

Soil form	WRB Reference Soil Group	Possible prefix qualifiers	Possible suffix qualifiers
Eluvic-prismacutanic (Estcourt Es)	Solonetz	Vertic, *Gleyic*, Stagnic	Glossalbic, Albic, Abruptic
	Planosols	Solodic, Vertic, Haplic	Albic, Sodic
Eluvic-pedocutanic (Klapmuts Km)	Luvisols	Cutanic, Albic, *Vertic*	*Ruptic*, *Rhodic*, Chromic
	Lixisols (Klapmuts soils with argic CEC <24 cmol$_c$/kg and base saturation >50%)	Cutanic	Albic, Chromic
Prismacutanic (Sterkspruit Ss)	Solonetz	Vertic, Haplic	Abruptic
	Planosol	Vertic, Haplic	Chromic
Pedocutanic-cumulic-hydromorphic (Sepane Se)	Luvisols	Cutanic, Vertic, Gleyic, Calcic, Haplic	
	Lixisols (Argic CEC <24 cmol$_c$/kg and base saturation >50%)	Cutanic, Stagnic, Calcic, Haplic	
Pedocutanic-cu-mulic-aeromorphic (Valsrivier Va)	Luvisols	Cutanic, Vertic, Calcic, Haplic	Chromic
	Lixisols (Argic CEC <24 cmol$_c$/kg and base saturation >50%)	Cutanic, Calcic, Haplic	Rhodic, Chromic
Pedocutanic-lithic (Swartland Sw)	Luvisols	Cutanic, Leptic, Calcic, Haplic	Rhodic, Chromic
	Lixisols (Argic CEC <24 cmol$_c$/kg and base saturation >50%)	Cutanic, Leptic, Calcic, Haplic	Rhodic, Chromic

USDA Soil Taxonomy

Eluvic-prismacutanic (Estcourt Es): Alfisols (Natrixeralfs; Natrustalfs)
Eluvic-pedocutanic (Klapmuts Km): Alfisols (Rhodustalfs; Haplustalfs; Rhodoxeralfs; Haploxeralfs)
Prismacutanic (Sterkspruit Ss): Alfisols (Natrixeralfs; Natrustalfs); Aridisols (Natrargids)
Pedocutanic-cumulic-hydromorphic (Sepane Se): Alfisols (Endoaqualfs – Aeric, Typic; Haplustalfs – Aquic, Typic)
Pedocutanic-cumulic-aeromorphic (Valsrivier Va): Alfisols (Rhodustalfs – Typic; Haplustalfs – Typic); Aridisols (Haplargids)
Pedocutanic-lithic (Swartland Sw): Alfisols (Rhodustalfs – Typic; Haplustalfs – Typic); Aridisols (Haplargids)

Figure 2.7.4 Archetype duplex soil, derived from deeply weathered aeolian sediments on the coastal platform inland from George, Western Cape, with properties of both A and B horizons that are transitional between diagnostic criteria for form and family (i.e. between a bleached orthic A and an E, between prismacutanic and pedocutanic B, and between sombric and asombric properties in the B; see Table 2.7.1).

continuity. In some instances this is rendered more obvious by the presence of a stoneline (see box) at the interface of the A or E with the underlying B, in which the stones may or may not be of some rock type other than that of the weathered rock beneath the solum. Undoubtedly many (some argue even most) duplex soils owe their horizonation at least partially to a binary origin of this kind, in which a colluvial or aeolian layer has been deposited above material which has developed *in situ* through weathering of the rock or sediment that underlies the soil. Nevertheless most duplex soils show clear evidence of cutanic character, both macroscopically, in the form of darker pore in-fillings and coatings on grain and ped surfaces, and microscopically in the form of layered deposits of clay with a high degree of optical birefringence (Figure 2.7.5). This indicates that lessivage (from Fr. *lessivé* – to wash out) has taken place. Lessivage involves the removal of dispersed colloidal substances in the upper part of the solum (eluviation) and their deposition after downward migration in aqueous suspension (illuviation).

Duplex soils develop from a wide range of parent materials under diverse climatic conditions. There are noteworthy exceptions, however: seldom, if ever, are they encountered on basic and ultrabasic igneous rocks with high clay-forming potential, and they are not found where extreme chemical weathering has occurred in warm, humid climates. In the former case the coarse fraction, made up of resistant minerals such as quartz, is insufficient to allow a marked textural contrast to develop. In the latter, oxides of Fe and Al form stable aggregates with kaolinitic clay. Lessivage will consequently not occur to any marked degree. For these reasons the environments in which duplex soils are most expected are those with sedimentary or the more siliceous igneous rocks (including granites), weathering under sub-humid to semi-arid climates. Some of these rocks may also be richer in Na than basic igneous rocks, which further contributes to dispersibility and lessivage.

Factors which help mobilise soil colloids include a high surface charge on the colloid particles themselves, a low electrolyte concentration in the soil solution and high hydration energy relative to valence associated with the exchangeable cation suite. Thus mineralogy and humus content are important, as are both pH (since this can affect surface charge) and the proportion of exchangeable Na and Mg (Emerson and Bakker, 1973) relative to other cations, especially Ca. Adsorbed anions such as silicate, phosphate, bicarbonate and those of natural organic acids may enhance dispersion (Frenkel *et al.* 1992). There must also be sufficient energy to disaggregate soil particles (e.g. by raindrop impact or the viscous drag of soil water flowing rapidly through macropores). Consequently the main cause of dispersive behaviour in a given soil is not readily identified, and even though exchangeable Na is one of the more conspicuous parameters associated with clay dispersion in soils, its importance relative to other factors, especially ionic strength of the soil solution, has probably been exaggerated in the literature on soil genesis.

The accumulation of clay to form a B horizon is just as fascinating. Besides simple physical filtration of suspended colloids there are chemical factors that promote flocculation and enhance the filtration process, especially the concentrations of electrolyte and exchangeable Ca. Wetting and drying cycles, wetting front penetration and lithological discontinuity are also potentially involved. Once clay has begun to accumulate the process becomes self-enhancing. Absorption of water from the clay suspension percolating into dry subsoil induces deposition of oriented clay platelets on pore walls and ped surfaces. Disruption through shrink-swell and bioturbation ensures that only a small proportion of accumulated clay retains the layered orientation exhibited in Figure 2.7.5 and which manifests itself as shiny surfaces when viewed in the field.

Although the mobile colloids involved in duplex soil formation are chiefly phyllosilicate clay minerals, they usually contain sufficient humic substances and/or Fe oxides to impart a characteristically darker colour to the surfaces of peds than that of the bulk soil matrix in both B and A horizons (see sombric and rhodic families in Table 2.7.1). As might be expected, when detailed particle size analysis is conducted a higher proportion of fine clay (< 0.1 μm) is commonly encountered in the B horizon. Migrating solids are not confined to this size threshold, however, and may even exceed that of silt and include very fine sand, since the boundary between plasmic (mobile) and skeletal (immobile) particles is about 100 μm (Mills et al., 2006).

There are two other mechanisms of duplex soil formation that could operate in parallel with the dispersion-flocculation/filtration model described above. The first of these (Paton et al, 1995) is the winnowing effect associated with excavation by mesofauna of finer, more clayey material which is then removed by erosion. Since the fines come mainly from the topsoil, textural contrast develops relative to the subsoil.

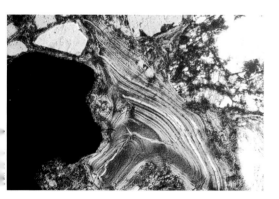

Figure 2.7.5 Light microscope view through a thin section of an argillan showing successive depositional layers of oriented clay deposited in the pore space between quartz grains.

The second process, known as ferrolysis (Brinkman, 1970) involves pH fluctuations associated with redox cycles. Bicarbonate and ferrous iron are produced and partially leached during wet phases while silicate clays are progressively destroyed by the residue of net acidity that is produced by oxidation during dry phases. A bleached, more coarsely textured surface horizon is the result. The process is more intense in the topsoil where organic matter is most available to fuel the reduction process. Once textural contrast (i.e. duplex character) has developed, the process is intensified because saturation is prolonged. Bleaching and, eventually, the full development of an E horizon, including bleached caps on columns of the prismacutanic B, are the most conspicuous evidence that such a process is commonly associated with many duplex soils (see eluvic forms and achromic families in Table 2.7.1). In some cases the E has developed without complete removal of Fe resulting in pale yellow (hypoxanthic) rather than grey colours when moist (albic families – Table 2.7.1). In the eluvic duplex forms and in the achromic (bleached orthic A) families of the other duplex forms, lateral discharge of water perched above the prismacutanic or pedocutanic B can be expected.

An explanation for the development of blocky or prismatic structure in the B horizon of duplex soils is not easily arrived at. Prismatic (or columnar) structure is most commonly encountered when the transition from the overlying orthic A or E horizon is abrupt. Discontinuity in soil cohesion across the A/B transition is presumably involved. Often there is secondary blocky structure in which the prismatic or columnar peds tend to cleave naturally into blocky peds, typically with clay coatings on their exteriors that are less marked than those which occur on the vertical walls of the primary peds. Shrinking and swelling has a predominantly horizontal direction in the prismacutanic B. Coarser, more angular blocky structure in the pedocutanic B (macropedal families, Table 2.7.1) appears more frequently to have an exchangeable cation suite similar to that which is generally found in prismacutanic B horizons (i.e. higher Na or Mg relative to Ca). Slickensides are sometimes found in the lower B horizon attesting to the presence of smectitic clay and a high shrink-swell potential in many duplex soils. The lower B is

STONELINES

A discussion of stonelines could be inserted at any point in this book because they are widely evident in many types of South African soils. In duplex soils, however, they often mark the transition between A or E and B horizons and a more detailed consideration of their origin seems appropriate here.

Because stonelines are often evident in alluvial strata as a result of flood events of different intensity, a similar kind of sorting is sometimes inferred in cases where colluviation (down-slope transport) is evident. According to this view, less mobile coarse fragments are deposited as a deeper lag concentrate covered by more readily transported fine particles.

A more widely accepted explanation of stoneline formation involves the activity of soil fauna (see Chapter 3). Ants, termites, earthworms and other small, burrowing invertebrates (mesofauna) continually bring fine earth to the surface. Gravel and stones, if present, remain behind because they are too large to be moved. Most biological activity is concentrated near the soil surface since this is where most of the plant material (roots and litter) occurs on which the animals feed. Consequently the coarse fragments become concentrated towards the lower limit of faunal activity; the high concentration of these stone fragments also tends to inhibit deeper faunal activity, which has the effect of accentuating the contrast between the fine-earth topsoil and the stoneline at its base. Downward migration of clay associated with the formation of duplex morphology would conceivably also constitute a depth impediment to the activity of mesofauna, since burrowing would be easier in the overlying sand or loam than in the underlying clay. Hence the stoneline would be expected to coincide with the (textural) transition between A (or E) and B horizons. Observation suggests that this is invariably the case. A similar situation would prevail where the subsoil is less workable either because of cementation (e.g. in silicic, calcic or plinthic soils) or due to insufficient rock weathering (in lithic soils).

The activity of mesofauna could possibly also promote duplex soil formation, since clay migration during lessivage is more likely to occur through wider pores and cracks (>20 µm) than through

often also gleyed (especially in the hydromorphic Sepane form but in some of the other forms too) and may contain concretionary calcium carbonate (calcic families, Table 2.7.1) as well as soluble salts. Predominantly red coloured B horizons (rhodic families) may not necessarily reflect more stable structure and better physical properties, since sometimes this colour is inherited from the parent material (e.g. red and purple mudstones in the central Orange River Basin, near Aliwal North).

When viewed from above, the cleavage planes separating peds in the prismacutanic B intersect to form to a polygonal pattern, reminiscent of mud cracks in a dried lake bed. Rounding of the tops of prismatic polygons to form columnar structure is typically found in more mature eluvic forms and represents an advanced stage of clay removal (by dispersion and ferrolysis) resulting in tongues of bleached quartz grains (skeletans) extending downward into the B. It is tempting to think that bleached columnar structure, such as that in Figure 2.7.3b, represents the zenith of duplex formation.

Although duplex soils can be found in all topographic positions, they are most likely to be encountered on concave lower slopes and old river terraces. The pedocutanic-lithic Swartland form (Table 2.7.1) is often associated with soils of the lithic group (Section 2.13) and is more commonly found on convex upper and mid-slope positions.

Use

Two overriding concerns in using duplex soils are erodibility (Figures 2.7.7a) and the impediment presented by the B horizon to water and plant roots (Figure 2.7.7b). Under the right circumstances these soils can be very productive for annual crops (Figure 2.7.7c), but it is usually recommended that intensive use be avoided and that they be managed with great care.

The main cause of erosion is clay dispersion

fine pores in the bulk soil (Nortcliff, 1992). Hence a kind of feedback could be envisaged, with clay migration being accentuated by – and progressively limiting the depth of – mesofaunal activity.

Besides useful clues which stonelines provide about the lithological nature of the upper soil mantle and about the balance between competing biological forces of burial and exhumation of coarse fragments (see Chapter 3 for examples, such as that of elephants uprooting trees), stonelines are a treasure trove for archaeologists since they commonly contain stone tools (Figure 2.7.6) which once must have lain on the soil surface. To the extent that their age is known, these buried artifacts allow us to estimate how long it took for stonelines to form at the base of sorted, fine-earth topsoils. Road cuttings and dongas are ideal places to hunt for such artifacts.

Figure 2.7.6 Examples of stonelines: a. in a duplex soil exposed by erosion, containing bleached Fe concretions, dolerite fragments and a stone-age artifact, formed in mainly shale-derived colluvium; and b. at the transition from the A to the B horizon of a lithic soil developed in gabbro saprolite, both near Kokstad, KwaZulu-Natal.

which is especially prevalent in duplex soils. This gives rise to surface crusting, which reduces the infiltration of rainwater and intensifies surface runoff. Gully erosion can become especially severe in the cumulic forms derived from deep pedisediments on concave footslopes once the main solum is breached and highly unstable subsoil clay is exposed. Slaking and spalling of the subsoil leads to undercutting and eventual collapse of the topsoil. Duplex soils on level topography such as that of river and coastal plain terraces do not carry the same erosion risk and in some regions these are intensively utilised under irrigation or for winter wheat and vegetable production. In the Swartland wheat belt of the Western Cape severe erosion problems on duplex and lithic soils in the early half of the twentieth century were eventually contained by intensive conservation methods such as contour banks. Productivity is now more sustainable (Figure 2.7.7c). Duplex soils in the summer rainfall regions are possibly at greater risk of degradation because of higher rainfall intensity and greater susceptibility of grassland and savanna vegetation to periodic droughts and overgrazing. In the Tugela Basin of KwaZulu-Natal, van der Eyk et al. (1969) rated soils of the Estcourt form as being suitable for no more intensive use than permanent pastures.

Duplex soils often have a shallow effective depth because the B horizon is inhospitable for plant root development (Figure 2.7.7b). Afforestation of these soils is usually avoided not only because of root penetration problems but also because duplex morphology indicates climatic conditions too dry for productive forestry. A wetness hazard is also associated with the eluvic forms and with the achromic (bleached orthic A) families (Table 2.7.1). Ridge and furrow cultivation is widely practised to improve drainage, and subsurface drains (ideally stone-filled French drains) may also be needed. On the other hand rice cultivation under flood irrigation in Swaziland has capitalised on the naturally poor drainage of duplex soils.

Figure 2.7.7 Land use considerations on duplex soils: a. partly rehabilitated badlands on duplex soils in the Mzinhlava River valley near Kokstad; deeply gullied land in the foreground once extended above the fence and was filled in with a bulldozer. It now supports irrigated grass-legume pasture; b. tree root confinement above a pedocutanic B; c. harvesting wheat on the contour in rolling duplex soil terrain near Riebeek West in the Swartland.

The structural instability of clays associated with duplex soils means that earth structures such as contour banks and dam walls will fail unless well protected. A sustainable vegetation cover is sometimes not effective enough, however, because subsurface erosion (piping) may occur. Other engineering-related properties of duplex soils that are conveyed (or will be, when revisions are introduced) by classification at form and family level are the nature of the deeper regolith and whether or not this is consolidated (lithic vs. cumulic forms, Table 2.7.1), excessively wet (hydromorphic families) or expansive (vertic qualifier). Besides showing crusting tendencies many orthic A horizons of duplex soils are hard setting; this can greatly impede tillage when the soil is too dry.

Chemical properties of duplex soils most likely to need attention are sodicity (and sometimes salinity); surface applications of gypsum are widely used for amelioration. The surface horizon of duplex soils usually has a near-neutral pH but some sandier variants are sometimes surprisingly acidic in the bleached topsoil and may require small amounts of lime to rectify this. Trace element deficiencies are not uncommon especially in the achromic and eluvic variants (e.g. Cu in soils of the George district). Such deficiencies are more likely to need correction for shallow rooted cereal and vegetable crops than for deeper rooted perennials. Denitrification of fertiliser nitrate is likely to be more severe in eluvic and achromic forms and split N applications may be judicious depending on crop requirements. Maintenance of organic matter levels, which are naturally low anyway, must always be a priority when these soils are cultivated. In an ecological context, there appears to be some control on vegetation type as exemplified by the association of Mopane veld with duplex (solonetzic) soils in various parts of southern Africa.

In conclusion, the duplex soil group occurs widely in South Africa and presents a variety of serious management challenges to farmers, foresters and engineers. These soils should receive major attention in future pedological and agronomic research. Correlation with soil groups in international classification will enable researchers to build on the extensive work already done on such soils elsewhere.

2.8 PODZOLIC SOILS

Introduction

Podzolic soils are so central to pedology that it is probably better to adhere to convention and simply refer to them as podzols (from Ru. *pod*, under; and *zol*, ash, a reference to the light coloured E horizon). They are the most familiar of all soils because typically they show dramatic horizonation. They are recognised by all soil classifications although the terminology differs. Invariably they are associated with siliceous parent materials having a low clay forming potential. Soluble humic substances in such environments have little to hold them back and are leached out of the regolith, imparting a dark colour to drainage water. A portion of these humic substances may combine with aluminium and iron, however, and the resultant metal humate complex precipitates to form a B horizon. In some cases this is cemented or contains a thin cemented layer known as a placic pan. The podzol B (and placic pan when present) is the defining criterion for podzols.

Although removal of some organic matter from the A horizon and its deposition in the B is the hallmark of podzols, we prefer to recognise it for classification only when it is marked and when other processes such as lessivage and the enrichment of oxides by redox and weathering reactions are relatively subdued. Dark colours in the B horizon of duplex soils (sombric and some arhodic families, Section 2.7) and subsurface carbon enrichment in some plinthic and oxidic soils (Sections 2.9 and 2.10) are indications that podzolisation nevertheless often occurs in parallel with other soil forming processes.

Soils of the podzolic group are most common on sandy parent materials in the higher rainfall parts of the western and south-eastern Cape (Figure 2.8.1). Rare occurrences are also encountered on the coastal plain of northern KwaZulu-Natal and on the Drakensberg escarpment of Mpumalanga Province. A typical podzol landscape is shown in Figure

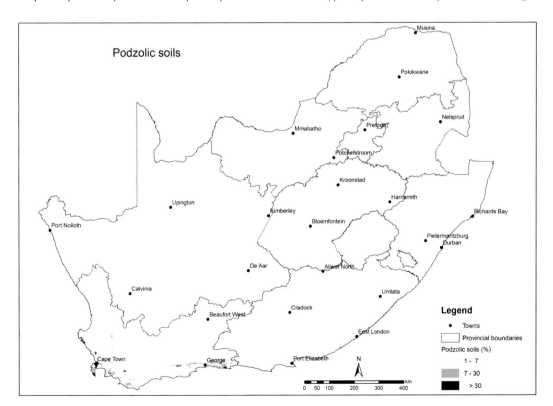

Figure 2.8.1 Podzolic soils in South Africa (abundance classes refer to estimated percentages within land types)

2.8.2 although even here a strongly developed podzol might be elusive, and rewarding to find.

Properties
General
The podzol B horizon is distinguished from other B horizons by an accumulation of illuvial organic matter and sesquioxides but without the textural contrast that defines the duplex group and the luvic families of other soil groups. The podzol B horizon can display a range of morphologies. Bleaching and eluviation to form an E horizon are typically associated with the mineral soil above the B horizon although residual humus can sometimes mask the visual evidence of eluviation in which case chemical analysis is the only sure way to establish podzolic character.

Morphological
Usually the most striking feature of podzolic soils is the colour contrast shown by the soil horizons (Figures 2.8.3 and 2.8.4). The colour of the orthic A ranges from pale grey in podzols with very low clay content and in drier conditions, to dark in higher rainfall areas. It is almost black in many instances. Although it is generally not thick (<300 mm) it may be very thick in more hydromorphic positions. It is underlain either by a bleached, almost white E horizon or, as in Figure 2.8.3f, directly by the podzol B. This latter has a number of variants i.e., a soft, very dark grey to black horizon with little sesquioxic hardening that is often wet; a dark coloured, vesicular ortstein hardpan that can normally be broken by hand; a soft reddish horizon with or without sesquioxic hardening; and a reddish or yellowish horizon with podzolic accumulation in the form of distinct to diffuse, dark coloured mottles and/or tongues (Soil Classification Working Group, 1991). In all cases a defining feature is that the horizon is darker or more chromatic than both the overlying E horizon, if present, and the material immediately below it. Most commonly, the E–B transition in non-hydromorphic podzols is wavy whereas hydromorphic podzols have a smooth transition associated with a fluctuating water table. With combustion the colour of the orthic A changes to nearly white, similar to that of the underlying E.

When moist the horizon has a loose to slightly firm consistence, unless cemented, and is non-plastic. Two soil forms (Table 2.8.1) have one or more diagnostic placic pans associated with the podzol B. This is a thin (usually 2–10 mm thick), wavy, continuous or discontinuous, black to dark reddish brown sheet cemented by organic matter with or without Fe and Mn (Figures 2.8.3c,f and 2.8.4a,b). It has a shiny, conchoidal fracture. Two other forms have families identified on the basis of ortstein hardening (densic) in the podzol B. This represents cementation to a point of brittleness when moist, even though the material may be broken readily by hand (Figure 2.8.4e).

A feature of some podzols, notably those under coniferous forests is a thick build-up of undecomposed organic matter on the surface. This so-called 'mor' collects under aerobic conditions and should not be confused with the organic material of organic soils that is a result of anaerobic conditions (Section 2.1). Many podzols are coarse textured, a feature best expressed in the underlying parent material which is often sandstone or coarse colluvium.

Figure 2.8.2 Siliceous parent materials, high rainfall and fynbos vegetation maximise the opportunity for finding podzolic soils (Hottentots Holland Nature Reserve, Western Cape).

Figure 2.8.3 Examples of podzolic soils in the western and southern Cape: a-c: Albic-cumulic-aeromorphic (Cc) forms, a. in quartzitic colluvium near Witelsbos; b. in old dune on the Knysna coastal plain; c. from sandstone colluvium near Porterville; d. Albic-lithic (Hh) form in sandstone saprolite near Storms River; e. Albic-cumulic-hydromorphic (Lt) form in cover sands on the Cape Flats; and f. Lithic (Gk) form above sandstone saprolite near George.

Figure 2.8.4 Further examples of podzolic soils in the western and southern Cape: a. and b. both examples the placic (Jb) form with an abrupt irregular E–B transition via a placic pan, near George; c. Albic-cumulic-hydromorphic (Lt) form near Oyster Bay; d. bisequal profile (Lt form) with gleyed clay near Humansdorp; e. Albic-cumulic-aeromorphic (Cc) form with firm podzol B near Knysna; and f. undulating E–B transition in Cc form near Grabouw.

The depth of the podzol B can vary enormously from <1 m to many metres. The so-called 'giant' podzols (with E horizons up to 20 m deep) are found predominantly in the tropics where they are formed on deep, alluvial sands. In South Africa the cumulic Fernwood form (Section 1.11) can have a very thick E horizon that may owe its origin at least in part to the podzolisation process (e.g. at Kwambonambi, Kwazulu-Natal where the illuvial iron-rich layer is found about 8 m below the surface). On the Cape Flats in the Western Cape, with deep aeolian sands as parent material, the E is usually thicker than 1 m (Figure 2.8.3e) and several relict, buried podzol B horizons can be found in the deep subsoil.

Chemical and physical

The most salient chemical property of podzols is the accumulation of metal-organic complexes in the podzol B and placic pan. The criteria for classification require the use of acid oxalate and alkaline pyrophosphate solutions to extract, in the laboratory, fractions of Al, Fe and C that best characterise the podzolisation process (Soil Classification Working Group, 1991). Although not part of the defined criteria, a sodium fluoride colorimetric test can be useful in the field as a semi-quantitative indicator of amorphous alumina concentration, which is often higher in the podzol B than in horizons above or beneath it.

Podzols are very acidic soils when judged in terms of pH, with values as low as 3.5 being found in humus-rich surface horizons. The quantity of acidity is generally low, however, since podzols are sandy and consequently weakly buffered. The acidity often moderates with depth so that the B horizon may have a pH of about 5.5 – a value critical to one theory of podzol formation (see below). This is not a uniform attribute, however, and podzolised profiles near Knysna in the Southern Cape have been found with a pH of 3.3 in the B and a pH of 5.1 in the orthic A.

Cation exchange capacity and plant nutrient levels are extremely low as a result of a high degree of leaching, low clay content, predominance in the clay fraction of low activity clays, base-poor vegetation and initially coarse, acid parent material. Co-accumulation of certain trace elements with Al, Fe and C in the B horizon is often found to have occurred.

The physical properties of podzols are a function of their coarse sandy texture and their hydrological characteristics. Their texture means that they are weakly structured and often freely draining, especially in upslope positions. However, even these can show signs of wetness in the upper solum if the illuvial horizon has become relatively impermeable as a result of ortstein hardening or the formation of placic horizons, in which case a perched water table may occur in the E horizon (Figure 2.8.4b). Hydromorphic podzols (groundwater podzols or aquods – Table 2.8.2) are mostly found in lower landscape positions.

Although the bulk mineralogy is nearly always dominated by quartz, the range in pH that occurs through the profile results in a considerable spectrum of clay-size minerals. In the upper solum where weathering is most intense, clays tend to be broken down so that only small amounts of the most resistant minerals, mainly kaolin and quartz, survive, although smectite has also been found in some E horizons. In the B horizon, iron oxides accumulate and if conditions are suitable imogolite and allophane recrystallise. In the only detailed study of the mineralogy of the podzols of Cape Province, Hawker (1986) and Hawker et al. (1992) analysed the silt and clay of 17 samples from three podzols located as a slope sequence (two well drained and one poorly drained member) near George. The silt fraction had a very high quartz:feldspar ratio and contained gibbsite in some of the lower horizons. The quartz content in the clay fraction increased almost linearly with profile development. The kaolinite content was always lowest in the E horizon or placic pan while pedogenic chlorite increased towards the surface in two profiles and decreased in another. Gibbsite was absent from all three A horizons but was common in the saprolite or lowest horizon in two profiles. None of the soils contained smectite or imogolite while goethite and ferrihydrite, but not hematite, were present in selected samples. The goethite in the placic pan contained about 21 mole % aluminium (Hawker and Fitzpatrick, 1989).

Classification

Podzolic soils are known as Spodosols in the USDA Soil Taxonomy (Soil Survey Staff, 1999; 2003) since they have the diagnostic spodic horizon (from Gr.

spodos – wood ash); as Podzols in the WRB (IUSS Working Group WRB, 2006) with a characteristic podzic horizon; while in South Africa they are distinguished on the basis of having a podzol B horizon. The definition is slightly different in each case. A further diagnostic criterion is the presence of a placic pan in some soils of this group. Local classification and international correlation are summarised in Tables 2.8.1 and 2.8.2.

Genesis

The region where most podzols are found in South Africa (Figure 2.8.1) largely coincides with the acid fynbos in which plants of the *Ericaceae, Proteacae* and *Restioacae* make up a dominant component. Although soils that show indications of having been affected to some extent by podzolisation occur elsewhere in South Africa, true podzols are rare and isolated e.g. eastern Mpumalanga; the Drakensberg mountains (on sandstone parent rocks) and on old dune sands in coastal KwaZulu-Natal. In such cases the vegetation is quite different from that of the fynbos but nevertheless contains a significant component of what might be termed acidic vegetation – i.e. species adapted to growing on dystrophic soils, having an inherently low concentration of bases in

Table 2.8.1 Forms of podzolic soils and their constituent families. Nomenclature is shown alongside form names, abbreviations and family codes as defined by the Soil Classification Working Group (1991).

Eluvic-placic (Tsitsikamma Ts)

1000 aeromorphic	2000 hydromorphic

Eluvic-cumulic-hydromorphic (Lamotte Lt)

1100 adensic, friable	1200 adensic, firm
2100 densic, friable	2200 densic, firm

Eluvic-cumulic-aeromorphic (Concordia Cc)

1000 friable	2000 firm

Eluvic-lithic (Houwhoek Hh)

1100 orthosaprolitic, aeromorphic	1200 orthosaprolitic, hydromorphic
2100 hypersaprolitic, aeromorphic	2200 hypersaprolitic, hydromorphic

Placic (Jonkersberg Jb)

1000 aeromorphic	2000 hydromorphic

Cumulic-hydromorphic (Witfontein Wf)

1100 adensic, friable	1200 adensic, firm
2100 densic, friable	2200 densic, firm

Cumulic-aeromorphic (Pinegrove Pg)

1000 friable	2000 firm

Lithic (Groenkop Gk)

1100 orthosaprolitic, aeromorphic	1200 orthosaprolitic, hydromorphic
2100 hypersaprolitic, aeromorphic	2200 hypersaprolitic, hydromorphic

Table 2.8.2 Approximate placement of podzolic forms in international classification systems (IUSS Working Group WRB 2006; Soil Survey Staff, 2003)

WRB (Suffix qualifiers have not been identified)

Soil form	WRB Reference Soil Group	Possible prefix qualifiers
Eluvic-placic (Tsitsikamma Ts)	Podzols	Placic, Albic, Gleyic, Haplic
Eluvic-cumulic-hydromorphic (Lamotte Lt)	Podzols	Ortsteinic, Albic, Gleyic, Haplic
Eluvic-cumulic-aeromorphic (Concordia Cc)	Podzols	Albic
Eluvic-lithic (Houwhoek Hh)	Podzols	Albic, Gleyic, Leptic, Haplic
Placic (Jonkersberg Jb)	Podzols	Placic, Gleyic, Haplic
Cumulic-hydromorphic (Witfontein Wf)	Podzols	Ortsteinic, Gleyic, Haplic
Cumulic-aeromorphic (Pinegrove Pg)	Podzols	
Lithic (Groenkop Gk)	Podzols	Gleyic, Leptic, Haplic

USDA Soil Taxonomy

All forms: Spodosols (mostly Aquods and Orthods)

the biomass and thus producing a high concentration of organic acids upon decomposition. Often such species are rich in phenols and animals find them less palatable; the Miombo woodland of central Africa being a prime example (Figure 2.8.5).

The limited geographic spread of podzols highlights the dual importance of climate (high rainfall) and parent material (siliceous) in their formation. The role of vegetation type is not easily resolved from that of climate or parent material because, as mentioned above, highly leached soils in areas of high rainfall are conducive only to certain types of vegetation. Nevertheless certain plants do promote podzolisation more strongly: the egg-cup podzol at the base of the Kauri tree (*Agathis australis*) in New Zealand is probably the best known example.

The manner in which the podzol B develops is the subject of intense investigation and discussion. Despite a fairly wide consensus about the main processes involved there have been some interesting debates over alternative or additional explanations (Farmer et al., 1980; Farmer, 1981; Buurman and van Reeuwijk, 1984). A good summary of the factors involved in mobilising metals and humic substances in the surface horizon and their subsequent immobilisation at depth to form a B horizon is given in a recent review by Sauer et al. (2007):

'The main theories on mobilisation and transport of organic matter, Fe, Al, and Si are (1) metal-organic migration, (2) metal reduction, and (3) inorganic sol migration. Immobilisation theories include precipitation or polymerisation due to increasing pH/abundance of base cations with depth, mechanical filtering in soil pores, oxidation of metal-organic complexes, biodegradation of the organic part, decreasing C-to-metal ratios during translocation, adsorption to soil particles, and flocculation at the point of zero charge.'

Also of great interest are the catenary sequences that occur in podzolised landscapes. Besides the broad division of podzols into those formed residually from sandstone on mountain slopes (e.g.

Figure 2.8.5 Podzol in Miombo woodland in central Zambia, derived from Kalahari sands under an annual rainfall of more than 1 200 mm: a. the abrupt, wavy transition from the E horizon manifests itself as parallel troughs in the podzol B horizon; b. typical vegetation in the region.

Figures 2.8.3 d and 2.8.3f) and those which have developed in deep colluvial sands on foot-slopes and decalcified aeolian coastal sands, there are further, more subtle distinctions that can be recognised on the basis of topographic position, such as the deep trough formed by the E horizon in the Figure 2.8.4f, which probably reflects a down-slope concentration of water movement, giving rise to a locally enhanced depth of eluviation. Solutes discharged from such sites give rise to hydromorphic podzols such as those in Figures 2.8.3e and 2.8.4c, in which a fluctuating water table, periodically recharged from up-slope positions and replete with soluble metal humates, feeds the formation of a podzol B in the phreatic zone. Careful field studies have quantified such processes and the term lateral podzolisation has been proposed to encapsulate them (Sommer et al., 2001).

Not all of the mobile organic matter is immobilised. Podzolised catchments usually give rise to dark drainage water (Figure 2.8.6). According to Midgley and Schafer (1992) the main feature of catchments that generate clear (as opposed to dark) water within the fynbos region is a preponderance of soils with red and yellow B horizons, suggesting that if there is enough Fe (and Al), from parent material, the humic substances will be filtered out before discharging from the regolith. On the west coast where podzolised coastal sands occur there is no dark drainage water. In general these podzols occur under a drier climate, approaching the lower limit of rainfall under which podzols are found. The deep regolith probably therefore contains sufficient calcium and/or has a high enough pH for confinement of the humic substances. A higher proportion of base flow in the coastal sands than in the mountain podzol zone may also be a factor in reducing the occurrence of dark surface waters.

Ortstein hardening appears to be associated with wetter podzols, suggesting that redox processes are at work similar to those that operate in plinthic soils (Section 2.9). The occurrence of placic

Figure 2.8.6 Water rich in humic acids draining podzolised terrain: a. near Grabouw and b. river mouth near Knysna, Western Cape.

pans (Figure 2.8.4a and b) is less predictable. Their common occurrence at the E–B transition suggests that redox reactions may be associated with their formation, e.g. oxidation of ferrous iron which then co-precipitates with humic substances; or possibly adsorption of ferrous iron on organic surfaces in the upper B horizon during the wet season followed by oxidation during summer, eventually leading to cementation which in turn intensifies the overlying reducing conditions as a result of a perched water table developing during wet phases. The commonly wavy, convoluted appearance of the placic pan may be due to the sandy texture of the matrix (which minimises the probability of preferential flow paths developing through shrinkage) since similar morphology is often exhibited by clay lamellae in sandy soils (see Section 2.12).

Just as organic matter in the B horizon may attenuate and concentrate Fe to form a placic pan, so the existence of an Fe rich material may serve to immobilise and concentrate organic matter. We see evidence of this near Kleinmond in the Western Cape, where a very prominent podzol B horizon has formed in the upper zone of a relict hard plinthic (laterite) horizon. Plinthic soils are common in the Western Cape and many of the plinthic horizons have some podzol character. Conversely there are podzol B horizons (especially those with ortstein hardening) that have much in common, morphologically and mineralogically, with plinthic horizons (Figure 2.8.7). Similarly the black duplex soils in the Langkloof area (which is still within the fynbos

Figure 2.8.7 Vesicular podzol B with ortstein hardening on the shores of the Steenbras Dam near Gordon's Bay. Lake water has removed the softer material to expose the plinthite-like skeleton.

biome) doubtless owe some of their sombric character (dark cutans on the surfaces of prismatic peds) to podzolisation.

According to Sauer et al. (2007), the podzolisation process can respond quite rapidly to vegetation changes, with incipient podzol formation taking 100–500 y and mature podzols forming within 1 000 to 6 000 y. In old (<100 y) pine plantations near Graskop, Mpumalanga and on the coastal plain near George in the Western Cape, a thin layer of distinctly bleached sand grains can often be discerned beneath the mor layer of decomposing pine needles.

Use

Land use on podzols is constrained largely by the climate under which they occur and their sandy texture. Forestry is common (Figure 2.8.8a). Dry summers, however, mean that irrigation is needed for agricultural intensification. Water repellency is common after dry spells. Acidity invariably requires lime but less for neutralisation than might be expected because of the low buffer capacity of the surface horizons. Phosphorus and other nutrients, including trace elements, are often severely deficient. Because nutrient retention is low, fertigation (i.e. the application of nutrients through the irrigation water) is a desirable practice. Land preparation for orchards and vineyards often includes deep cultivation to mix the A and E as well as the B horizon as far as possible (to depths of as much as 1 m). Trace element deficiencies in livestock are also apparent on grazed pastures. Certain specialty crops such as blueberries and proteas do particularly well on podzols.

The unique character of the fynbos biome owes much to the podzolised nature of the soil mantle. The strong emphasis on conserving biodiversity (Figure 2.8.8b) could well see the effect of subtle differences between podzols (in terms of depth, wetness, permeability and nutrient status) on plant species distribution being recognised to an increasing extent. Classification of vineyard terroir (Figure 2.8.8c) might one day highlight podzolic character as one of the factors giving rise to distinctive wine.

Figure 2.8.8 Land use on podzols: a. Conserved fynbos; b. forestry in the Outeniqua foothills near Knysna; c. vineyard on podzolic soil (Lamotte estate, Franschhoek).

2.9 PLINTHIC SOILS

Introduction

Where iron oxides are found segregated and concentrated in soil in the form of mottling and cementation, pedologists call this plinthite (Gk. *plinthos* – brick). The strong pigmenting effect of Fe oxides and their cementation either as hard nodules or as a variegated, vesicular hardpan, make for easy identification provided there is no marked evidence of organic matter having been involved (see podzolic soils in the previous section). The presence of a soft or hard plinthic B horizon defines the plinthic soil group. It is usually indicative of a fluctuating water table and conveys valuable information about seasonal soil water status. The distribution and abundance of plinthic soils across South Africa (Figure 2.9.1) seems to confirm the general observation that they are largely absent from regions of extremely low or high rainfall.

If it is weakly developed, with only mottles and incipient cementation, a soft plinthic B horizon is not easily distinguished from some E or G horizons (Section 2.11) or other materials with signs of wetness. When well developed, plinthite is broadly equivalent to what geologists call laterite (L. *later* – brick; Buchanan, 1807). To South African road builders it is known as *ouklip* or *ngubane* while the term *murram* is used in East Africa. Because of its durability, hard plinthic material is persistent and at some localities is considered to be relict, having formed under different conditions from those prevailing at present. A distinction between modern and paleo-plinthite is not always easy, especially since the connotation of wetness may still apply in view of the restricted permeability of plinthic hardpans. In this section we explore one of the most widely encountered pedogenic materials and discover why name changes – from laterite to plinthite or ferricrete – do not necessarily reduce the diversity of opinions as to how such materials form and the environmental history they reflect.

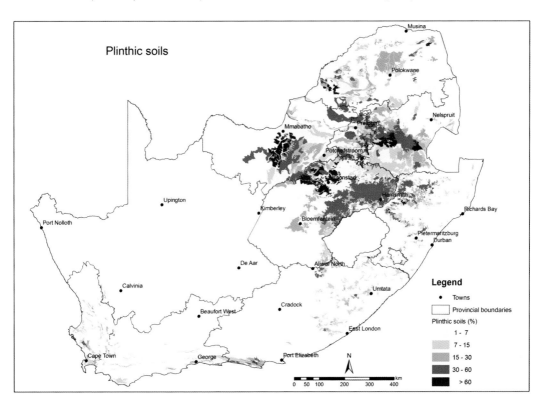

Figure 2.9.1 Plinthic soils in South Africa (abundance classes refer to estimated percentages within land types)

Properties
General
Plinthic soils consist of an orthic A horizon which grades into a soft or hard plinthic B horizon either directly, or indirectly via a red apedal B, yellow-brown apedal B or E horizon. In many respects, therefore, the properties of plinthic soils are very similar to those of either oxidic soils (Section 2.10) and/or gleyic soils (Section 2.11). The distinction between hard and soft plinthic material not only indicates a different degree of pedogenic expression but also implies different practical considerations of land use. The most important of these is that plinthite indicates 'periodic saturation with water, for example between the limits of fluctuation of a water table' (Soil Classification Working Group, 1991). Also important are the degree of hardening and the depth of the plinthic horizon within the soil profile, since both will determine whether the horizon functions as a barrier to plant root growth.

Morphological
The morphological recognition of plinthite is either easy or difficult, depending on which definition and which parts of a particular definition are being applied. Examples of the wide range in morphology are given in Figures 2.9.2 and 2.9.3. Although this is not the place for a detailed critique of plinthite definitions, the opportunity will be taken of pointing out some deficiencies in both local and international classification which surely will be ironed out eventually. The discussion on genesis later in this section will attempt an explanation for the divergent views on what constitutes plinthite.

The Soil Classification Working Group (1991) definition of a soft plinthic B requires 'many (more than 10% by volume of the horizon) distinct reddish brown, yellowish brown and/or black mottles, with or without hardening to form sesquioxidic concretions' and 'grey colours caused by gleying, either in the horizon itself or immediately beneath it'. Even though there may have been some cementation, the material can be cut with a spade when wet. A hard plinthic B horizon is defined as consisting of 'an indurated zone of accumulation of iron and manganese oxides which cannot be cut with a spade, even when wet'.

Contrast these with the criteria (morphological only) that are diagnostic for plinthic and petroplinthic horizons in the WRB classification (IUSS Working Group WRB, 2006):

> The plinthic horizon 'has 25% by volume or more of an iron-rich, humus-poor mixture of kaolinitic clay with quartz and other diluents, which changes irreversibly to a hard mass or to irregular aggregates on exposure to repeated wetting and drying with free access to oxygen'. The petroplinthic horizon 'has cementation by iron to the extent that dry fragments do not slake in water and it cannot be penetrated by roots except along vertical fractures (which have a horizontal spacing of 10 cm or more and which do not occupy more than 20 percent by volume of the layer)'.

The differences in these definitions simply reflect the perspectives of pedologists working in different parts of the world during different periods, going back to the original description of laterite in India by Buchanan in 1807. Suffice it to say here that:

(i) The volumetric percentage of soil occupied by mottles (and the extent to which some portion of them has already hardened) is a minor issue and can easily be agreed upon internationally since the threshold is inevitably an arbitrary one, being designed primarily to exclude expressions of plinthic character that are deemed too weak to be of significance;

(ii) Contrary to what is implied in the South African definition, grey colours in or below the horizon are not an essential attribute of 'active' plinthite. As explained in the section on genesis, iron oxides, especially goethite, may be quite stable under intermittently reducing conditions, particularly when they contain a substantial degree of aluminium substituting for iron in their crystal structure;

(iii) Hardening upon exposure to repeated wetting and drying is not only impossible to determine in the field but is a property largely based on myth, derived from both mineralogical error and circumstantial field observation rather than sound evidence. Although irreversible hardening upon repeated wetting and drying may well have been verified on occasions

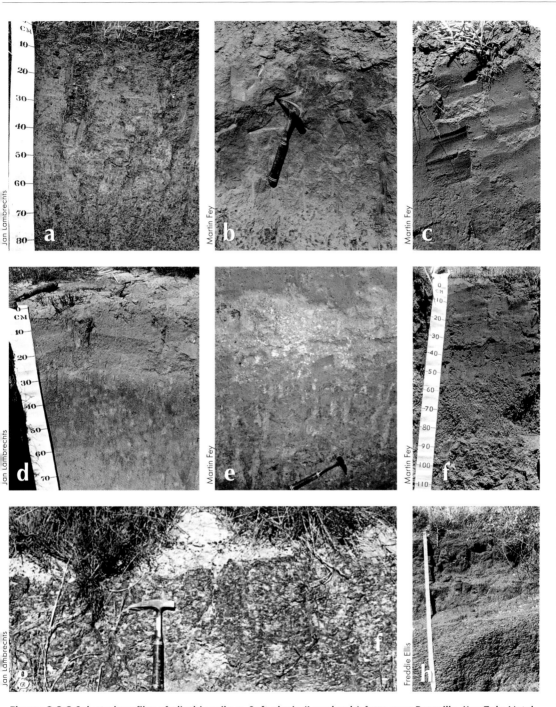

Figure 2.9.2 Selected profiles of plinthic soils: a. Soft-eluvic (Longlands) form near Bergville, KwaZulu-Natal; b. Soft-rhodic (Bainsvlei) form at Elsenburg, Western Cape; c. Soft-xanthic (Avalon) form near Wartburg, KwaZulu-Natal; d. Soft (Westleigh) form in the Koue Bokkeveld, Western Cape; e. Soft-eluvic (Longlands) form near Komga, Eastern Cape; f. Hard-eluvic (Wasbank) form near Wartburg, KwaZulu-Natal; g. Hard (Dresden) form near Bredasdorp, Western Cape; h. Hard-xanthic (Glencoe) form near Naboomspruit/Mookgopong, Limpopo Province.

Figure 2.9.3 Further variants of plinthic morphology: a. Hard rhodic form near Lichtenburg, Northwest Province; b. Hard plinthic B (close-up) at the base of profile in a; c. Hard plinthite at the foot of the Maskamberg near Vanrhynsdorp, Western Cape; d. Manganiferous plinthite in red subsoil near Karkloof, KwaZulu-Natal; e. Soft plinthite with 'zebroid' mottling derived from hornfels in rainforest, Kopinang Basin, Guyana. The red zones are partly cemented and contain hematite, gibbsite and kaolinite while yellow zones contain kaolinitic clay with goethite; f. Hard plinthic horizon near Hermanus, Western Cape.

since the phenomenon was first described by Buchanan, there are simply no reports of this having been done routinely enough to establish it as a property warranting inclusion as a defining characteristic;
(iv) While manganese undoubtedly accumulates along with iron in some plinthic materials (Figure 2.9.3d), its inclusion is not essential; and
(v) The criterion of using a spade to distinguish soft from hard is probably more sensible than inferring penetrability by plant roots.

Excluded from the definition of plinthic material are discrete iron concretions, even when concentrated in stonelines by other processes such as bioturbation. Such accumulations of iron are not considered to contribute to plinthic character. (Concretions and non-hardening mottles are accommodated to some extent, however, in the ferric and paraplinthic qualifiers of the World Reference Base – IUSS Working Group WRB, 2006).

Other difficulties in identifying plinthic horizons occur when they are soft and mottled with relatively few concretions; in such cases a G horizon is usually indicated if there is substantial luvic character i.e. a marked increase in clay content relative to the overlying horizon.

Chemical and physical

The physical properties of plinthic soils are probably in most cases dominated by the plinthic horizon itself, both in terms of water status and, especially when cementation is advanced, root penetration. When the plinthic horizon is shallow the consequence of slow permeability of the plinthic horizon (or the material below it which may have been the initial cause of plinthite development through creating a perched water table) is most likely to be one of excessive wetness during and following periods of heavy rain. On the other hand a deep plinthic horizon may function as a water barrier. Plinthic horizons even when soft do not appear to be well colonised by roots. This could be due to intermittently waterlogged (anoxic) conditions. The extent of such waterlogging is amenable to quantification based on soil colour and morphology. Van Huyssteen et al. (1997) developed an equation relating the duration of free water to the hue, value and chroma of red, yellow and grey (E) soil horizons. Field studies by van Huyssteen (2004) showed that measured wetness, in terms of both duration and intensity, correlates nicely with morphological indicators such as mottling.

Differences in the degree of wetness of plinthic soils are also indicated by the colour of the overlying horizon. Thus soil families in the following sequence would represent a decreasing degree of wetness:

Albic-eluvic (grey E)	Hypoxanthic-eluvic (yellow E)	Xanthic (yellow B)	Rhodic (red B)

Although the chemical characteristics of plinthic soils are somewhat overshadowed by their physical properties in terms of edaphic importance, certain generalisations are worth noting. Plinthic soils are usually non-calcareous. (Some, especially the dystrophic families listed in Table 2.9.1, are strongly acidic and deficient in bases).

The international definition of plinthite implies that the clay fraction is dominated by kaolinite, which would imply a low CEC, but there has been insufficient collation of data to confirm this locally. Similarly it is implied that the Fe in plinthite is largely crystalline in contrast to that of the podzol B and placic pan (Section 2.8) in which a high proportion of Fe is in an amorphous and organically complexed state. The iron minerals in South African plinthic materials have been studied in detail (Fitzpatrick, 1978), and most usually consist of goethite with accessory hematite in redder variants. Concretions are sometimes magnetic, in the form of maghemite. Manganese oxides may sometimes exist alongside those of iron although rarely as the dominant mineral (Figure 2.9.3d).

The redox potential of plinthic soils is probably their most interesting chemical feature and there is enormous scope for investigating soil management implications. Intermittent wetness affects plants directly through an oxygen deficit and indirectly by reducing the availability of nitrogen and sometimes causing manganese toxicity. The organic matter content of plinthic horizons is generally low.

Classification

Soil forms and families of the plinthic group are listed in Table 2.9.1. International correlation (bearing in mind the divergent views discussed earlier on what constitutes plinthite) is given in Table 2.9.2.

Genesis

Plinthic soils are widely distributed in South Africa but appear to be largely absent in the most arid and humid regions (for example they are seldom if ever encountered either in the Karoo or in the highest

Table 2.9.1 Forms of plinthic soils and their constituent families (Nomenclature is shown alongside form names, abbreviations and family codes as defined by the Soil Classification Working Group, 1991). Terms in italics refer to the orthic A horizon.

Soft-eluvic (Longlands Lo)

1000 albic	2000 hypoxanthic

Hard-eluvic (Wasbank Wa)

1000 albic	2000 hypoxanthic

Soft (Westleigh We)

1000 haplic	2000 luvic

Hard (Dresden Dr)

1000 *chromic*	2000 *achromic*

Soft-xanthic (Avalon Av)

1100 dystrophic, haplic	1200 dystrophic, luvic
2100 mesotrophic, haplic	2200 mesotrophic, luvic
3100 eutrophic, haplic	3200 eutrophic, luvic

Hard-xanthic (Glencoe Gc)

1100 dystrophic, haplic	1200 dystrophic, luvic
2100 mesotrophic, haplic	2200 mesotrophic, luvic
3100 eutrophic, haplic	3200 eutrophic, luvic

Soft-rhodic (Bainsvlei Bv)

1100 dystrophic, haplic	1200 dystrophic, luvic
2100 mesotrophic, haplic	2200 mesotrophic, luvic
3100 eutrophic, haplic	3200 eutrophic, luvic

Hard-rhodic (Lichtenburg Li)*

1100 dystrophic, haplic	1200 dystrophic, luvic
2100 mesotrophic, haplic	2200 mesotrophic, luvic
3100 eutrophic, haplic	3200 eutrophic, luvic

*Evidence of wetness in the hard plinthite has been proposed as an additional family criterion.

(**Note:** The hard-rhodic variant has recently been proposed. Previously, hard plinthite was considered relict when overlain by a red apedal B horizon. There is evidence to suggest that this may not always be the case (P le Roux, personal communication).

Table 2.9.2 Rough placement of plinthic forms in international classification systems (IUSS Working Group WRB, 2006; Soil Survey Staff, 2003)

WRB
Note: Red and yellow mottles in RSA soft plinthic horizons do not harden on exposure. According to WRB plinthic horizon definition RSA soft plinthic do not qualify as WRB plinthic horizon.

Soil form	WRB Reference Soil Group	Possible prefix qualifiers	Possible suffix qualifiers
Soft-eluvic (Longlands Lo)	Plinthosols	Pisolithic, Stagnic, Haplic	Albic, Manganiferric, Ferric, Dystric, Eutric
	Stagnosols	Plinthic, Endogleyic	Albic, Manganiferric, Ferric, Dystric, Eutric
	Acrisol	Plinthic, Gleyic, Haplic	Albic, Manganiferric, Ferric
	Arenosol	Albic, Endogleyic, Plinthic, Haplic	Dystric, Eutric
Hard-eluvic (Wasbank Wa)	Plinthosols	Petric, Stagnic, Haplic	Albic, Dystric, Eutric
Soft (Westleigh We)	Plinthosols	Acric, Haplic	Manganiferric, Ferric, Dystric, Eutric
	Stagnosols	Plinthic, Endogleyic	Manganiferric, Ferric, Dystric, Eutric
	Acrisol	Plinthic, Gleyic, Haplic	Manganiferric, Ferric
	Lixisol	Gleyic, Plinthic, Haplic	Manganiferric, Ferric
	Arenosol	Endogleyic, Plinthic, Haplic	Dystric, Eutric
Hard (Dresden Dr)	Plinthosols	Petric	
Soft-xanthic (Avalon Av)	Plinthosols	Pisolithic, Acric	Dystric, Eutric
	Ferralsol	Plinthic, Haplic	Dystric, Eutric, Xanthic
	Acrisol	Plinthic, Gleyic, Haplic	Manganiferric, Ferric, Chromic
	Lixisol	Gleyic, Plinthic, Haplic	Manganiferric, Ferric, Chromic
	Arenosol	Endogleyic, Plinthic, Haplic	Dystric, Eutric
Hard-xanthic (Glencoe Gc)	Plinthosols	Petric, Acric, Haplic	Dystric, Eutric
	Acrisol	Petroplinthic, Gleyic, Haplic	Manganiferric, Ferric, Chromic
	Lixisol	Gleyic, Petroplinthic, Haplic	Manganiferric, Ferric, Chromic
	Arenosol	Endogleyic, Petroplinthic, Haplic	Dystric, Eutric
Soft-rhodic (Bainsvlei Bv)	Plinthosols	Pisolithic, Acric, Haplic	Dystric, Eutric
	Ferralsol	Plinthic, Haplic	Dystric, Eutric, Xanthic
	Acrisol	Plinthic, Gleyic, Haplic	Manganiferric, Ferric, Rhodic, Chromic
	Lixisol	Gleyic, Plinthic, Haplic	Manganiferric, Ferric, Rhodic, Chromic
	Arenosol	Endogleyic, Plinthic, Haplic	Dystric, Eutric
Hard-rhodic (Lichtenburg Li)	Plintosols	Petric, Acric, Haplic	Dystric, Eutric
	Acrisol	Petroplinthic, Gleyic, Haplic	Manganiferric, Ferric, Rhodic, Chromic
	Lixisol	Gleyic, Petroplinthic, Haplic	Manganiferric, Ferric, Rhodic, Chromic
	Arenosol	Endogleyic, Petroplinthic, Haplic	Dystric, Eutric

> **USDA Soil Taxonomy**
>
> **Soft-eluvic** (Longlands Lo): Ultisols (Typic Plinthaquults); Alfisols (Typic Plinthaqualfs)
> **Hard-eluvic** (Wasbank Wa): Ultisols (Typic Plinthaquults); Alfisols (Typic Plinthaqualfs)
> **Soft** (Westleigh We): Ultisols (Typic Plinthaquults); Alfisols (Typic Plinthaqualfs)
> **Hard** (Dresden Dr): Entisols (Udorthents – Lithic; Typic; Ustorthents – Lithic; Typic)
> **Soft-xanthic** (Avalon Av): Oxisols (Udox; Ustox); Ultisols (Udults; Ustults); Alfisols (Udalfs; Ustalfs) – plinthic variants of all of these
> **Hard-xanthic** (Glencoe Gc): Oxisols (Udox; Ustox); Ultisols (Udults; Ustults); Alfisols (Udalfs; Ustalfs) – plinthic variants of all of these
> **Soft-rhodic** (Bainsvlei Bv): Oxisols (Udox; Ustox); Ultisols (Udults; Ustults); Alfisols (Udalfs; Ustalfs) – plinthic variants of all of these
> **Hard-rhodic** (Lichtenburg Li): Oxisols (Udox; Ustox); Ultisols (Udults; Ustults); Alfisols (Udalfs; Ustalfs) – plinthic variants of all of these

rainfall regions of KwaZulu-Natal and Mpumalanga). This climatic relationship has recently been investigated by le Roux (2002). Association with a sub-humid as opposed to a humid climate suggests that a pronounced dry season is a prerequisite for plinthite formation, coupled with a sufficiently wet season to induce saturation with water and mobilisation of Fe in the reduced form. The association of plinthite with soils of the humic group is negligible for this reason and exceptions are widely considered (Maud, 1965; King 1972) to represent relict laterite associated with old land surfaces. This may seem anomalous given the extensive occurrence of plinthite in supposedly modern soils of tropical rainforests.

Broadly speaking there are two different kinds of plinthite forming processes:

(i) Residual iron enrichment, predominantly through weathering and removal of silica and bases, in which case resistant oxides of aluminium, titanium and zirconium will become co-enriched with Fe, giving rise to a range of materials referred to generally as laterite or, if Al enrichment is strong enough, bauxite. In this case the term plinthite is appropriately confined to those materials in which the iron has become segregated into mottles and cement (vesicular or pisolitic) through the action of a fluctuating water table, either contemporaneously or subsequently. Typically this kind of plinthite is associated with areas of relatively impeded drainage on old land surfaces:

(ii) Imported iron enrichment, in groundwater after being mobilised by reduction and then concentrated by oxidation within that portion of the intermittently aerated zone that most frequently hosts an influx of dissolved ferrous iron (Figure 2.9.4). In this case the only other element which may exhibit significant co-accumulation with Fe is manganese, depending on its concentration in the parent material from which the Fe was derived.

Most plinthic soils in South Africa have formed by the second process. The plinthite (old laterite) associated with humic soils (Section 2.2; see also Figure 2.9.4) is probably a result of the first. The distinction may be useful in resolving some of the discrepancies between international and local definitions of plinthite mentioned above. Plinthite formed by the second process is usually conspicuous for its association with positions in the landscape which afford reception of Fe-rich water through seepage from surrounding areas i.e. gentle, usually concave, lower slopes.

Although plinthite forms in the regolith derived from most rock types, there seems to be, in any one region, a much stronger tendency for it to form on sedimentary than on igneous (especially basic) rocks. This could be related to a shallower depth of weathering and therefore lower permeability of shale and sandstone strata, with concomitantly greater lateral discharge of groundwater into basins.

Redox transformations of Fe, Mn and other soil constituents are intriguing. Some of the phenomena associated with the genesis of plinthic and other hydromorphic soils can be summarised as follows

Figure 2.9.4 Examples of paleo-plinthite (laterite), both of which are on the early Tertiary (African) erosion surface and derived from Table Mountain sandstone: a. beneath a humic A horizon, near Fawnleas, KwaZulu-Natal and b. on the Gifberg plateau near Vanrhynsdorp, Western Cape.

(further details can be found in Blume, 1967; Fey, 1982 and Schwertmann, 1991):

- Ferrous iron moves both passively, in groundwater, and actively, by diffusion, the latter in response to redox gradients created by the oxidative poise of metal oxide foci relative to an intermittently reduced matrix. Over time, Fe (and Mn) thus assumes increasingly localised concentrations as nodules or in a vesicular or reticulate pattern in environments of fluctuating redox potential.
- Iron concretions within a red or yellow matrix are not necessarily relict and transported but may well have formed *in situ* and still be undergoing Fe accumulation.
- Specialised bacteria increase the rate of both oxidation and reduction.
- Yellow goethite is generally more resistant to reductive dissolution than red hematite, because of its commonly higher content of substituent Al (Fey, 1982; Schwertmann, 1991). This helps to explain chromatic segregation within mottled zones, between horizons in soil profiles and across catenas (see box).
- Organic matter fuels the reduction process. The role of waterlogging is merely to allow bacterial reduction to proceed faster by retarding the diffusion of oxygen. Topsoils are therefore more intensely reduced than yellow or red subsoils during spells of wet weather, despite the fact that the subsoil can be saturated with water for longer periods.
- Much of the Fe in a body of plinthite may come from the topsoil of adjacent, more elevated terrain and not only from Fe dissolution in groundwater.
- Reduction fueled by organic matter consumes organic carbon and volatilises nitrogen, the loss of which further inhibits humus accumulation. Hence topsoils in plinthic landscapes are often a light grey colour since the conditions favour destruction of both Fe oxides and humus.
- Eh (Redox potential) and pH are complementary. This means that for any given level of aeration, acid conditions are inherently more reducing than alkaline ones. Conversely a higher pH is more conducive to oxidation. Horizontal and vertical pH gradients will therefore also affect Fe and Mn segregation.
- As discussed in Section 2.11, the process of ferrolysis (Brinkman, 1970) helps to explain why bleached zones (and especially E horizons) are often more acidic and depleted of clay.
- Many yellow-brown apedal B horizons are

THE RED-YELLOW-GREY PLINTHIC CATENA

Since Milne (1935) first introduced it to describe the toposequence of soils across a valley as though it were a chain suspended between interfluvial crests, the term 'catena' has firmly established itself in the language of pedology. The word is used in South Africa to describe a widely observed toposequence consisting of red soils on well drained crests, grading via yellow soils on mid-slopes to grey soils in poorly drained bottomlands. More often than not a plinthic horizon is found in profiles of the yellow and grey members. Sometimes plinthite occurs throughout the sequence.

It is tempting to draw a comparison between the catenary sequence in Figure 2.9.5a and that which carries the same distinctive colour transition in Figure 2.9.5b (over a distance of centimeters rather than kilometers). However, although in both cases the segregation of Fe results from a variation in redox potential such that Fe oxides are allowed to become concentrated under well oxidised conditions and are stripped out by reduction under poorly aerated conditions (with aluminous goethite persisting for longer than hematite; see Section 2.2), there are also important differences. In the former case, leaching of ferrous iron in solution under gravity is the main mode of transfer, which is lateral and downwards. In the latter, diffusion of ferrous iron over a concentration gradient is the mechanism by which Fe is removed from less oxidised zones and deposited when it encounters better oxidation. Gradients not only in water and oxygen but also in acidity and organic matter content (the latter fuels microbial catalysis of Fe reactions) provide the spatial framework for iron segregation, both within peds or pedons and across drainage basins. For plinthite to develop in such situations, temporal fluctuation (cyclic wetting and drying) seems to be the *sine qua non*.

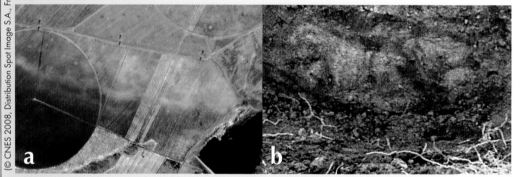

Figure 2.9.5 A matter of scale: a. satellite image of cultivated lands on the Mpumalanga Highveld revealing the red-yellow-grey plinthic catena, straddled by irrigation centre-pivots and power lines; b. an analogous 'micro-catena' associated with iron segregations in plinthite weathered from sandstone saprolite (enlargement from lower left of Figure 2.9.4a; the scale is provided by grass roots). The mineralogical source of pigmentation and the redox processes which give rise to these patterns at such different scales are essentially the same.

simply weaker (or more resilient) expressions of E horizon character.

- Mn behaves quite similarly to Fe except that it is more readily reduced, more mobile under acid conditions and is present at a much lower concentration than Fe in most parent materials. Some plinthic materials are rich in Mn (Figure 2.9.3d).

Use

Except where the soft or hard plinthic B horizon occurs at sufficient depth beneath a red or yellow-brown apedal B (rhodic and xanthic forms – Table 2.9.1), poor drainage will render plinthic soils only marginal at best for the production of most crops besides vegetables and pastures. The remaining soils will present problems with rooting depth and periodic waterlogging that are proportional to the depth of the plinthic horizon and the degree of hardening which it has undergone. Artificial drainage can mitigate the wetness hazard. Ripping has been employed to break up hard plinthic horizons and improve rooting depth. Soils such as Avalon and Bainsvlei are prized by maize farmers on the Highveld because the upper solum drains freely while the plinthic horizon dams water within the lower part of the profile which can be tapped by maize roots during dry spells (Figure 2.9.6). Similar considerations apply to soils of the oxidic group that show signs of wetness beneath the B horizon (Section 2.10). Plinthic soils in which the orthic A grades directly to a plinthic B or to an E horizon can be expected to be wetter on average than those with a red or yellow-brown apedal B horizon.

Apart from the soil water status associated with the plinthic B horizon, the properties of xanthic and rhodic forms of plinthic soils are similar, from a land use point of view, to those of their counterparts in the oxidic soil group (Section 2.10). For example soil acidity (including subsoil acidity in dystrophic families of xanthic and rhodic forms – Table 2.9.1) can be a problem on some plinthic soils and lime application is often needed. Gypsum has been used – when applied at the surface and allowed to leach into the subsoil of a Plinthic Paleudult (Avalon form) – to alleviate Al toxicity and improve the rooting depth of crops such as maize (Farina et al., 2000; Figure 2.9.6).

The surface horizon of plinthic soils is typically poorly structured – or at least the structure becomes degraded quite easily through cultivation. Low organic matter content and a lack of iron oxides (especially in the eluvic forms) can lead to poor water infiltration and hard setting problems (especially in the more loamy or clayey soils) which call for judicious tillage and careful management of crop residues.

Because of the variation in wetness tolerance of different deciduous fruit types in the winter rainfall region, pears (very tolerant) are successfully grown on most drained plinthic soils (even the eluvic, albic Wasbank and Longlands forms), while apples (less tolerant) are planted on drained xanthic and rhodic plinthic forms with yellow-brown and red apedal B horizons. Peaches and nectarines that are very sensitive to wetness are only planted on plinthic soils

Figure 2.9.6 Maize roots extending into the acid subsoil of a soft-xanthic plinthic soil (Avalon form) ameliorated with gypsum at Geluksburg, northern KwaZulu-Natal. A water table coincides with the plinthic B horizon during summer and represents a reservoir for the crop provided deep rooting is possible.

when the depth to the plinthic B, beneath a red or yellow-brown B, is > 1.0 m.

In many parts of the summer rainfall region, the transition from an area of high, dependable and well distributed rainfall (e.g. the midlands mist-belt in KwaZulu-Natal) to one of lower, less dependable and more erratically distributed rainfall considered marginal for dryland cropping (e.g. the tall grassveld) is typically marked by the appearance of plinthic soils in the landscape. This can be an important consideration in valuing agricultural land where rainfall records are inadequate. That plinthic soil is not regarded as being a good choice for forest plantations may be due as much to this climatic association as it is to considerations of poor internal drainage.

Plinthic material is used as road metal (Netterberg, 1985). In the Cape peninsula, *ouklip* that was originally abundant (its use for road making was noted by Charles Darwin) has long ago been dug up for this purpose. The advantage of using it is the combination of ease of excavation and durability once it is broken into gravel. Transport networks over much of Africa will no doubt continue to be stabilised with this valuable material.

> ## In Praise of Laterite
>
> *If it form the one landscape that we, the inconstant ones,*
> *Are consistently homesick for, this is chiefly*
> *Because it resists dissolution. Mark these flattened hilltops*
> *With their surface fragrance of* **Artemesia**, *their autumn tinge of* **rooigras**
> *And, beneath, a vesicular carapace of red and brown cement;*
> *Hear the springs that spurt out everywhere with a chuckle,*
> *Bearing their secret load of silica and bases to feed the oceans of the world*
> *And leave behind their* **ouklip** *residue that paves the African surface*
> *And protects the underground waterways, letting them open intermittently*
> *To host the tree fern and blue swallow. Examine this region*
> *Of long distances and definite places: marvel at the equilibration*
> *Of oxygen, water, igneous rock and microbes, a metamorphosis*
> *Of raw earth to ripe* **mgubane**, *driven through countless millennia*
> *By warmth and sunlight.*
>
> (Martin Fey, with apologies to W H Auden, In Praise of Limestone)

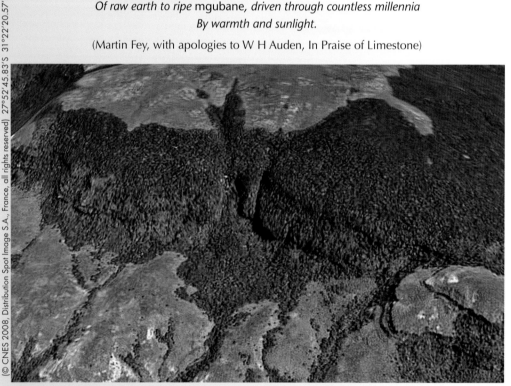

Figure 2.9.7 Laterite capped, grassland plateau above the Ngome Forest in northern Kwazulu-Natal

2.10 OXIDIC SOILS

Introduction

Oxidic soils have a B horizon that is uniformly coloured with red and/or yellow oxides of iron and, as distinct from humic soils (Section 2.2), have an orthic A horizon. Soils with a dorbank or carbonate horizon beneath a red apedal B belong in the silicic and calcic groups, respectively (Sections 2.5 and 2.6). Signs of wetness are permitted at the base of the solum provided these do not constitute a diagnostic plinthic horizon (Section 2.9). A wide range in degree of weathering is possible and these soils therefore exhibit a broad geographic distribution (Figure 2.10.1), which accounts for the marvellous redness of the landscape that confronts travellers through much of South Africa. (The most popular of all SA soils is the Hutton form e.g. Figure 2.10.2). The concept underlying the group is one of relative maturity (although not all members would conform to that description) coupled with free drainage and aeration in the upper solum. These soils should not be confused with oxisols which represent a narrower class of extremely weathered soils within the oxidic, plinthic or humic soil groups.

Properties

General

Uniformity of B horizon colour is an overriding feature of oxidic soils. Cutans and other signs of colour variegation are permitted provided they are of the same hue as the matrix with minimal contrast in value and chroma. Clay enrichment and structural development relative to the overlying orthic A horizon are also permitted provided the transition is gradual. Mottling in the form of red and yellow segregations of Fe or black segregations of Mn, commonly with grey or paler yellow matrix colours, is not permitted within the horizon but may occur in the material beneath it.

Morphological

Some examples of profile morphology are shown

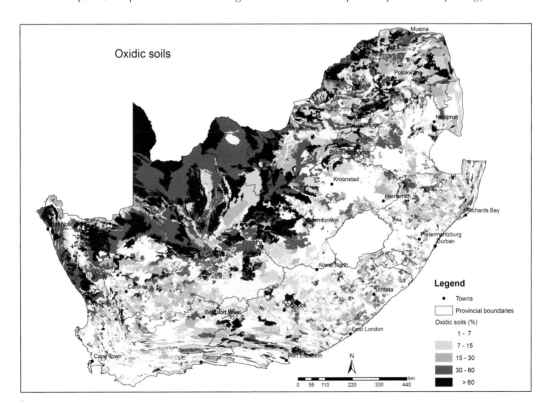

Figure 2.10.1 Oxidic soils in South Africa (abundance classes refer to estimated percentages within land types)

Figure 2.10.2 Thick, colluvial, rhodic, oxidic regolith derived from dolerite on the Singisi cuttings (N2 route) near Kokstad, KwaZulu-Natal

in Figures 2.10.2 (apedal variants), 2.10.3 (hydromorphic variants) and 2.10.4 (red structured or pedorhodic form). The range of red and yellow-brown colours is strictly defined (Soil Classification Working Group, 1991). Apedal B horizons must have structure that is weaker than moderate blocky or prismatic in the moist state (see box under Genesis). In borderline cases the CEC is considered (see below).

Chemical and physical

The apedal soils in this group are characterised by a relatively low CEC (< 11 $cmol_c$ kg^{-1} clay determined by the pH 7 ammonium acetate method) reflecting oxidic mineralogy in association with a predominantly kaolinitic clay mineral assemblage (sometimes referred to as kandic or low activity clays). This criterion is only invoked in borderline cases of structure development, however, which means that lighter textured variants may still contain a significant component of 2:1 phyllosilicates (e.g. mica or illite) in the clay fraction. The degree of leaching varies across the full spectrum from dystrophic (high acid saturation) to eutrophic (fully base saturated) or even calcareous, provided the latter manifests itself only in the form of discrete nodules in a non-calcareous matrix. By contrast, humic soils with apedal B horizons (Section 2.2) tend to be restricted to areas of high rainfall and therefore are seldom fully saturated with bases and are generally confined to a narrower range of strong weathering. Oxidic soils have an orthic A horizon which means that, if the topsoil has the degree of leaching required for a humic horizon, the organic carbon content is always less than 1.8%.

The red structured B horizon (pedorhodic Shortlands form) is usually more clayey and/or contains more smectitic (high activity) clay than its apedal counterparts, although kaolinite dominance of the clay fraction is still common. Consequently CEC, water retention and shrink-swell potential are better developed, with extreme variants being difficult to distinguish from red members of the vertic soil group (Section 2.3).

Because of the strong pigmenting effect of iron oxides (especially hematite) the Fe content of oxidic soils may be as little as one percent and properties usually linked to a high Fe content, such as P fixation and anion exchange capacity, are therefore not always strongly expressed in soils of this group; in fact, inferences based on oxide mineralogy are probably more applicable to the humic soils in Section 2.2. The main connotation associated with oxidic soils is the relatively free drainage and well aerated condition of the B horizon. Yellower colours (from goethite) are generally associated with cooler, moister conditions than those linked with red colours (hematite). As indicated in Section 2.2, the combination of yellow and red colours within the same pedon (the xanthirhodic form) can have special significance with respect to redox status and its modification due to the presence of organic matter as a microbial substrate in the upper solum.

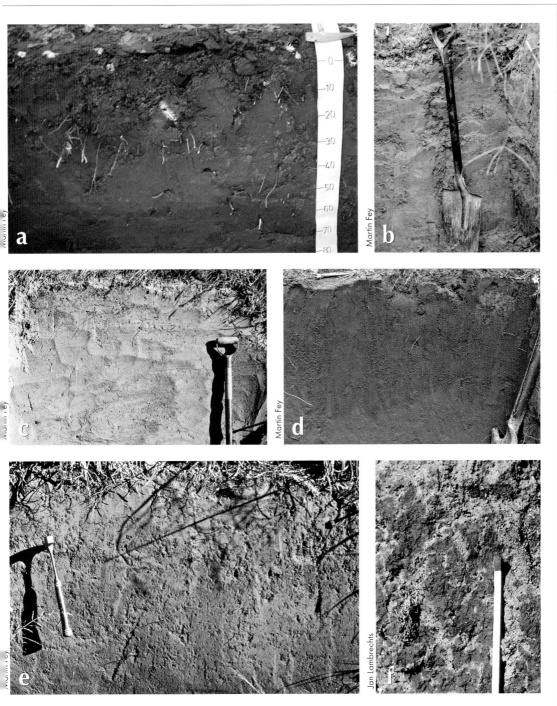

Figure 2.10.3 Oxidic soils with uniformly pigmented, apedal subsurface horizons: a. Rhodic (Hutton) derived from Kalahari sands in vineyard near Upington, Northern Cape; b. Xanthic (Clovelly) derived from sandstone, near Greytown, KwaZulu-Natal; c. Xanthic (Clovelly) from dune sand near Humansdorp; d. Rhodic (Hutton) in aeolian sand north of Vanrhynsdorp, Western Cape; e. Xanthirhodic (Griffin) derived from shale-dolerite colluvium near Richmond, KwaZulu-Natal); f. Differential yellowing along root channels in the B1/B2 transition in a xanthirhodic (Griffin) form near Sabie, Mpumalanga.

Figure 2.10.4 Diagnostic signs of wetness in xanthic oxidic soils (Pinedene form) in the form of: a. Gleyed saprolite and b. Mn oxide glaebules beneath a yellow-brown apedal B horizon (both near Knysna, southern Cape).

A mottled transition from the yellow to red B such as that illustrated in Figure 2.10.3f may also imply hydromorphy, as discussed under Genesis.

Classification

Forms and families of oxidic soils are summarised in Table 2.10.1 and international correlation is presented in Table 2.10.2. It is as well to recall that other soils with 'rhodic' and 'xanthic' character are already accommodated at earlier stages in applying the key to soil groups. The soils in question are found in the humic, vertic, silicic, calcic, duplex and

Figure 2.10.5 Examples of the pedorhodic (Shortlands) form of oxidic soils with a red structured B horizon, both derived from dolerite: a. near Eston and b. near Albert Falls, KwaZulu-Natal.

Table 2.10.1 Forms of oxidic soils and their constituent families (Nomenclature is shown alongside form names, abbreviations and family codes as defined by the Soil Classification Working Group, 1991)

Xanthic-hydromorphic (Pinedene Pn)

1100 dystrophic, haplic	1200 dystrophic, luvic
2100 mesotrophic, haplic	2200 mesotrophic, luvic
3100 eutrophic, haplic	3200 eutrophic, luvic

Xanthirhodic (Griffin Gf)

1100 dystrophic, haplic	1200 dystrophic, luvic
2100 mesotrophic, haplic	2200 mesotrophic, luvic

Xanthic (Clovelly Cv)

1100 dystrophic, haplic	1200 dystrophic, luvic
2100 mesotrophic, haplic	2200 mesotrophic, luvic
3100 eutrophic, haplic	3200 eutrophic, luvic

Rhodic-hydromorphic (Bloemdal Bd)

1100 dystrophic, haplic	1200 dystrophic, luvic
2100 mesotrophic, haplic	2200 mesotrophic, luvic
3100 eutrophic, haplic	3200 eutrophic, luvic

Rhodic (Hutton Hu)

1100 dystrophic, haplic	1200 dystrophic, luvic
2100 mesotrophic, haplic	2200 mesotrophic, luvic
3100 eutrophic, haplic	3200 eutrophic, luvic

Pedorhodic (Shortlands Sd)

1110 mesotrophic, haplic, micropedal	1120 mesotrophic, haplic, macropedal
1210 mesotrophic, luvic, micropedal	1220 mesotrophic, luvic, macropedal
2110 eutrophic, haplic, micropedal	2120 eutrophic, haplic, macropedal
2210 eutrophic, luvic, micropedal	2220 eutrophic, luvic, macropedal
3110 calcic, haplic, micropedal	3120 calcic, haplic, macropedal
3210 calcic, luvic, micropedal	3220 calcic, luvic, macropedal

plinthic groups. Other soils with 'oxidic' character key out later in the cumulic group (Section 2.12) where colour variegation is sufficiently marked or there is insufficient evidence of soil formation having occurred in recent sediments.

Another aspect that requires comment in relation to the family criteria in Table 2.10.1 is the omission of texture which featured prominently in defining soil series in the earlier binomial classification (MacVicar et al., 1977). The omission of soil series in the second edition was ostensibly to encourage scientists to re-define soil series (Soil Classification Working Group, 1991, p. 3), but it is unlikely that the information needed for this to happen will become

Table 2.10.2 Approximate placement of oxidic forms in international classification systems (IUSS Working Group WRB, 2006; Soil Survey Staff, 2003)

WRB

Soil form	WRB Reference Soil Group	Possible prefix qualifiers	Possible suffix qualifiers
Xanthic-hydromorphic (Pinedene Pn)	Ferralsols	Posic, Acric, Haplic	Dystric, Eutric, Xanthic
	Acrisols	Gleyic, Haplic	Chromic
	Lixisol	Gleyic, Haplic	Chromic
	Arenosols	Hypoluvic, Rubic, Endogleyic, Ferralic, Haplic	Dystric, Eutric
	Cambisols	Endogleyic, Ferralic	Dystric, Eutric, Chromic
Xanthirhodic (Griffin Gf)	Ferralsols	Posic, Acric, Haplic	Dystric, Eutric, Rhodix, Xanthic
	Acrisols	Haplic	Chromic
	Lixisol	Haplic	Rhodic, Chromic
	Arenosols	Hypoluvic, Rubic, Ferralic, Haplic	Dystric, Eutric
	Cambisols	Ferralic	Dystric, Eutric, Rhodic, Chromic
Xanthic (Clovelly Cv)	Ferralsols	Posic, Acric, Haplic	Dystric, Eutric, Xanthic
	Acrisols	Haplic	Chromic
	Lixisols	Haplic	Chromic
	Arenosols	Hypoluvic, Rubic, Ferralic, Haplic	Dystric, Eutric
	Cambisols	Ferralic	Dystric, Eutric, Chromic
Rhodic-hydromorphic (Bloemdal Bd)	Ferralsols	Posic, Acric, Haplic	Dystric, Eutric, Rhodic
	Acrisols	Gleyic, Haplic	Rhodic
	Lixisols	Gleyic, Haplic	Rhodic
	Arenosols	Hypoluvic, Rubic, Endogleyic, Ferralic, Haplic	Dystric, Eutric
	Cambisols	Endogleyic, Ferralic	Dystric, Eutric, Rhodic
Rhodic (Hutton Hu)	Ferralsols	Posic, Acric, Haplic	Dystric, Eutric, Rhodic
	Acrisols	Haplic	Rhodic
	Lixisols	Haplic	Rhodic
	Arenosols	Hypoluvic, Rubic, Ferralic, Haplic	Dystric, Eutric
	Cambisols	Ferralic	Dystric, Eutric, Rhodic
Pedorhodic (Shortlands Sd)	Nitisols	Ferralic, Acric, Haplic	Dystric, Eutric, Rhodic
	Alisols	Nitic, Haplic	Rhodic
	Acrisols	Nitic, Haplic	Rhodic
	Luvisols	Nitic, Haplic	Rhodic
	Lixisols	Nitic, Haplic	Rhodic
	Cambisols	Ferralic, Haplic	Calcaric, Dystric, Eutric, Rhodic

available in the foreseeable future. However, as proposed in Chapter 1, the naming routine adapted from that used in international classification allows the textural class to be appended as a suffix qualifier to the soil name. For example, the old Balmoral series of the first edition, which has given way to a broader family in the second, can be restored as a dystrophic, haplic (or luvic), rhodic, oxidic clay.

USDA Soil Taxonomy

Xanthic-hydromorphic (Pinedene Pn): Oxisols (Aquox; Ustox; Udox); Ultisols (Aquults; Udults; Ustults); Alfisols (Aqualfs; Ustalfs; Udalfs)
Xanthirhodic (Griffin Gf): Oxisols (Ustox; Udox); Ultisols (Udults; Ustults); Alfisols (Ustalfs; Udalfs)
Xanthic (Clovelly Cv): Oxisols (Ustox; Udox); Ultisols (Udults; Ustults); Alfisols (Ustalfs; Udalfs)
Rhodic-hydromorphic (Bloemdal Bd): Oxisols (Aquox; Ustox; Udox); Ultisols (Aquults; Udults; Ustults); Alfisols (Aqualfs; Ustalfs; Udalfs)
Rhodic (Hutton Hu): Oxisols (Ustox; Udox); Ultisols (Udults; Ustults); Alfisols (Ustalfs; Udalfs);
Pedorhodic (Shortlands Sd): Ultisols (Paleudults; Rhodudults; Hapludults; Paleustults; Rhodustults; Haplustults); Alfisols (Rhodustalfs; Haplustalfs; Rhodudalfs; Hapludalfs)

Genesis

Soils with red and yellow-brown apedal B horizons develop from most parent materials under a wide range of climatic conditions. As a general rule, however, red colours are more prevalent under warmer, drier circumstances and/or in soils derived from more ferruginous (e.g. mafic) parent materials; conversely, yellow colours signify moister, cooler, more acidic environments and/or a parent material lower in weatherable, iron-containing minerals. The factors that influence hematite (red) or goethite (yellow) formation when ferric iron oxides precipitate during weathering and oxidation are numerous and include pH, temperature, organic complex formation, rate of Fe supply, redox potential, soluble base cation concentration and the presence of interfering or substituting elements such as Al, Si and Ti (Schwertmann and Taylor, 1977). The case of a yellow-brown apedal B horizon overlying a red apedal B horizon is therefore fascinating and has been discussed under humic soils in Section 2.2, on the basis of which it can be concluded that the mottled transition which is occasionally encountered (Figure 2.10.3f) is probably hydromorphic notwithstanding earlier assertions to the contrary (de Villiers, 1964a). In a similar vein, the inference by de Villiers (1964b) of subliminal podsolisation in the xanthirhodic form (his Kranskop series) based on a consistent increase in Fe content from the yellow to red horizon, is more readily explained in terms of hydromorphy, although as Simonson (1959) would no doubt have argued in principle, both redox and podsolisation processes (along with lessivage and still others) are admissible in such situations.

Both yellow-brown and red apedal B horizons may grade either directly into parent material or saprolite, or into material showing pedogenic evidence of wetness either as plinthic material (Section 2.9) or as a mottled horizon with insufficient segregation of Fe or Mn to qualify as plinthite (e.g. Figure 2.10.4b). The wetness factor also manifests itself topographically, with red soils occupying the crests, yellow soils the midslopes and grey soils the footslopes and bottomlands of a drainage catena that is typical on the Highveld and elsewhere (see box in Section 2.9 relating to plinthic soils). For the same reason, evidence of wetness at depth is probably more common beneath a yellow-brown apedal B than a red apedal B. Aspects of soil colour related to wetness are further addressed in Sections 2.2, 2.9 and 2.11.

Uniformity of subsoil colour is probably best explained by pedoturbation (chiefly faunal and transportational), since residual profiles on weathered rock typically retain the variegation characteristic of saprolite even when intensely weathered to a soft kaolinitic mass. Time is doubtless an important prerequisite for pedoturbation to achieve homogenisation, although the period required can only be guessed at. Inevitably, however, apedal B horizons are more likely to be found on gentle rather than steep slopes and especially where colluvial aggradation has occurred.

Since oxidic soils possess an orthic A horizon their organic matter status will reflect to some extent the climatic conditions under which they are encountered. Areas of highest rainfall and coolest temperatures fall largely outside the distribution range and are occupied by soils of the humic group.

There is some evidence to suggest that the correlation between base status and climate is clearer when the former is included in an index of total cations including acidity (i.e. reflecting so-called effective CEC) and the latter is expressed in terms of a calculated hydrological index reflecting degree of leaching (Fey and Donkin, 1994). Such relationships are valuable not only in understanding soil formation but also in using soil properties as a guide to land use potential. The luvic property of some of the apedal soils suggests that oxidic mineralogy in some cases does not impart sufficient stability to micro-aggregates to prevent clay dispersion. Soils in this category would show affinities with alfisols and ultisols rather than oxisols (Table 2.10.2).

The term 'apedal' invites some explanation and justification (see box). In the case of light textured soils a weak structure is self evident, but many apedal B horizons contain 60 percent or more clay and yet are still effectively apedal and massive although still porous and friable. This description, it should be emphasised, only applies strictly when the soil is judged in terms of blocky or prismatic structure. In terms of granular structure such soils could be described as strong, fine, granular. Stable micro-aggregates of oxides and silicate clay are responsible for this condition and especially for the high and stable porosity, notwithstanding the clayey texture. The term apedal is therefore meaningful only when used in reference to the kind of structure by which the red structured B (e.g. strong, sub-angular blocky) and the pedo- or prismacutanic B horizons of duplex soils are recognised.

There seems to be increasing consensus that the apedal soils (clays in particular) in the summer rainfall regions derive their weak structure, at least to some extent, from the fact that the subsoil remains permanently moist as a result of a dormant grassland vegetation during the winter which, although much drier than the summer, is cool enough and with sufficient rainfall for water deficits to rarely, if ever, occur. Shrinkage of the soil mass is barely evident in freshly dug profiles (fine cracks extending into the upper B are a natural expression in Figure 2.10.6b) and only occurs after long exposure through excavation (e.g. road cuts) or when evergreen tree plantations (pine, wattle, eucalypts) are established. Oxidic soils elsewhere, that are even more weathered under a higher rainfall but with a warmer and more pronounced dry season and an evergreen forest canopy, are known to exhibit strong blocky structure, for example Ferralsols and Acrisols in East Africa and Kraznozems in Queensland.

Why do so many red structured B horizons in South Africa have kaolinite as essentially the only phyllosilicate in their clay fraction and yet possess a strong blocky structure? Such soils, belonging to the Shortlands form (pedorhodic), typically occur in the warmer, somewhat drier zones of savanna and thicket biomes than their apedal counterparts of the humid grasslands. According to the Soil Classification Working Group (1991) at least some smectite is believed to be present in the clay fraction in accounting for the strong structure in this soil form although mineralogical analysis often shows otherwise. Base status in these soils is typically high (at least mesotrophic). There is often a connection between the pedorhodic form and melanic and vertic soils to form red-black toposequences (Sections 2.3 and 2.4) in which the black member with a vertic or melanic horizon typically occurs at a lower, less drained elevation in the catena.

Use

Because of the wide range in climatic conditions over which oxidic soils are found, there are few generalisations that can be drawn about land use practices and problems. In general the tillage of oxidic soils (Figure 2.10.7) is much easier, and erosion less prevalent, than with many other soil groups because the oxides provide a micro-aggregating effect which reduces the dispersibility of fine particles and its attendant problems of sealing and hard-setting.

Since CEC of the mineral colloids is low, a loss of organic matter through cultivation may have especially adverse effects on soil quality. In higher rainfall areas soil acidity and phosphate fixation are research priorities. The more leached and weathered apedal variants may be deficient in a number of nutrients (base cations and even trace elements such as zinc and boron). In range and wildlife management this has important consequences for grazing quality (sourveld). The dystrophic families of the apedal forms will exhibit the greatest problems of acidity and infertility. Subsoil acidity is quite widespread in the summer rainfall region and deep lime incorporation (and to a lesser extent surface application of gypsum) has proved effective in reducing susceptibility to mid-summer drought (see also Section 2.9). The eutrophic, apedal forms and the pedorhodic Shortlands form do not present such

GENESIS AND IMPLICATIONS OF APEDAL STRUCTURE

Apedal red clay is friable, porous and massive (i.e. it shows no planes of natural cleavage). Why?

- It is a flocculated assemblage of kaolinite and iron oxides in close to coulombically equivalent amounts (i.e. near to a point of zero charge). The clay therefore does not disperse and block pores. Nor does it swell markedly.

- With little seasonal drought (dry season is cool and grass is dormant), it remains hydrated more or less permanently, so does not shrink and is more easily penetrated by burrowing fauna which continue to create pore space.

Result: When it does dry out, it shrinks and cracks dramatically. The shrinkage is irreversible, producing large columnar clods of soil on exposed profiles and wide cracks in soil of forest plantations. The exposure is stable and vertical repose is preserved, but large trees threaten buildings through drying.

Pedal red clay is firm, dense and blocky (i.e. it consists of peds clearly demarcated by cleavage planes). Why?

- Being further from a point of zero charge (i.e. having a substantial net negative charge) as a result of either a more smectitic silicate clay and/or a lower content of iron oxides relative to silicate clay and/or a higher pH, the clay fraction is more readily dispersed and blocks pores. It also coats fabric surfaces separated by shrinkage cracks, creating cutans and providing greater permanence to planes of weakness (ped exteriors).

- It is repeatedly desiccated, which provides capillary suction for the collapse of pores. Loss of pore space means a denser and less hospitable fabric for burrowing fauna, so regeneration of porosity is slower.

- The more negatively charged clay also has a greater tendency to swell. This further reduces porosity under confinement.

Result: When the soil is exposed, shrinkage loosens the blocky peds and subsequent shrink-swell cycles cause large peds to break up into finer peds, leading to self-mulching (mellowing) and a shallow angle of repose on the sides of erosion gullies and excavated cuttings. Erodibility is enhanced.

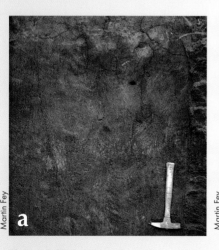 

Figure 2.10.6 Contrasting origin and consequences of structure (pedality) in two oxidic clays: a. Red apedal B (the old Balmoral series) in the Midlands Mistbelt near Donnybrook and b. Red structured B (enlarged; grass root for scale) in Tall Grassveld savanna near Kokstad, KwaZulu-Natal.

Figure 2.10.7 Cultivated oxidic soil near Vredendal, Western Cape

problems, are highly productive when irrigated and are typically associated with sweet grazing.

Hydromorphic forms of the oxidic group are likely to provide the same benefits for deeper rooted crops as those associated with plinthic soils (Section 2.9). Conversely the drainage of non-hydromorphic forms may be excessive and therefore counterproductive in dryland crop production where growing season water deficits are common, such as over much of the Highveld.

Relatively free drainage and structural stability imply that for irrigated and dryland agriculture, and forestry, these soils will be among the best available in any given region of South Africa. As already mentioned the erosion hazard is low on most soils of the oxidic group, with the pedorhodic Shortlands form as well as eutrophic families of the other forms (i.e. those associated with a drier climate) probably being the least stable. In the luvic families the textural contrast between top- and subsoil implies that moderate problems of compaction under cultivation and some degree of impeded drainage will be encountered. It is also worth noting that the yellower apedal soils are in some situations moister (and therefore better for dryland cropping) but in others wetter (i.e. more prone to waterlogging) than their red counterparts. In the next section the more extreme types of natural wetness in soil are explored.

CONSTANTIA: A FORM ON ITS OWN

The Constantia form is defined (Soil Classification Working Group, 1991) as consisting of an orthic A – E – yellow-brown apedal B horizon sequence. Four families within the form are defined on the basis of whether or not they are luvic and whether or not the material beneath the yellow-brown apedal B horizon has podzolic character. Following the precedent of scrapping the old Shepstone form (MacVicar et al., 1977) it is recommended that the Constantia form be similarly disbanded and its families allocated to the podzolic group (eluvic-cumulic-aeromorphic form (Concordia)) and the cumulic group (eluvic-neocutanic form (Vilafontes)). The oxidic group (the only non-humic, non-plinthic group with a yellow-brown apedal B horizon) is not a suitable home for Constantia because the E horizon potentially contradicts the implications of free drainage and aeration in the upper solum that constitute one of the hallmarks of the oxidic group. The profile in Figure 2.10.3c comes close to qualifying for the Constantia form because strongly achromic character is evident in the upper solum. Its locality suggests a probable tendency towards one of the families of the Constantia form that have podzolic character.

2.11 GLEYIC SOILS

Introduction

In the previous section we encountered a group of soils for which the defining characteristic is the accumulation of iron oxides, sufficient to produce a B horizon with strong and uniform pigmentation and implying that oxidation, in a predominantly aerated soil environment, is the key genetic process as well as the one most relevant to soil quality. 'Gley' specifies the obverse, in which reduction is the key process in a predominantly anaerobic environment. This happens most intensely in wetland soils that are found in low-lying parts of the landscape such as vleis and pans, producing a groundwater gley. In any region of the soil mantle, however, it can happen that water periodically infiltrates more rapidly than it drains away, either because of impervious subsoil or due to a wet climate. The result has been referred to, historically, as stagnogley.

Perhaps because the distribution of gleyic soils in South Africa (Figure 2.11.1) is so much more limited (happily) than that of oxidic soils (Figure 2.10.1), local pedology has not yet incorporated all the subtle nuances of gley into formal classification. The G horizon, when it occurs beneath an orthic A or E horizon, defines the gleyic soils. It includes, however, all subsurface horizons beneath either an organic O, a vertic, melanic or orthic A or an E horizon, that are 'saturated with water for long periods' and are 'dominated by grey, low chroma matrix colours, often with blue or green tints, with or without mottling' (Soil Classification Working Group, 1991).

Thus there are soil forms in the organic, vertic and melanic groups which have a G horizon, and there are soil families in still other soil groups that have saprolitic, unconsolidated or unspecified materials with 'signs of wetness' which exhibit the properties of a G horizon and might even have

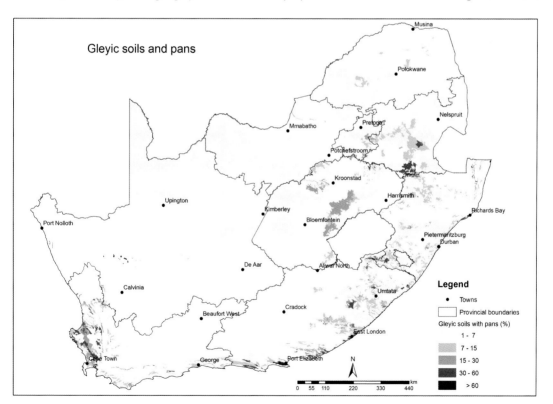

Figure 2.11.1 Gleyic soils, including pans, in South Africa (abundance classes refer to estimated percentages within land types; it should be noted, however, that because gleyic soils often occupy a very limited area along watercourses their indicated virtual absence over much of South Africa is potentially misleading).

qualified as a G had the position in the profile been different. The discussion of both genesis and land use of gleyic soils has relevance to all these other soils even though they are not dealt with explicitly in this section.

Many E horizons originate through gleying – the E was formerly called a perched gley in a forerunner to the present classification (Van der Eyk et al. (1969) – but the dominant process in the E is one of eluviation of clay (such as in some duplex soils), iron (some plinthic soils) and/or metal-humic compounds (some podzolic soils). Sometimes ferrolysis is a dominant process (see Sections 2.7 and 2.9 for further discussion). An E horizon (or its surface equivalent, the achromic or bleached orthic A) may also occur in in the duplex, plinthic, cumulic and lithic soil groups. The temptation to employ the E horizon, either alone or jointly with the G horizon, for creating a soil group has therefore been avoided. Nevertheless some of the implications of there being an E horizon in addition to the G horizon, both genetically and in relation to use, are also addressed in this section.

Properties
General
The overriding property of gleyic soils is the gley (mostly greyish) colour, which is a manifestation of their tendency to be extremely wet. This happens either directly beneath an orthic A, or via an E horizon which itself is wet at least intermittently. In the following discussion of properties the orthic and eluvic forms (Table 2.12.1) are dealt with separately, bearing in mind that in some respects it is inappropriate to consider them either separately from one other or from other soils which also tend towards excessive wetness without having a diagnostic E or G horizon (e.g. shallow plinthic or lithic soils with an achromic A horizon). Selected profiles of gleyic soils are given in Figure 2.11.2.

Morphological
The G horizon is naturally saturated with water for long periods and is dominated by grey colours especially on macro-void and ped surfaces, with or without mottling. It has firmer consistence than the overlying horizon. It does not qualify as a pedocutanic, prismacutanic, lithocutanic or plinthic B horizon, all of which may in some instances have evidence of strong gleying (Soil Classification Working Group, 1991).

The WRB specifies gleyic properties (from the Russian word gley, mucky soil mass) more demandingly, in terms of having a gleyic colour pattern (either oximorphic or reductomorphic) either in 50% of the soil mass or in 100% of the soil mass below a surface horizon (IUSS Working Group WRB, 2006). Oximorphic properties refer to ferric mottling in a grey matrix as opposed to predominantly (95%) neutral or bluish to greenish colours in the case of reductomorphic properties. The normal gley morphology in the WRB refers to ferric iron coloration at the top of the gley horizon and towards ped exteriors, which contradicts the description above of grey colours being especially evident on ped surfaces. Some examples of the various colours discussed above are shown in Figure 2.11.3.

The E horizon which occurs between the orthic A and the G in the eluvic (Kroonstad) form has defined grey matrix colours (Soil Classification Working Group, 1991) and may contain, although not necessarily, discernible mottling or streaking due to localised segregation of iron oxides. It tends to be loose or friable when moist and is non-plastic and may become very hard and brittle when dry, depending on texture. It is weakly structured and, relative to overlying or underlying horizons, shows evidence of having been depleted of colloidal matter (iron oxides, clay and organic matter). It may be very thick (Figure 2.11.2e, f and g) and even pale yellow (hypoxanthic) when moist (Figure 2.11.2f).

Chemical and physical
If the G horizon is close to the surface (orthic, Katspruit form) the soil is typically wet throughout much of the year. The deeper G horizon beneath an E in the eluvic, Kroonstad form implies that the upper part of the profile will exhibit saturation only intermittently and properties are likely to be similar to those of the eluvic forms of other soil groups (especially duplex and plinthic). The orthic form is especially typical of bottomlands or vleis which invariably have a better reserve of plant nutrients and a higher pH, CEC and organic matter content than soils of surrounding uplands. Such soils usually display a high degree of stickiness and plasticity. Calcareousness is used as a criterion for differentiating families.

The leached nature of the eluvic (Kroonstad)

Figure 2.11.2 Examples of gleyic soils: a, b, c and d are variants of the orthic (Katspruit) form, all in the vicinity of George and Knysna, Western Cape; e, f and g are variants of the eluvic (Kroonstad) form: e. albic family, near Alexandria, Eastern Cape; f. hypoxanthic family, in the Koue Bokkeveld, Western Cape; and g. albic family, with a thick orthic A, near Humansdorp, Eastern Cape.

Figure 2.11.3 Variants of gley morphology: a. classic blue-green colours of partly reduced iron compounds in salt pan sediment near Darling, Western Cape. White crystals of halite and black sulfide minerals are also evident. The green oxidises to brown a few hours after exposure to air; b. strong prismatic structure with gleyed cutans on ped surfaces, Experimental Farm near George; and c. rusty mottles along root channels in a grey matrix, on the farm Longdown near Vyeboom, Western Cape.

form of gleyic soils and the possible involvement of ferrolysis in their genesis imply that the upper solum is likely to be quite depleted of nutrients and, especially in lighter textured variants, to have a low CEC. Typically the pH of the A and E horizons is acidic, sometimes severely so. By contrast the usually more clayey B horizon is likely to have a higher pH and CEC and to contain a larger reserve of plant nutrients. Probably the most important characteristic of the eluvic form is intermittent saturation with water and a predominantly lateral discharge of water through the A and E horizons during these wet periods. Frequent wetting and drying can be expected to have a strong influence on the extent of N loss by denitrification and in limiting the quantity of humus that can accumulate. As mentioned in Section 2.9, the intensity and duration of saturation with water can be correlated with gley morphology (Van Huyssteen, 2004).

Gleys have no particularly characteristic mineralogy, which is usually dominated by those minerals imported in the alluvial or colluvial deposit which makes up the parent material. Nevertheless there are secondary minerals that sometimes are hallmarks of intense gleying, including pyrite, black sulfides (especially clay-sized minerals such as greigite and mackinawite), hydromagnetite and green rust (see Genesis below). Sulfidic oozes, as they are sometimes called, are readily identified by the smell when treated with HCl and their rapid paling with hydrogen peroxide. Green-blue colours (Figure 2.11.3a) are similarly unstable and do not last long when exposed to air.

Classification

Although only two soil forms key out in the gleyic group (Table 2.11.1), mention has already been made of other soils with a G horizon, in the organic, vertic and melanic groups. International correlation is given in Table 2.11.2.

Genesis

The orthic form (Katspruit) is probably the most widely encountered wetland (vlei) soil in South Africa. The fact that organic wetland soils are much rarer is probably a result of temperatures being too high. Reduction of iron under anaerobic conditions decolorises the soil if the iron is removed by gradual leaching, or imparts a bluish or greenish colour to

Table 2.11.1 Forms of gleyic soils and their constituent families (Nomenclature is shown alongside form names, abbreviations and family codes as defined by the Soil Classification Working Group, 1991)

Orthic (Katspruit Ka)

1000 acalcic	2000 calcic

Eluvic (Kroonstad Kd)

1000 albic	2000 hypoxanthic

Table 2.11.2 Approximate placement of gleyic forms in international classification systems (IUSS Working Group WRB, 2006; Soil Survey Staff, 2003)

WRB (qualifier terms in italics are less certain)

Soil form	WRB Reference Soil Group	Possible prefix qualifiers	Possible suffix qualifiers
Orthic (Katspruit Ka)	Gleysols	Acric, Haplic	*Thionic*, Abruptic, Calcaric, Eutric
Eluvic (Kroonstad Kd)	Planosols	Endogleyic, Haplic	*Albic, Dystric, Eutric*
	Stagnosols	Vertic, Luvic, Haplic	*Albic*

USDA Soil Taxonomy

Orthic (Katspruit Ka): Ultisols; Alfisols; Entisols – aquic variants of all
Eluvic (Kroonstad Kd): Ultisols; Alfisols; Entisols – aquic variants of all

the soil if some of the iron remains (Figure 2.11.3a). A ferrous-ferric hydroxide (green rust) is responsible for these blue-green colours. The clay fraction is typically dominated by kaolinite in higher rainfall areas where discharge of water matches evaporation. In areas of lower rainfall and higher temperatures, evaporative concentration conserves silica and bases and creates conditions suitable for smectite neoformation, in which case vertic or melanic soils (Sections 2.3 and 2.4) are more likely to be found, although examples do exist of soils with an orthic A and vertic properties, especially slickensides, in the underlying G horizon.

Not all G horizons are uniformly reduced. Many have grey colours but show little indication of any remaining reduced free iron and may contain a fairly high proportion of ferric mottling (red, yellow and/or brown). Ferrihydrite, goethite and hematite are common Fe minerals associated with mottling in gleys. Mottles of jarosite and schwertmannite (pale yellow) in very acid vleis and of lepidocrocite (orange) in cool, highland vleis may be encountered, although rarely. As is the case with plinthic horizons, manganese oxides may also accumulate as black mottles and concretions. Rusty root channels (Figure 2.11.3c) and increased mottling near the upper boundary of the gley are additional indicators of localised redox gradients in these soils.

The eluvic gleyic (Kroonstad) soils (together with their plinthic and duplex counterparts, and soils with a bleached orthic A) are found in almost every region of South Africa, although occurrence is much rarer in areas of highest rainfall such as in the KwaZulu-Natal midlands and Mpumalanga escarpment where the degree of weathering has ensured sufficiently stable structure and free drainage to prevent the formation of a water table in the solum. As was indicated to be the case for the plinthic group, these eluvic gleyic soils are particularly common in sub-humid regions where there is still sufficient

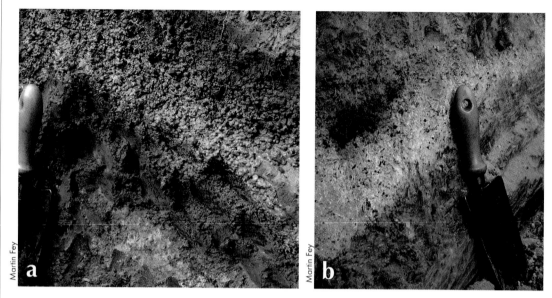

Figure 2.11.4 Differentially wetted E horizon after rain, consisting mostly of pisolitic Fe concretions in a fine sandy loam: a. saturated and past its liquid limit and b. dry and hard, one metre away in the same pit, overlying a mottled G horizon derived from shale colluvium, near Kokstad, KwaZulu-Natal.

rainfall to produce periodic subsurface saturation with water, and especially on parent materials with a lower iron content such as shales and sandstones and acidic, igneous rocks such as granites. Basic igneous rocks rarely allow for the formation of E horizons because their high clay forming potential and/or higher Fe content minimises the tendency for clay migration needed to build up the strong textural contrast between upper and lower parts of the solum. For further discussion of the significance of E horizons associated with wetness above a less permeable B horizon the reader is referred to sections 2.7 (duplex) and 2.9 (plinthic soils).

Field observations during wet phases are valuable in revealing hydrological subtleties such as preferential, lateral flow paths within E horizons (Figure 2.11.4).

Use

Land use for the gleyic (Katspruit) form is similar to that for the organic group (Section 2.1) except that problems of oxidative degradation are not as pronounced. Artificial drainage is imperative for planted crops. Trees tolerant of wetness such as poplars (KwaZulu-Natal) and pears in the winter rainfall region can be grown quite successfully on these bottomland soils. Otherwise the conservation of wetlands for both hydrological and ecological reasons (Figures 2.11.5 and 2.11.6) is probably the most common use to which gleyic soils are put. In the KwaZulu-Natal midlands some vleis with Katspruit soils yield high quality kaolinitic clay used for local pottery.

Land use considerations on the eluvic forms of gleyic soils are similar to those on plinthic soils and deeper variants of duplex soils (Sections 2.7 and 2.9). Besides intermittent wetness, the need for drainage and the likelihood of unreliable rainfall, crusting and structural deterioration are common except on members with very sandy texture. Precautionary tillage practices are therefore advisable. Cases have been recorded of maize doing poorly on eluvic (Kroonstad) soils in years of normal rainfall but spectacularly out-yielding maize on adjacent, better drained soils in drought years. Families with a yellower (hypoxanthic) E can also be expected to exhibit somewhat less intense wetness than families with a grey (albic) E. For fruit and grape production in the winter rainfall region these soils are deep tilled in such a manner that the A and E horizons are mixed to create a more uniform texture and organic matter profile, for better root development.

Figure 2.11.5 Land use on gleyic soils: a. bottomlands with river meanders adjoining cultivated fields used for livestock grazing near Swartberg, KwaZulu-Natal; b. conserved wetland habitat near Harrismith in the Free State.

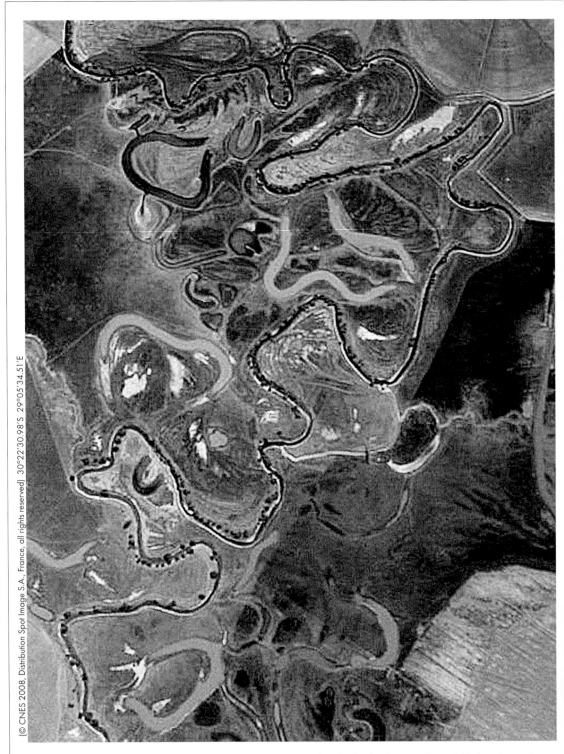

Figure 2.11.6 An alluvial floodplain outside the town of Cedarville in the Eastern Cape. This is a dramatic representation of the interface between gleyic and cumulic soils.

2.12 CUMULIC SOILS

Introduction

In this and the remaining two sections we examine soils that are immature. Either derivation from their parent material is more recent or, at least, the expression of soil forming processes is weaker. Although the idea of maturity in soil formation is challenging and controversial, all classification systems recognised immature soils in terms of parent material having only *recently* been subjected to processes of soil formation (hence Entisols in Soil Taxonomy) or showing some signs of soil formation but with such weak expression that soil formation is best described as *incipient* (hence Inceptisols in Soil Taxonomy). The concept is useful because it allows the diagnostic criteria for most soil classes to be confined to strong, readily recognised properties. This reduces ambiguity.

The three groups created to accommodate incipient pedogenesis are based on the nature of the parent material. Soils of the cumulic group (this section) have formed in unconsolidated natural deposits; those of the lithic group (next section) in rock or saprolite that is still geologically recognisable, and those of the anthropic group (final section) in deposits of material produced by human activity.

Unconsolidated materials accumulate by a variety of geomorphic processes. The commonest of these are colluvial (down-slope) or alluvial (down-valley) deposition from flowing water, mass movement (through landslides or soil creep), and aeolian (wind-blown) deposition. The distribution of cumulic soils is therefore widespread (Figure 2.12.1).

Properties

General

The nine forms of cumulic soils (Table 2.12.1) are differentiated on the basis of the nature and degree of alteration of subsurface materials. These may be:

1. negligibly altered, as in the arenic (Namib) and fluvic (Dundee) forms;

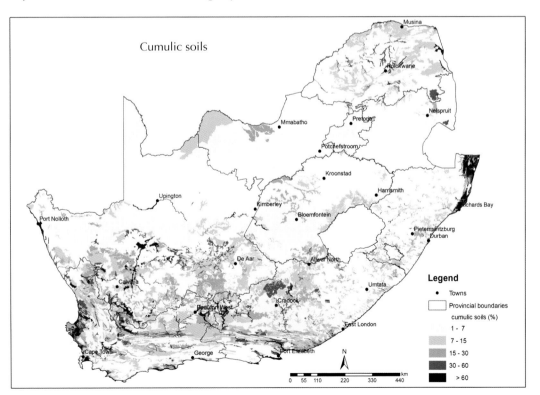

Figure 2.12.1 Cumulic soils in South Africa (abundance classes refer to estimated percentages within land types)

2. weakly altered in a luvic direction (i.e. an increase in clay content with depth) but insufficiently to qualify as duplex soils, as in the neocutanic (Oakleaf, Tukulu and Vilafontes) forms;
3. weakly altered in the direction of carbonate accumulation but insufficiently to qualify as calcic soils, as in the neocalcic (Augrabies, Montagu and Kinkelbos) forms;
4. characterised by the presence of an E horizon, either alone in the eluvic (Fernwood) form or above a neocutanic or neocarbonate B horizon (eluvic Vilafontes and Kinkelbos forms, respectively); or
5. characterised by signs of wetness beneath a neocutanic or neocarbonate B horizon, as in the hydromorphic forms (Tukulu and Montagu, respectively).

Additional features such as surface bleaching, subsurface reddening, weak yellow (hypoxanthic) colour or clay lamellae in the E horizon, calcareousness, clay illuviation and signs of wetness are used in defining families within some of the forms. Although all such features are weakly expressed they provide a basis for interpreting genesis and anticipating soil behaviour, correlating with some soils of earlier groups for which the same features are employed, in more strongly expressed form, as diagnostic criteria.

Morphological

Identification of arenic (Namib) and fluvic (Dundee) forms in the cumulic group is relatively straightforward, with regic sand and stratified alluvium, respectively, carrying clear signs of recent aeolian or fluvial deposition (Soil Classification Working Group, 1991). The geomorphic context of the site often provides useful confirmation of recent deposition while clues are also sometimes provided in the form of textural stratification in fluvial deposits and cross-bedding and dune features in aeolian deposits. Two fluvic examples are shown in Figures 2.12.2.

Besides the arenic (Namib) form, deep cover sands may also be eluvic (Fernwood form), with a thick E horizon, some examples of which are shown in Figure 2.12.3. Dark (umbric) or bleached (achromic) topsoil is one differentiating criterion for families. Regarding additional family criteria, the thick E horizon is most recognisably albic (grey), though sometimes hypoxanthic (pale yellow); and haplic (uniform), though sometimes lamellic (having darker, wavy sheets of clay known as lamellae; Figure 2.12.3).

The neocutanic (Tukulu, Oakleaf, and Vilafontes) forms (Figure 2.12.4) are not always so easily recognised. In terms of structure development the neocutanic B horizon is defined identically to the apedal B horizon (red or yellow-brown) and the only clear difference is in terms of colour variega-

Figure 2.12.2 The fluvic (Dundee) form of cumulic soils: a. in vineyard on the Olifants River floodplain, Vredendal, Western Cape; b. more distinctively stratified, near De Rust, Klein Karoo.

Figure 2.12.3 Four examples of families in the eluvic (Fernwood) form of cumulic soils: a. *umbric*, albic, haplic, in a dambo (seasonal wetland) derived from Kalahari sand, north-west Zambia; b. *achromic*, albic, lamellic, from recent aeolian sand near Mbumbulu, KwaZulu-Natal; c. *achromic*, hypoxanthic, haplic, and d. *achromic*, hypoxanthic, lamellic, both developed on slopes of old sand dunes under high rainfall near Knysna, Western Cape.

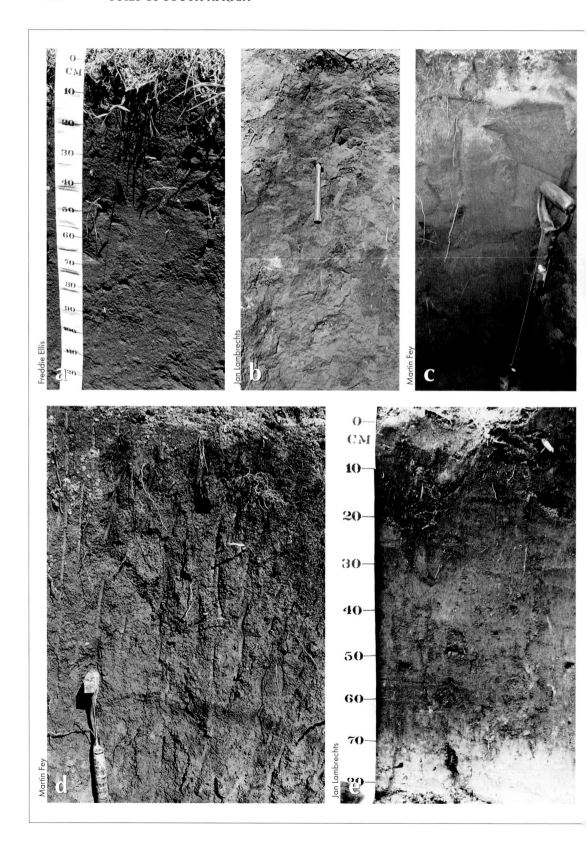

Figure 2.12.4 Miscellaneous neocutanic and neocarbonate forms of cumulic soils: a. neocutanic (Oakleaf) with humus-rich orthic A, near Sabie, Mpumalanga; b. neocalcic-hydromorphic (Montagu), Keisievallei near Montagu, Western Cape; c. eluvic-neocutanic (Vilafontes) near Jeffrey's Bay, Eastern Cape; d. neocutanic-hydromorphic Tukulu form (luvic, arhodic, chromic family) developed in granitic colluvium near Stellenbosch, Western Cape; e. neocutanic-hydromorphic (Tukulu) in the Koue Bokkeveld, Western Cape; f. neocalcic (Augrabies) near Groot-Brakrivier, Western Cape; g. neocutanic (Oakleaf) with weak silica cementation in terrace gravels near Aus, Namibia.

tion, with the definition of the latter allowing for some variegation in the form of dark coloured worm casts and illuvial cutans (Soil Classification Working Group, 1991), provided the dominant matrix colour is red or yellow within defined ranges. Often the transitional AB and BA horizons between the A and a red or yellow-brown apedal B horizon show sufficient variegation in colour to suggest (incorrectly) neocutanic character. Problems are inevitable when this transition is so diffuse that it extends well into the solum, effectively permeating almost the full depth of the red or yellow-brown subsoil. Distinction then becomes a matter of determining the permissible maximum thickness of a transition before it must be demarcated as a horizon in its own right. The proportion of the horizon occupied by uniformly coloured matrix material and (usually) darker infillings also needs to be specified. Understandably, pedologists with perspectives formed in different regions of the country do not always agree on these morphological thresholds. When weak brown or greyish colours dominate the matrix then such problems are not encountered. Bleaching of the overlying A horizon can be a useful indicator of neocutanic character in the B. If sufficient accumulation of iron oxides has occurred to impart strong red or yellow-brown colours then enough maturity is suggested for classification in the oxidic group. Future refinement of the definition is likely.

The neocalcic (Augrabies, Kinkelbos and Montagu) forms have essentially the same character as the neocutanic forms but with the additional distinction of having a neocarbonate B, with sufficient accumulation of carbonate to produce visible effervescence when tested with HCl though insufficient to qualify as a soft or hardpan carbonate horizon (Soil Classification Working Group, 1991).

Families within some cumulic forms are recognised on the basis of bleaching in the surface horizon (chromic when absent or achromic when present), red colours predominating or not in the B horizon (rhodic, arhodic), some clay illuviation having taken place or not (luvic, haplic), evidence of wetness or lack of it (hydromorphic, aeromorphic) and, in the case of the arenic and fluvic forms, the presence or absence of secondary carbonates (calcic, acalcic). The haplic/lamellic, albic/hypoxanthic and umbric/achromic distinctions within the eluvic

(Fernwood) form have already been mentioned.
Physical and chemical
Given the wide range in morphological properties just described it stands to reason that no generalisations are possible about physical characteristics other than the fact that the parent material is unconsolidated, which has both agricultural and engineering significance. Despite the absence of weathered rock there may still be potential for compaction, especially in the luvic variants. At the other extreme the arenic Namib form has minimal water holding capacity and is susceptible to erosion by wind. Intermittent perching of a water table could be inferred in some eluvic and achromic variants. Better subsurface aeration is implied in rhodic families while either poor aeration or an additional source of water for plants at depth is indicated in hydromorphic forms. The manifestation of these various traits in soils of the inceptic group is not necessarily less intense than it is in other soil groups which have the same properties more strongly expressed, since weak expression may merely be due to there having been insufficient time for morphological response.

Chemical properties of cumulic soils are also too diverse to permit more than a few generalised statements to be made, such as about expected periodic reducing conditions in the wetter variants (described above), soil reaction in the neocalcic forms and minimal buffering and nutrient storage capacity in the arenic and eluvic forms. As a rule, nutrient status can be expected to be better than in more mature oxidic, plinthic or podzolic soils, but exceptions are possible especially when the parent material has been transported from sites that have been subjected to considerable weathering. In the Western Cape, oxidic soils and associated red and yellow Tukulu and Oakleaf soils in high rainfall zones are similar to one another in terms of pH and concentration of extractable bases. Generally it is useful, where maps allow it, to make inferences about base status and fertility of cumulic soils by association with oxidic soils (for which the degree of leaching is a family differentiating characteristic) that may occur within the same land type.

Classification
The nine forms of cumulic soils and their families are listed in Table 2.12.1. Their counterparts in international classification are listed in Table 2.12.2.

Genesis
The fluvic Dundee form is recognised more easily than any other soil by its position in the landscape. It will be confined by nature to floodplains and specifically to those parts of an alluvial toe slope that have flooded sufficiently regularly in recent times to be rejuvenated with successive layers of stream sediment. Families with red colours (rhodic) merely

Table 2.12.1 Forms of cumulic soils and their constituent families (Nomenclature is shown alongside form names, abbreviations and family codes as defined by the Soil Classification Working Group, 1991). Terms in italics refer to the orthic A horizon.

Neocutanic (Oakleaf Oa)

1110 *chromic*, arhodic, haplic	1120 *chromic*, arhodic, luvic
1210 *chromic*, rhodic, haplic	1220 *chromic*, rhodic, luvic
2110 *achromic*, arhodic, haplic	2120 *achromic*, arhodic, luvic
2210 *achromic*, rhodic, haplic	2220 *achromic*, rhodic, luvic

Eluvic-neocutanic (Vilafontes Vf)

1110 *albic*, arhodic, haplic	1120 *albic*, arhodic, luvic
1210 *albic*, rhodic, haplic	1220 *albic*, rhodic, luvic
2110 *hypoxanthic*, arhodic, haplic	2120 *hypoxanthic*, arhodic, luvic
2210 *hypoxanthic*, rhodic, haplic	2220 *hypoxanthic*, rhodic, luvic

Neocutanic-hydromorphic (Tukulu Tu)

1110 *chromic*, arhodic, haplic	1120 *chromic*, arhodic, luvic
1210 *chromic*, rhodic, haplic	1220 *chromic*, rhodic, luvic
2110 *achromic*, arhodic, haplic	2120 *achromic*, arhodic, luvic
2210 *achromic*, rhodic, haplic	2220 *achromic*, rhodic, luvic

Neocalcic (Augrabies Ag)

1110 *chromic*, arhodic, haplic	1120 *chromic*, arhodic, luvic
1210 *chromic*, rhodic, haplic	1220 *chromic*, rhodic, luvic
2110 *achromic*, arhodic, haplic	2120 *achromic*, arhodic, luvic
2210 *achromic*, rhodic, haplic	2220 *achromic*, rhodic, luvic

Eluvic-neocalcic (Kinkelbos Kk)

1110 *albic*, arhodic, haplic	1120 *albic*, arhodic, luvic
1210 *albic*, rhodic, haplic	1220 *albic*, rhodic, luvic
2110 *hypoxanthic*, arhodic, haplic	2120 *hypoxanthic*, arhodic, luvic
2210 *hypoxanthic*, rhodic, haplic	2220 *hypoxanthic*, rhodic, luvic

Neocalcic-hydromorphic (Montagu Mu)

1110 *chromic*, arhodic, haplic	1120 *chromic*, arhodic, luvic
1210 *chromic*, rhodic, haplic	1220 *chromic*, rhodic, luvic
2110 *achromic*, arhodic, haplic	2120 *achromic*, arhodic, luvic
2210 *achromic*, rhodic, haplic	2220 *achromic*, rhodic, luvic

Eluvic (Fernwood Fw)

1110 *achromic*, albic, haplic	1120 *achromic*, albic, lamellic
1210 *achromic*, hypoxanthic, haplic	1220 *achromic*, hypoxanthic, lamellic
2110 *umbric*, albic, haplic	2120 *umbric*, albic, lamellic
2210 *umbric*, hypoxanthic, haplic	2220 *umbric*, hypoxanthic, lamellic

Fluvic (Dundee Du)

1110 arhodic, aeromorphic, acalcic	1120 arhodic, aeromorphic, calcic
1210 arhodic, hydromorphic, acalcic	1220 arhodic, hydromorphic, calcic
2110 rhodic, aeromorphic, acalcic	2120 rhodic, aeromorphic, calcic
2210 rhodic, hydromorphic, acalcic	2220 rhodic, hydromorphic, calcic

Arenic (Namib Nb)

1100 arhodic, acalcic	1200 arhodic, calcic
2100 rhodic, acalcic	2200 rhodic, calcic

Table 2.12.2 Approximate placement of cumulic forms in international classification systems (IUSS Working Group WRB, 2006; Soil Survey Staff, 2003)

WRB

Soil form	WRB Reference Soil Group	Possible prefix qualifiers	Possible suffix qualifiers
Neocutanic (Oakleaf Oa)	Acrisols	Cutanic, Haplic	Rhodic, Chromic
	Lixisols	Cutanic, Haplic	Rhodic, Chromic
	Arenosols	Haplic	Dystric, Eutric
	Cambisols	Haplic	Dystric, Eutric, Rhodic, Chromic
Eluvic-neocutanic (Vilafontes Vf)	Acrisols	Cutanic, Gleyic, Haplic	Albic, Rhodic, Chromic
	Lixisols	Cutanic, Gleyic, Haplic	Albic, Rhodic, Chromic
	Arenosols	Albic	Dystric, Eutric
	Cambisols	Haplic	Dystric, Eutric, Rhodic, Chromic
Neocutanic-hydromorphic (Tukulu Tu)	Acrisols	Cutanic, Gleyic, Haplic	Rhodic, Chromic
	Lixisols	Cutanic, Gleyic, Haplic	Rhodic, Chromic
	Arenosols	Endogleyic; Haplic	Dystric, Eutric
	Cambisols	Endogleyic, Haplic	Dystric, Eutric, Rhodic
Neocalcic (Augrabies Ag)	Luvisols	Cutanic, Calcic, Haplic	Chromic, Rhodic
	Lixisols	Cutanic, Haplic	Rhodic, Chromic
	Arenosols		Calcaric
	Cambisols	Haplic	Calcaric, Rhodic, Chromic
Eluvic-neocalcic (Kinkelbos Kk)	Acrisols	Cutanic, Haplic	Albic, Rhodic, Chromic
	Lixisols	Cutanic, Calcic, Haplic	Albic, Rhodic, Chromic
	Arenosols	Albic	Calcaric, Dystric, Eutric
	Cambisols	Haplic	Calcaric, Eutric, Rhodic, Chromic
Neocalcic-hydromorphic (Montagu Mu)	Luvisols	Gleyic	Rhodic, Chromic
	Lixisols	Cutanic, Gleyic Calcic, Haplic	Rhodic, Chromic
	Arenosols	Endogleyic, Haplic	Calcaric
	Cambisols	Endogleyic, Haplic	Calcaric, Rhodic, Chromic
Eluvic (Fernwood Fw)	Arenosols	Lamellic, Hyperalbic, Haplic	
Fluvic (Dundee Du)	Fluvisols	Gleyic, Haplic, Stagnic	Calcaric
Arenic (Namib Nb)	Arenosols	Rubic, Haplic	Calcaric

USDA Soil Taxonomy

Neocutanic (Oakleaf Oa): Inceptisols (Ustepts; Xerepts; Udepts)
Eluvic-neocutanic (Vilafontes Vf): Ultisols; Alfisols; Inceptisols
Neocutanic-hydromorphic (Tukulu Tu): Inceptisols (Aquepts; Ustepts; Xerepts; Udepts)
Neocalcic (Augrabies Ag): Aridisols; Inceptisols (Ustepts; Xerepts)
Eluvic-neocalcic (Kinkelbos Kk): Alfisols; Inceptisols
Neocalcic-hydromorphic (Montagu Mu): Aridisols; Inceptisols (Ustepts; Xerepts)
Eluvic (Fernwood Fw): Inceptisols; Entisols
Fluvic (Dundee Du): Inceptisols (Aquepts; Ustepts; Xerepts; Udepts) – Fluventic variants
Arenic (Namib Nb): Inceptisols (Ustepts; Xerepts; Udepts); Entisols

convey an indication that aeration is sufficient to preserve iron oxides whereas arhodic families may or may not be well aerated. Arhodic families that developed from low-iron containing parent material (e.g. Table Mountain Sandstones) could however be well aerated and non-hydromorphic. Hydromorphy is evident at depth in some families while secondary carbonate may also have accumulated. None of these incipient pedogenic indicators are sufficiently developed, however, to qualify the stratified alluvium as a diagnostic horizon. Texture and organic matter status also vary widely. The majority of soils on current floodplains probably do not qualify as fluvic because, despite occasional or even frequent flooding, the accumulation of sediment may be gradual and fine stratification is rapidly obscured through pedoturbation. Clear stratification represents a visible record of past flooding events. Interpretation of the strata using mineralogical, palynological and geochemical data in combination with geomorphological catchment analysis could in theory reveal much about the periodicity and intensity of major floods.

Regic sands are found widely in littoral and arid regions of the country. The origin of the arenic Namib form is typically aeolian and residual cross bedding may be evident. The soil may be part of an active dune system or, if stabilised by vegetation, lacks clear evidence of sufficient secondary clay formation and enrichment to bridge sand grains and provide coherence. Horizon differentiation is confined to darkening by organic matter. When red or yellow colours occur these are normally related to the parent sand rather than to secondary weathering. Besides obvious regional features such as the sands of the Namib and Kalahari and numerous coastal dune fields, regic sands are also associated with deflationary pans (lunette dunes) or as local accumulations on the leeward side of river channels.

Soil formation on coastal sands in KwaZulu-Natal has received some noteworthy attention from MacVicar et al. (1985) and, more recently with special reference to geochronology, from Botha and Porat (2007). The time scale over which soil formation occurs is shorter in unconsolidated sands because of more rapid weathering and most soils in regions where a sand blanket has been deposited have advanced beyond the arenic stage to at least the eluvic or neocutanic and even to the oxidic or podzolic stage.

The distinguishing feature of the eluvic Fernwood form is its thick E horizon. Although in some cases this E is associated with podzolisation, gleying and/or ferrolysis linked to a water table, in many cases it appears to result simply from intense lessivage (as discussed under duplex soils in Section 2.7) involving the removal of dispersed clay to leave a relatively bleached matrix of sand. Instances can quite frequently be found where the clay has not entirely been removed but instead has segregated into lamellae (Figure 2.12.3b and d). An explanation for how these lamellae develop (Van Reeuwijk and de Villiers, 1985) has been to some extent contradicted by Schaetzl (2001) who presented evidence suggesting that some pre-existing, banded, textural differentiation in aeolian sands could initiate the process of lamellae development. The wavy, convoluted appearance of the lamellae has been suggested, in relation to podzolisation in Section 2.8, as being due to the sandy texture of the matrix which minimises the probability of preferential flow paths developing through shrinkage.

The concept of incipient pedogenesis in unconsolidated materials was initially introduced in South African soil classification as the Oakleaf form. Intensive mapping of irrigable land revealed a consistent chronosequence associated with old alluvial terraces – deep red soils occurring on the oldest terrace and deep, but not yet reddened soils on younger terraces. Generally even the oldest red soils showed marked colour variegation due to faunal activity and clay illuviation, as well as weak structural development. The diagnostic neocutanic horizon corresponded approximately in concept to the cambic subsurface horizon in USDA Soil Taxonomy. This step served to focus attention on non-uniformity of soil colour in weakly structured or apedal materials and it soon became apparent that this criterion applies to extensive bodies of soil in youthful landscape positions, including sloping colluvial sites. Further differentiation became necessary, based on calcareousness and underlying hydromorphy, degree of clay illuviation, redness of the B horizon, bleaching of the orthic A horizon and presence of an E horizon above the B. Orthic bleaching and the presence of an E above a neocutanic B do not necessarily reflect hydromorphy, since they may also be the result of incipient podzolisation (Section 2.8) or eluviation of clay with humus (Section 2.7). Similar interfaces can be identified with oxidic soils (Section 2.10) in relation to colour uniformity and with plinthic soils (Section 2.9) based on degree of iron accumulation and segregation into mottles and concretions. (A future 'neoplinthic' variant could thus be envisaged should the need arise, analogous to the neocalcic concept representing immature variants of the calcic group in Section 2.6. Even 'neosilicic' variants of the silicic group in Section 2.5 are conceivable.)

It becomes clear that the genesis of neocutanic and neocalcic forms in the cumulic group embraces the full range of possibilities within unconsolidated materials, too weakly expressed to fit the definition of the appropriate diagnostic horizon. Consequently all landscapes will contain representatives, associated typically with alluvial or colluvial, lower (seldom convex) slope positions and sometimes with stabilised aeolian sediments. Commoner occurrence can in general be expected in drier climates, with the neocalcic forms (Montagu, Kinkelbos, Augrabies) being confined to arid regions.

Use

Soils in the cumulic subgroup are deep and generally highly suited to cultivation. Problems related to tillage, permeability and root penetration are usually minor although they may need to be addressed especially in the luvic families. Nitrogen and other nutrients may be seriously deficient in eluvic forms and achromic families. The hydromorphic families have signs of wetness beneath the neocutanic or neocarbonate horizon which may require more careful management in irrigated soils but which generally favours dryland farming with deeper-rooted crops. The arenic form requires very careful management because of proneness to wind erosion as well as poor water and nutrient retention (Figures 2.12.5a and c). The fluvic form and alluvial members of the neocutanic and neocalcic forms constitute a large proportion of irrigated soils in South Africa (Figures 2.12.5b and d). Consequently, textural and structural subtleties not accommodated in their definition down to family level need to be carefully considered. Rhodic families are prized by farmers practising irrigation. Especially the fluvic, and to a lesser extent the arenic, forms can be deep tilled to mix the texturally different layers and create a more uniform profile for good root development under perennial crops such as grapes and deciduous fruit.

From an engineering standpoint the cumulic group conveys important information to civil engineers about depth to bedrock. Instability of the arenic form when disturbed has been mentioned, as has the periodic flooding of the fluvic form. There are numerous cases of flooding in residential areas in South Africa (e.g. Ladysmith in KwaZulu-Natal) where inability to recognise the significance of alluvial stratification has proved very costly.

Recognisable indications of incipient pedogenesis in cumulic soils need to be considered in determining not only their suitability for various uses but also their ecological significance. Thus a cumulic soil that is genetically related to one or more of the other soil groups needs to be treated accordingly. This principle applies equally to lithic soils – the upland counterparts of cumulic soils (Figure 2.12.6) which are covered in the next section.

Figure 2.12.5 Examples of land use on cumulic soils: a. small-scale farming on eluvic Fernwood soil in northern KwaZulu-Natal; b. table grape vineyard under flood irrigation on the lower Orange River floodplain near Upington, Northern Cape; c. cover sands on the Maputaland coastal plain in northern KwaZulu-Natal are highly susceptible to wind erosion if the natural bush vegetation is removed; d. floodplain under sugarcane on a complex of fluvic and neocutanic forms of cumulic soils near Port Shepstone, KwaZulu-Natal.

Figure 2.12.6 The most widely encountered juxtaposition of soil groups in the South African landscape is that of lithic and cumulic soils. Their respective origins through erosional and depositional processes could not be more dramatic than in this scene just outside Khubus at the entrance to the Richtersveld National Park.

2.13 LITHIC SOILS

Introduction

Lithic and cumulic soils are complementary, the former characterising convex crests and steep slopes where natural erosion keeps pace with weathering and the latter being typical of gentler, concave footslopes and valley basins where deposition keeps pace with soil formation. Even though not entirely accurate, this simple view can be a useful generalisation.

In the lithic soil group (Gk *lithos*, stone) we find rock, or its weathered but intact derivative, saprolite, dominating the solum. Three forms of lithic soils may be recognised: the orthic (Mispah) form has an orthic A overlying hard rock; the glossic (Glenrosa) form has an orthic A which 'tongues' into saprolite, forming a lithocutanic B, and the eluvic-glossic (Cartref) form has similar tonguing beneath an E horizon. Partly because of aridity and partly due to broken terrain with steep slopes, both of which are widespread in South Africa, natural erosion commonly occurs more rapidly than weathering and the resultant lithic soils dominate much of the landscape (Figure 2.13.1). Rock- or saprolite-dominated soils also appear in the first four of the soil groups that have organic, humic, vertic or melanic topsoil horizons. Furthermore, shallower variants of other soil groups such as the duplex and podzolic soils have saprolite beneath the B horizon. All such soils are closely related to and usefully compared with the lithic group in terms of their properties, genesis and use.

Rock or saprolite close to the soil surface does not necessarily imply that there is less to interest the pedologist. The partial transformation of weathered rock to an incipient B horizon can have fascinating ramifications in terms of colour, texture, structure and consistence, many of which are unique to particular rock types. Transformation from primary to secondary minerals is often best studied in lithic soils because the topotactic nature of such reactions, in which clay and oxide minerals fill voids created by the dissolution of primary silicates, has not yet been obliterated by bioturbation.

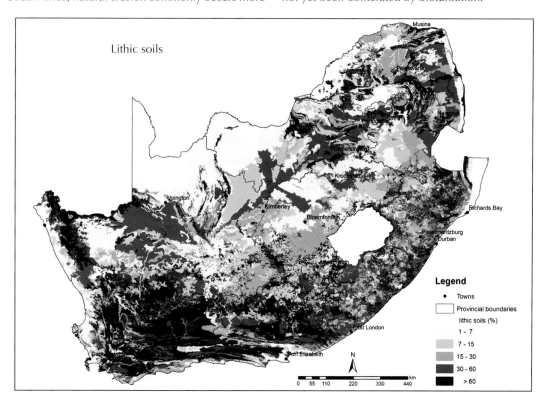

Figure 2.13.1 Lithic soils in South Africa (abundance classes refer to estimated percentages within land types)

Properties

General

The overriding feature of lithic soils is their clear affinity with underlying parent rock, either because of the latter's presence directly beneath an orthic A horizon or because the lithocutanic B horizon exhibits clear morphological evidence of such affinity. Identification is therefore essentially visual and is made easier through geological experience. A selection of lithic soil profiles is illustrated in Figures 2.13.2 to 2.13.5. Their classification is explained later in Table 2.13.1.

Morphological

Diagnostic hard rock, defining the orthic (Mispah) form, cannot be cut with a spade when wet. The lithocutanic B (distinguished from hard rock by not only consistence and degree of weathering but also tonguing and cutanic character) may itself be 'hard or not hard' (Soil Classification Working Group 1991). To be called hard in this context, however, at least 70% by volume of the horizon must be rock material and this must have a defined hard consistence irrespective of wetness. To avoid confusion here between these different definitions of 'hard' we use the terms hypersaprolitic and orthosaprolitic, respectively, instead of hard and not hard, as family criteria in the glossic (Glenrosa) and eluvic-glossic (Cartref) forms, i.e. those with a lithocutanic B (Table 2.13.1). Cutanic character may take the form of tongues of topsoil material extending into saprolite or of argillans which are more clearly indicative of clay movement (lessivage – see duplex soils, Section 2.7). In the lithocutanic B, horizontally discontinuous pockets of well formed peds may also be found, and in such cases the interface with a pedocutanic or prismacutanic B horizon (Section 2.7) is often difficult to determine (this is very much a question of scale of observation).

Other morphological features that are given prominence in the lithic soils are the presence of an E above the lithocutanic B, bleaching of the A, calcareousness (in the A or B, depending on form), and evidence of marked wetness. Whether or not the E, when present, is albic or hypoxanthic is an additional family criterion (Table 2.13.1). The orthic A horizon of lithic soils can encompass a very wide range of colour, texture, structure and consistence which is not necessarily related to the underlying rock, saprolite or lithocutanic B. Because of the steep topography with which lithic soils are often associated, the upper solum is frequently distinct, lithologically, from the B horizon and underlying parent rock. Quite commonly the vertical discontinuity is marked by a stoneline within or at the base of the A or E horizon, the contents of which alert us to the binary origin of the profile.

Figure 2.13.2 Glossic (Glenrosa) form of lithic soils developed in basic igneous rock (gabbro of the Insizwa Complex) near Kokstad, KwaZulu-Natal: a. tonguing into joint planes between blocks of saprolite (*chromic, orthosaprolitic, aeromorphic, acalcic* family); b. gradual transition from jointed saprolite into spheroidally weathered rock.

LITHIC SOILS 137

Figure 2.13.3 Glossic (Glenrosa) form of lithic soils developed in granite: a. near Hammarsdale, KwaZulu-Natal, with partial transformation of feldspar to kaolinite and of biotite to secondary Fe oxides along the margins of dark tongues of topsoil; b. unusual calcic variant (*chromic, hypersaprolitic, aeromorphic, calcic* family) developed in association with termite activity (the carbonate occurred as fine powdery coatings on peds and rock fragments) in the Valley of a Thousand Hills, KwaZulu-Natal.

Physical and chemical

One can no more generalise about the physical and chemical properties of lithic soils than about the properties of their parent materials. Perhaps the most interesting aspect of physical properties is that of spatial heterogeneity associated with differential weathering, illuviation and biotic disturbance, especially along joint or bedding planes. This gives rise to preferential pathways of water movement that are of special interest in assessing both soil water status and groundwater recharge. The pattern of plant root development is also strongly influenced by such features. Incipient soil formation and proximity of readily recognised rock material close to the surface would in most situations suggest a high base status and reserve of weatherable minerals but this is not necessarily the case (see the section on genesis).

Classification

The local classification and international correlation of lithic soils is summarised in Tables 2.13.1 and 2.13.2.

Genesis

In discussing the genesis of lithic soils the best place to begin is in the landscape. Carson and Kirkby (1972) remark that a description of geomorphic evolution through erosion by rainwater can be reduced to the simple generalisation that all hill slope profiles consist of an up-slope convexity and a concave base. These are separated from each other by a neutral main slope, the length and inclination of which determine the overall hill slope profile.

Lithic soils are pre-eminently associated with convexity as a result of divergent water flow which determines that soil is removed. Conversely water flow is convergent in concave zones, where soil material that was removed higher up is deposited. If the neutral main slope is shallow enough – and the vegetation cover dense enough – then some of the eroded material will be deposited in transit across the main slope. If the main slope is very steep, however, erosion will keep pace with deposition and lithic soils will extend from the convex crest to the foot of the main slope where it merges with

Figure 2.13.4 Profile of the albic, orthosaprolitic family in the eluvic-glossic (Cartref) form of lithic soils, near Grahamstown, Eastern Cape. The albic E horizon and dark lithocutanic B formed in weathered shale show a degree of development approaching that of duplex soils. Fragments of quartzite in the A and E horizons indicate binary origin, through colluviation.

the concave base. On the main slope all intermediate positions are possible between the extremes of bare rock and deep colluvial sediment. On convex crests, however, the fluvial creation of cumulic soils (Section 2.12) is essentially impossible; conversely lithic soils are seldom if ever found in concave basins. Operation of the time factor in the evolution of hill slopes is not easily quantified since erosion

 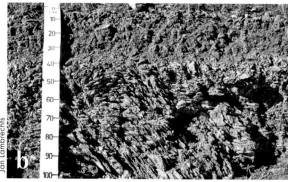

Figure 2.13.5 Lithic soils on shales: a. horizontally bedded Ecca shale near Kokstad, KwaZulu-Natal has resulted in the orthic (Mispah) form; b. diagonally dipping Bokkeveld shale near Caledon, Western Cape, has resulted in the glossic (Glenrosa) form, with achromic bleaching of the orthic A horizon.

Table 2.13.1 Forms of lithic soils and their constituent families (Nomenclature is shown alongside form names, abbreviations and family codes as defined by the Soil Classification Working Group 1991). Terms in italics refer to the orthic A horizon.

Orthic (Mispah Ms)

1100 *chromic, acalcic*	1200 *chromic, calcic*
2100 *achromic, acalcic*	2200 *achromic, calcic*

Glossic (Glenrosa Gs)

1111 *chromic, orthosaprolitic, aeromorphic, acalcic*	1121 *chromic, orthosaprolitic, hydromorphic, acalcic*
1112 *chromic, orthosaprolitic, aeromorphic, calcic*	1122 *chromic, orthosaprolitic, hydromorphic, calcic*
1211 *chromic, hypersaprolitic, aeromorphic, acalcic*	1221 *chromic, hypersaprolitic, hydromorphic, acalcic*
1212 *chromic, hypersaprolitic, aeromorphic, calcic*	1222 *chromic, hypersaprolitic, hydromorphic, calcic*
2111 *achromic, orthosaprolitic, aeromorphic, acalcic*	2121 *achromic, orthosaprolitic, hydromorphic, acalcic*
2112 *achromic, orthosaprolitic, aeromorphic, calcic*	2122 *achromic, orthosaprolitic, hydromorphic, calcic*
2211 *achromic, hypersaprolitic, aeromorphic, acalcic*	2221 *achromic, hypersaprolitic, hydromorphic, acalcic*
2212 *achromic, hypersaprolitic, aeromorphic, calcic*	2222 *achromic, hypersaprolitic, hydromorphic, calcic*

Eluvic-glossic (Cartref Cf)

1100 albic, orthosaprolitic	1200 albic, hypersaprolitic
2100 hypoxanthic, orthosaprolitic	2200 hypoxanthic, hypersaprolitic

Table 2.13.2 Approximate placement of lithic forms in international classification systems (IUSS Working Group WRB, 2006; Soil Survey Staff, 2003)

WRB

Soil form	WRB Reference Soil Group	Possible prefix qualifiers	Possible suffix qualifiers
Orthic (Mispah Ms)	Leptosols	Lithic, Haplic	Calcaric
Glossic (Glenrosa Gs)	Leptosols	Hyperskeletic, Gleyic, Haplic	Calcaric, Dystric, Eutric
	Acrisols	Leptic, Gleyic, Haplic	Skeletic
	Lixisols	Leptic, Gleyic, Haplic	Skeletic
	Cambisols	Leptic, Endogleyic, Haplic	Calcaric
Eluvic-glossic (Cartref Cf)	Leptosols	Gleyic, Haplic	Dystric, Eutric
	Acrisols	Cutanic, Leptic, Gleyic, Haplic	Albic
	Lixisols	Cutanic, Leptic, Gleyic, Haplic	Albic
	Cambisols	Leptic	Dystric, Eutric

USDA Soil Taxonomy

Orthic (Mispah Ms): Entisols (Orthents – lithic variants)
Glossic (Glenrosa Gs): Inceptisols; Entisols
Eluvic-glossic (Cartref Cf): Inceptisols; Entisols

tends to be episodic in nature and prominent features may well have been created by a few events of exceptional intensity (Figure 2.13.6).

In general, lithic soils will be more widespread not only in more dissected terrain but also in more arid regions. Their dominance in Karoo land types can clearly be seen in Figure 2.13.1. Occurrence in more humid climates becomes increasingly limited to the steepest slopes or most convex crests. This relationship of lithic soil abundance with climate is probably governed largely by the covariant density of vegetation cover and the effect this has on erosion rate since, as illustrated in Figure 2.13.7a, the degree of weathering can be extreme in some lithic soils and weathering *per se* is not sufficient to obliterate saprolitic character.

Vegetation may have additional, more direct effects on lithic soil morphology. Besides plants intensifying glossic character through penetration of roots along planes of weakness in jointed saprolite, the phenomenon of tree-fall and ripping by large roots, a form of floral pedoturbation, represents one of the earliest stages in the disintegration of saprolite, allowing an acceleration of soil forming processes and the creation of a more hospitable biomantle (see Chapter 3). Examples of this are abundant in such diverse environments as the rainforest of Guyana and the calcrete terrain around Etosha Pan in Namibia (in the latter case the calcrete is broken and exhumed by roots of trees pushed over by foraging elephants). Such observations prompt speculation that the disorientation of rock fragments illustrated in Figure 2.13.7b has a similar origin even though this area is now largely treeless. Rare large Milkwood trees nevertheless remain from what may once have been extensive woodland.

Biotic influence also manifests itself indirectly in the redox transformations of hydromorphic lithic

Figure 2.13.6 Buried spurs of weathered shale, on a steep hillside near Kokstad, KwaZulu-Natal, separated by a paleo-channel filled (possibly by mass movement such as mudslides) with bouldery dolerite colluvium. The binary origin (dolerite/shale) of glossic profiles on the shale spurs is self-evident.

soils, with organic matter serving as the energy substrate for iron-reducing bacteria. In wet lithic soils, gleying and mottling features can be every bit as well developed (and photogenic: Figure 2.13.7c, d) as they are in gleyic or plinthic soils.

The lithocutanic B horizon shares with the neocutanic B of cumulic soils the distinction of being a forerunner to more strongly developed subsoil horizons, especially the pedocutanic and prismacutanic B horizons of duplex soils. Consequently processes linked to lessivage and the genesis of cutanic character, discussed in Section 2.7, are especially relevant to the genesis of lithic soils. Ultimately, however, it is worth following the advice of Simonson (1959), to the effect that we should consider all soil forming processes for the degree to which they may have influenced each soil we examine.

Use

Lithic soils often have a better quality and are far more useful agronomically than their generally shallow nature would suggest. Provided slope angle is not excessive, even the orthic (Mispah) form can be ripped and cultivated with heavy machinery and turned to profitable use if the economic returns are good, such as with deciduous fruits and vines in the Western Cape and on the *buitegronde* adjoining the lower Orange River (Figure 2.13.8a,b). Pedolo-

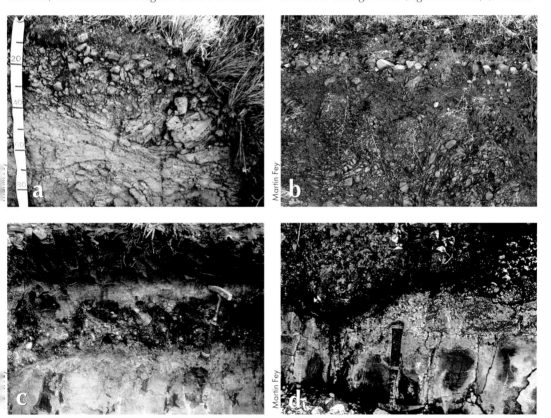

Figure 2.13.7 Special features of some lithic soils: a. bauxitic saprolite derived from dolerite, with gibbsite pseudomorphic after feldspar and the pattern of exfoliation still evident, occurs on the Ngome plateau, KwaZulu-Natal; most soils of this kind are likely to key out in the humic group and accordingly their special nature is discussed in Section 2.2; b. glossic (Glenrosa) profile on the coastal plain north of Cape Agulhas, where bioturbation by tree-fall is the putative explanation for the disturbed orientation of shale fragments (the stoneline is also biotic – see box, Section 2.7); c. and d. are two hydromorphic examples in sandstone saprolite near Storms River, Eastern Cape. Gleying along fracture planes is proximal to dark cutans in the first example and is considerably more advanced in the second, also nicely illustrating the red-yellow-grey sequence described in Section 2.9.

gists from up-country often remark in jest that their colleagues in the Western Cape do not know what real soils are like, which is actually a compliment to Cape farmers who have become adept at turning rock into soil. Such disturbed, stony soils require special care in irrigation scheduling because of their high permeability. Foresters have also learned to produce profitable timber on lithic soils throughout the higher rainfall regions of South Africa (Figure 2.13.8c) although drought damage in plantations can sometimes be serious.

The glossic variants (Glenrosa, Cartref) are obviously preferable to the hard, orthic form (Mispah) and, similarly, orthosaprolitic families of Glenrosa have a better physical quality than hypersaprolitic families. The eluvic-glossic (Cartref) form and the achromic (bleached) families of the Glenrosa form are generally more prone to intermittent wetness, and artificial drainage, even for dryland production of crops such as sugarcane in the eastern coastal regions, is often employed on soils of the old Cartref series (which, incidentally, is the most widely encountered soil series in the sugar belt, occupying about 7% of the area under production).

Where there are better soils to choose from, lithic soils are avoided for intensive use and left as unimproved veld for grazing (Figure 2.13.6). Livestock ranching and wildlife conservation are the most common types of land use on lithic soils, both in mountainous country (because of erodibility or mechanisation difficulties on steep slopes) and in arid areas except where irrigation water is locally available.

Lithic soils provide useful morphological indications of the nature of fractured rock aquifers. These are of interest to groundwater hydrologists because they require special consideration of preferential flow paths, when modelling groundwater movement and recharge in the assessment of aquifer vulnerability to pollution.

Figure 2.13.8 Examples of land use on lithic soils: a. Table grapes (background) are planted in shallow soil derived from weakly weathered granite-gneiss on higher ground adjacent to the Orange River floodplain near Upington, Northern Cape; alteration of the hyperthermic soil climate from arid to humid through irrigation has raised questions about salt generation through accelerated weathering; b. Vineyard on convex crest; c. forest plantation on steep mid-slope, near Stellenbosch, Western Cape.

2.14 ANTHROPIC SOILS

Introduction

Last, but not least, we consider those soils that have been so profoundly affected by human disturbance that their natural genetic character (i.e. their link to the natural factors of soil formation) has either largely been destroyed or, in some cases, has had insufficient time to express itself. There has been a spate of recent activity in this branch of pedology, both internationally and locally. The IUSS Working Group WRB (2006) caters for such soils in two groups: the Anthrosols, which are soils formed or strongly modified through long-term human activities (chiefly related to their agricultural use) and the Technosols, which are soils in recent deposits of artificial origin or soils mixed with alien products.

Properties and classification

In South Africa there is currently only one soil form (Witbank) that caters for the anthropic group (Soil Classification Working Group, 1991). However, a proposal for the recognition of six forms has recently been considered by the Soil Classification Working Group (Bloemfontein meeting, 2006). These six forms are presented in Table 2.14.1, along with a tentative descriptor term for each form. The diagnostic criteria are largely self explanatory and, because agreement is still to be reached on how to accommodate these additional forms in the South African classification, no further detail on the definition of forms and their subdivision into families will be presented here. The international classification of Anthrosols and Technosols is quite well developed (IUSS Working Group WRB, 2006) and probably caters for most of the anthropic soils likely to be encountered in South Africa.

Genesis and use

For present purposes it should suffice to provide some examples illustrating the challenges facing soil scientists in areas where the natural soil mantle has been converted to, or, having been destroyed, remains to be replaced by anthropogenic soil. In South Africa the most extensive areas of anthropic soils belong in the technic (Witbank) form, created as a result of the rehabilitation of mined land and especially where open-cast methods have been employed. Chief among these is coal mining mainly in Mpumalanga (Figure 2.14.1) with other noteworthy examples being diamonds and gypsum in Namaqualand (Figure 2.14.2), and heavy mineral sands on both the west and east coasts (Figure 2.14.3). As a result of various other mining, mineral processing or urban-industrial activities there are also numerous waste dumps which require rehabilitation such as gold mine tailings (Figure 2.14.4), sanitary landfills (Figure 2.14.5) and ash dumps from coal-fired power stations.

Open-cast coal mining (Figure 2.14.1a) has transformed enormous areas of land on the Highveld. Open pits are back-filled with rubble which is flattened then capped with 'topsoil' (stripped from about the top 0.5 m of soil and stockpiled prior to mining). Rehabilitation difficulties are associated mainly with soil compaction, and sustainability could depend on the re-establishment of a diverse population of soil fauna (see Chapter 3). Where rehabilitation is performed on coal discard dumps there is the additional problem of pyrite oxidation

Table 2.14.1 Some proposed forms of anthropic soils (Soil Classification Working Group, 2006)

Diagnostic materials	Soil form	Descriptor
Human-transported soils and materials	Witbank	Technic
Physically disturbed soils	Grabouw	Cultic
Soils strongly affected by chemical pollution	Industria	Chemic
Soils with strongly altered hydrological properties	Stilfontein	Hydric
Soil materials markedly affected by human activities in the past	Maropeng	Hortic
Soils or materials with multiple anthropogenic effects	Egoli	Urbic

Figure 2.14.1 Technic anthropic soils from coal mining activities: a. open pit colliery near Sasolburg, Gauteng; b. coal discard dump capped with soil and grassed, near Vryheid in KwaZulu-Natal, eroding as a result of poor vegetation establishment on the acid, saline footslope.

which can give rise to acid, saline seeps that kill the vegetation leading to erosion of the soil cover (Figure 2.14.1b). The 'new' soil in this case is a capping of mixed A and B horizons of the original soil overlying compacted mine spoil.

Restoration after diamond mining in Namaqualand (Figure 2.14.2a) is made doubly difficult by extreme aridity and the commonly saline, sodic

Figure 2.14.2 Anthropic soils in an arid setting: a. diamond mining in Namaqualand; b. gypsum mining near Vanrhynsdorp, Western Cape; c. biodiversity contrast between mined area after rehabilitation (upper left) and un-mined area on the right.

Figure 2.14.3 Restoration of dune topography and vegetation after mining heavy mineral sands, near Lake St Lucia, KwaZulu-Natal

nature of many of the spoil materials on the coastal plain. Modern mining methods include stockpiling of topsoil for restoration but the area still has many spoil dumps created two or three decades ago that were never capped with topsoil and revegetation has been very slow.

The anthropic soil being created by gypsum mining in Figure 2.14.2b is homogenised subsoil which includes crushed dorbank and gypsum. Here 'topsoil' (about 0.5 m depth) is separated prior to mining and replaced after levelling the spoil material. Rehab results are shown in Figure 2.14.2c. Francis et al. (2007) have emphasised the importance of natural soil horizons such as dorbank and calcrete in the functioning of these arid ecosystems. There could be great value in simulating the effects of these horizons and those of surface gravels (desert pavement – see box in Section 2.5), especially on soil water movement and storage, to ensure sustainable restoration.

Similar considerations apply to the rehabilitation of heavy mineral mines on the arid west coast near Vredendal, whereas on the east coast the rate of revegetation (shown in its initial stages in Figure 2.14.3) is extremely rapid because of favourable rainfall, and within ten years a forest cover can be recreated. The resultant anthropic soil is likely to be little different, within a few decades, from the neocutanic-cumulic soil which once occurred naturally on these forested dunes.

Gold mine tailings are another prominent parent material for anthropic soils (Figure 2.14.4a). In these unbuffered fine sands pyrite oxidation gives

Figure 2.14.4 Anthropic soils on gold mine tailings: a. mine dump formerly covered by Eucalyptus and scrub wattle and currently being reworked; b. anthropic fine sand with incipient horizonation developed in > 50 year-old tailings, both near Benoni, Gauteng.

rise to extreme acidity which needs to be corrected by liming before vegetation can successfully be established. The acidity accelerates chemical weathering which probably contributes to the horizonation apparent in Figure 2.14.4b. A study of soil

Figure 2.14.5 Anthropic soil in the making. Topsoil is used to cover solid waste daily at the Tyger Valley landfill, Western Cape.

forming processes on these old dumps could prove rewarding.

A form of anthropic soil familiar in most urban areas is created through sanitary land-filling of municipal refuse (Figure 2.14.5). Modern practice demands daily covering of freshly disposed waste with soil to reduce odour and the resultant anthropic soil is therefore layered with soil material and decomposing refuse.

These are just some of the more conspicuous examples of anthropic soils in South Africa, sufficient to make it clear that there is enormous scope for pedological studies to be conducted into the factors which make the restoration of disturbed land fully sustainable. An abiding theme that applies to technic-anthropic soils is the critical role of soil structure in determining infiltration and storage of water, as well as aeration and drainage. Without these physical properties being optimised no amount of chemical amelioration (organic or inorganic) will be sufficient to ensure rehabilitation. Since regular tillage is not an option on much restored land, the only way to ensure sustained physical quality is to encourage an active population of soil fauna. The relationship between soils and animals is therefore a vital one and is explored in the next chapter.

Namaqua psalms

These verses were inspired by a handful of souvenirs collected while visiting a mineral sands mine undergoing restoration on the west coast, near Brandsebaai. They link this final section of Chapter 2 with the next chapter.

I

What great history of moving waters
Do these smooth elliptic stones reveal?
Long ago washed down from arid rockscape
Of granite, chert and banded ironstone;
Washed up again and worked by waves on shore,
Rubbing with bright diamonds and each other,
Becoming fashioned into shape too beautiful
To imply a violence in their making,
Buried afterwards by sands of time,
Patiently awaiting other epoch.
Deep weathering the silica enriched,
Erosion of the coastal plain ensued.
Pallid zones stick out with silcrete cappings
Like half-clad breasts of land, God's work accrued.

II

Survey next this cluster of concretions,
This limestone nodule with its ripened breast:
Unfinished sculpture of a half-draped mistress
To feed a fertile mind's imagination;
Beside the broken cup of ironstone geode
With brownish colours of a German poet,
Smooth alluvial softer meerschaum fragment;
All three of these reveal a genesis.
Nestled among bright, loose grains of zircon
And ilmenite in maritime repose,
With vaporising winds reducing solutes
Evolved from brine blown inland to cement.
Thus nature's manufacturing proceeds
Through chemistry. Soil growth will not relent.

III

Sometimes people passed here running, hunting
With stones amenable to useful fracture,
Fashioned tools to cut and pierce their prey;
Left them lying, spent, awaiting termites
To undertake and then, a wait much longer,
As ancient greeting cards, to busier people
With much less time to muse and different tools
To make: refractory glazes, aircraft wings
And bright pigments, needing raw material
From the sands of time, exhuming them
And washing out the useful heavy grains,
Leaving quartz behind as regrowth substrate.
Without cover, sands will surely vanish
Just as the hunters vanished long ago.

IV

Here, among the scrambled, wind-blown wastes
Of last year's mining, lie also the fragments
Of fragile bone, holed shell and vygie seedhead
And dried-out stump of unnamed desert shrub,
All remains of recent living creatures.
They'll not wait so long with mineral patience;
They'll decompose with acids and enzymes
To feed new ecosystems that evolve
On sucked-out sands, recharging them in time
With plant roots, mycorrhyzae, rodents, snakes,
And termites – even aardvarks in the end.
Then granite koppies will look down once more
Across the coastal plain bedecked with sand;
New park on reworked earth in ancient land.

SOILS
OF
SOUTH AFRICA

Chapter 3

Animals in soil environments

Martin Fey, Antoni Milewski and Anthony Mills

Aardvark.
Source: Brehms Tierleben, Small Edition 1927
Artist unknown

INTRODUCTION

Although soil scientists are aware of the role that animals play in forming soil and the reciprocal role of soil in providing for animals, they tend to think of animals as being of somewhat less importance than geology, climate, topography and vegetation in shaping the character of soil over time. Since human disturbance has greatly diminished populations of wild animals interacting with soil we tend to underestimate the intensity of soil-animal relationships in undisturbed ecosystems. On the other hand there are some who consider the upper part of soil primarily as a biomantle (Johnson, 1990; Johnson et al., 2005).

South African soils are – or have been – affected by animals in many different taxa exhibiting an extraordinary range of body size, population and activity. Some animals that lack size often make up for it in numbers. Soil in turn may benefit animals in a variety of ways. It functions as shelter, breeding ground, waste dump, source of food, dietary supplement and means of bodily hygiene. In exploiting soil, animals change it. Mixing, segregation and nutrient cycling take place. The consequences for porosity, aeration, water storage, drainage, density, erodibility and nutrient status can be profoundly important, both ecologically and economically.

In deciding on material for this chapter we have chosen conspicuous examples and have not attempted to describe a full or balanced spectrum of animal-soil interactions. The objective is to heighten awareness of how fauna affect soil properties and how soil affects the welfare of animals. We begin our survey near the base of the food chain with the ubiquitous termites, followed by the animals that depend largely on termites for food. The same theme is applied to earthworms and their main burrowing predator, the golden mole. There are short accounts of various mesofauna, other burrowing animals and birds and, finally, large herbivores. The largest and smallest creatures are sometimes found to push soil formation in different directions. Hopefully, in what we have written to celebrate the biomantle, readers will find something to entertain, inform, inspire new research and serve as a reminder that there is not only life on earth, but in it too.

3.1 TERMITES

Although not unique to Africa, termites are one of the most distinctive fauna on the continent, not so much for what they look like but for what they produce. South Africa's landscape owes much of its character to their earth structures and in some regions termites exert a dominant effect on soil formation. Besides their tunnelling, construction of mounds, cycling of nutrients and alteration of the vegetation cover, many of their predators are burrowing animals and add further to what soil scientists refer to as faunal pedoturbation.

The genus *Macrotermes* builds the largest structures of any invertebrate and includes the largest termites (queens are up to 14 cm long and 3.5 cm wide; Carwardine, 1995; the soldier in Figure 3.1.1 is about 1 cm long). In South Africa it occurs as far south as Kimberley in the west and the Kei River in the east.

This genus builds not only spires, which are chimney systems located above the hives and reinforced with the equivalent of ceramic materials or mortar (Figure 3.1.2 compares a variety of spires from different genera), but also large mounds of earth extraneous to their nesting activities. In addition it builds foraging runways consisting of a crust of earth. *Macrotermes* also burrows deeper – as much as 70 m – than any other termite and is a powerful agent in returning eluviated clay, nutrients and groundwater to the surface. During the rainy season a constant battle is waged against erosion with spires frequently having to be patched with fresh earth (Turner et al., 2006). Pomeroy (1976) estimated that *Macrotermes natalensis* colonies in Uganda each move as much as a cubic metre of soil annually to the surface which could amount to about ten tons per hectare given that in many areas in tropical Africa there are more than ten *Macrotermes* colonies per hectare (Boyer, 1975). Lepage (1974) found that of the 2 t ha^{-1} of soil brought to the surface each year by *Macrotermes subhyalinus* at least a third was in the form of foraging runways.

Although *Macrotermes* builds its largest structures under the moist, tropical climate of central Africa, the earth mounds found in the Kruger National Park are extensive enough to support patches of forest trees (e.g. *Diospyros mespiliformis*).

These patches are taller than the surrounding vegetation and different in floristic composition (Figure 3.1.3). Vegetation on the mounds is often exempt from surrounding fires. Tree-clad mounds provide an attractive foraging substrate and refuge for a variety of animals. This serves further to intensify the partitioning of nutrients in the landscape that has already been achieved through the foraging of termites between the mounds. Although the food of fungus-culturing termites is completely mineralised

Figure 3.1.1 Termites in action: a. *Macrotermes* soldier; b. *Odontotermes* workers building spire; c. *Hodotermes* workers and soldiers at entrance to foraging hole.

Figure 3.1.2 Contrasting termite spires: a. in central Namibia (*Macrotermes*); b. north-west Zambia; c. Mabalingwe, Limpopo Province; d. in a wetland, southern Democratic Republic of Congo; e. northern Namibia (possibly *Odontotermes*); f. *Microhodotermes*, Knersvlakte, Namaqualand.

Figure 3.1.3 Woody vegetation associated with *Macrotermes* mounds: a. the Kruger National Park; b. Mozambique coastal plain.

by a rapid and thorough digestive process located in an aerobic culture of fungus (Basidiomycetes: *Termitomyces*) in the centre of the mound (Figure 3.1.4), organic matter continues to accumulate on mound soils because of the attraction of other animals and their faeces to the mounds, protection from fire, and the clonal self-propagation of many of the shrubs and trees that colonise the mounds. The growth of herbaceous annuals is typically better on the mounds of *Macrotermes* than between them and farmers in nearby countries such as Zambia and Zimbabwe use earth from the mounds to improve their crop lands. Similarly in the sugar belt of KwaZulu-Natal it is common to see more luxuriant cane growth on stulis (see below) than between them. This is not always the case, however, and near Likasi in the Democratic Republic of Congo, maize lands have been observed to be completely barren on patches of bulldozed termitaria. Soil analysis revealed severe acidity and infertility suggesting that the basal material from the mound consisted of sterile subsoil.

Macrotermes and other fungus-culturing termites (Termitidae: Macrotermitinae) such as

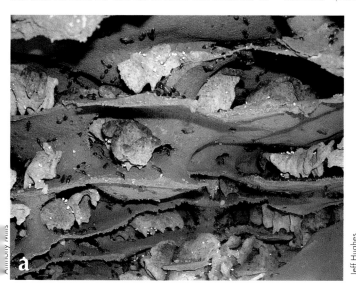

Figure 3.1.4 Subterranean fungus garden in *Macrotermes* mounds: a. in Namibia; b. in KwaZulu-Natal.

Odontotermes are exceptionally active probably owing to the capacity of their fungus cultures, which have no counterpart among termites in the Americas or Australia, to digest fibre by complete oxidation. However, *Macrotermes* is currently restricted to the warm and mesic parts of South Africa and even small species of fungus-culturing termites are today absent from the Mediterranean-type climate of the southwest Cape and from low-lying parts of the adjacent Karoo.

One of the puzzles of pedology in these areas is the existence of large earth mounds (known locally as heuweltjies or kraaltjies) which with the termite *Microhodotermes* is associated, although these have not yet been shown to have built the mounds. Heuweltjies (Figure 3.1.5) are comparable in mass to the large earth mounds of *Macrotermes* (Figure 3.1.3), by which they may possibly have been built during past relatively warm periods of summer rainfall extending to the southwestern Cape. It is noteworthy that the distribution of heuweltjies has an abrupt southerly limit at the latitude of Cape Town, leaving them absent from the Ruensveld in contrast to their commonness in the Swartland and Little Karoo. This anomaly is difficult to explain in terms of the current distribution of *Microhodotermes*, which remains common in the Ruensveld. However, it appears consistent with a former latitudinal limitation on termites of tropical affinity such as *Macrotermes*. Heuweltjies have a counterpart in KwaZulu-Natal where they are known is 'stulis' (Zulu: *isiduli*) illustrated in Figure 3.1.5b and f.

Whatever their origin, heuweltjies remain important in pedology and soil ecology because they are relatively nutrient-rich, continue to be inhabited by *Microhodotermes* even in dry climates, and form foci of activity for rodents and other digging animals. *Microhodotermes* belongs to a different family (Hodotermitidae) from *Macrotermes* and not only lacks a fungus culture, but relies on protozoan gut symbionts which are likely to digest cellulose slowly compared to the gut bacteria found in most termites. *Microhodotermes* is, however, a remarkably large and active termite, differing from most termites worldwide by being as much an herbivore as a detritivore. *Microhodotermes* often cuts down fresh green plant matter, particularly of the dwarf shrubs that dominate karoo vegetation. It is unusual among termites in bringing nutrient-rich faeces to the surface, where these may play a significant role in fertilising the plants that form staples for the termites and the larger herbivores of the region.

Heuweltjies often contain hardpans of secondary, pedogenic material (see Box) and may also develop into zones of hypersalinity. In the Kaokoveld of northern Namibia the distinctive circles of compacted red sands that remain bare of plants in perpetuity because of their extreme salinity are thought to be old, uninhabited *Macrotermes* mounds that have eroded into flat features over time (pers comm with Ken Tinley, 2006). Stonelines are another major feature in South African soils (see box, Section 2.7) that is attributed to termites (and other fauna) by some pedologists. The idea is that termites have, over long periods, moved fine particles to the surface leaving gravel and stones behind in a layer below the topsoil (Figure 3.1.5b). This effectively amounts to inverse sieving of the soil at the scale of the landscape.

The effect of termites on soil structure and infiltrability is to a large extent determined by soil type and chemistry of the site. The import of calcium from plant material into mounds will often decrease the proportion of exchangeable sodium and reduce dispersion, thereby potentially increasing infiltrability of dispersive soils (Ellis, 2002). However, mound earth tends to be made of finer particles and is consequently often less permeable than adjacent soil surfaces (de Bruyn & Conacher, 1990). Off the mound, termites can considerably boost soil macro-porosity through their construction of feeding channels and chambers. Elkins (1986), for example, found in the Chihuahuan desert that infiltration rates were significantly higher (88 mm h^{-1}) in the presence of termites than in their absence (51 mm h^{-1}). The effect of termites on landscape infiltrability is thus complex and requires analysis at several scales.

Microhodotermes currently inhabits heuweltjies but its hives do not appear to be restricted to them. Its colonies (recognised by small, hard, conical termitaria) frequently occur on the surface between these mounds, and its range extends beyond that of the heuweltjies, almost reaching Cape Agulhas. Although it remains possible that heuweltjies were constructed by *Microhodotermes* this notion is not supported by the activities of this genus

HEUWELTJIES AND PEDOGENIC CEMENTS

Heuweltjies in the Western Cape and Namaqualand commonly show definite geochemical differentiation within the mound, manifested as a secondary accumulation of calcium carbonate, silica, sepiolite, and even manganese oxides. In a single heuweltjie, the accumulation appears in the form of segregated layers or lenses of cemented calcrete, dorbank and/or sepiocrete (Ellis, 2002; Francis, 2007). Manganese oxide is less common and, when present, usually occurs in the form of a black horizon towards the base of the main calcrete or dorbank horizon. In the most arid parts of this region such horizons may be found throughout the landscape. In less arid parts (e.g. near Piketberg) such materials are found only in heuweltjies and not in the intervening soil mantle. Since the occurrence of such materials is typical of arid climates the simplest explanation for their persistence in heuweltjies but not in the intervening regolith is that the heuweltjies are more xeric than the surrounding soil. Quite simply, rainwater runoff from the mound surface wets the inter-mound area and the landscape is partitioned into arid heuweltjies and less arid intervening soil. The dorbank lenses in heuweltjies have been shown sometimes to have a flattened 'doughnut' shape with maximum thickness towards the periphery of the mound (Ellis, 2002). This could simply mean that the degree of leaching is least beneath the sloping banks of the mound where runoff is greatest, although it has also been suggested that solutes could be drawn by capillary action into the relatively arid mound from intervening moist soil before precipitating as silicate or carbonate cements. It is tempting to regard such features of heuweltjies as being biogenic. It is quite possible that termite activity, in concentrating nutrients over millennia (heuweltjie material has been dated as old as 24 000 years), has included Ca and Si (silica is a significant constituent of much plant material, especially grasses). However, the gradients in concentrations of water, carbon dioxide, and oxygen which have given rise to the accumulation and spatial segregation of carbonates, amorphous silica, sepiolite (magnesium silicate), and manganese oxide within heuweltjies are probably not unique, since such features often can be observed in soil mantles unaffected by heuweltjies (see earlier accounts of silicic and calcic soil groups in Chapter 2). Concentrations of carbon dioxide are likely to be greater in active termite mounds compared with normal soils, which would further promote the precipitation of calcium carbonate. The formation of secondary cements such as those described is therefore likely to have been mediated biogenically, even though the presence of these cements in heuweltjies is quite readily explained in terms of hydrological and geochemical processes. Interpretation is nevertheless complicated if we accept the theory that heuweltjies are paleo-features engineered by a different species of termite under different vegetation cover and climatic conditions than those prevailing today. Some possible evidence for this theory comes in the form of silica-cemented material such as that shown in Figure 3.1.5b. On the other hand Figure 3.1.5c shows evidence of secondary geochemical enrichment (in this case of Ca and Fe minerals) associated with a termite mound that is currently active.

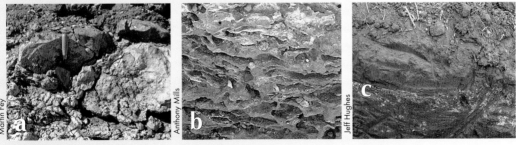

Figure 3.1.5 Cemented material in heuweltjies: a. secondary enrichment consisting of calcrete and sepiolite in a heuweltjie near Vanrhynsdorp, Western Cape; b. material of similar vesicular morphology to the fungus garden in Figure 3.1.4a, but cemented with silica, exhumed from a heuweltjie in the Western Cape and turned into a garden ornament; c. Secondary white calcite and orange iron oxide formed in a termite mound in northern KwaZulu-Natal.

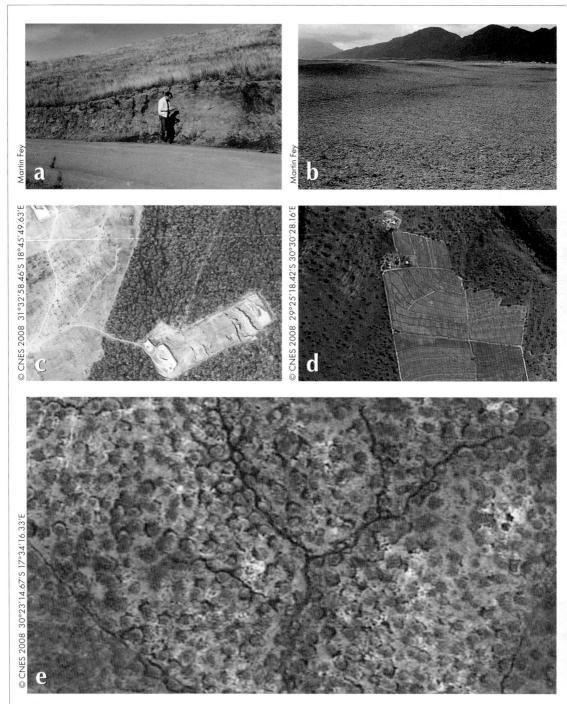

Figure 3.1.6 Large mounds are known as heuweltjies in the Western Cape and Namaqualand and as isiduli in KwaZulu-Natal: a. isiduli cross-section with stone line near Hammarsdale, KwaZulu-Natal; b. dark heuweltjies enriched with humus in wheat lands near Piketberg, Western Cape; c. At the gypsum mine north of Vanrhynsdorp, Namaqualand; d. Isiduli near Wartburg, KwaZulu-Natal; e. Modifying drainage lines near the road between Koingnas and Garies, Namaqualand; f. dense pattern associated with pans near Piketberg, Western Cape; g. peculiar amalgamation of heuweltjies near the Wolfberg, south of Springbok, Namaqualand. (© CNES 2008, Distribution Spot Image S.A., France, all rights reserved)

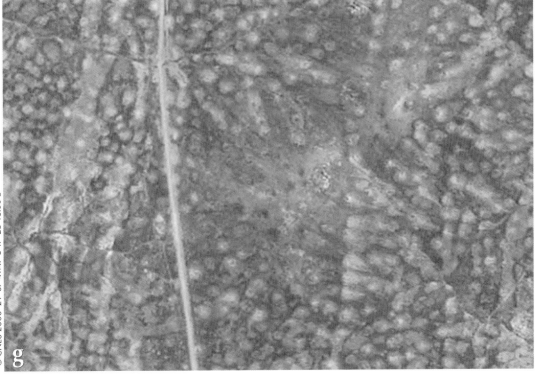

and members of the same family in other lands. *Microhodotermes* is native to North Africa, including Morocco, Tunisia, and Libya, and a related genus, *Anacanthotermes*, occurs in the Arabian Peninsula and Eurasia. Earth mounds are unrecorded in this vast area, which appears to be too far north for the previous occurrence of *Macrotermes*, the possible past builder of heuweltjies, even under the extremes of pluvial climate in the Pleistocene.

Microhodotermes is replaced northwards from the Cape region by another genus of Hodotermitidae, widespread in the bimodal and summer-rainfall areas of southern and eastern Africa, namely *Hodotermes*. Although *Hodotermes mossambicus* builds neither earth mounds nor any termitarium visible above ground, it is nonetheless important for soil ecology in South Africa. This is because *Hodotermes* is, even more than *Microhodotermes*, an exceptionally active termite analogous to a grazing herbivore, and regularly digs deeper than 7 m (pers comm, Mike Picker), bringing subsoil rapidly to the surface in a way likely to assist the rapid recycling of nutrients. Both *Microhodotermes* and *Hodotermes* can be observed foraging on the surface during the day which explains the dark pigmentation and visual perception possessed by *Hodotermes*. The latter has compound eyes in workers (unlike most other termites which only have compound eyes in alates). It is comparable to locusts (Orthoptera) rather than conventional termites in consuming green as well as dry grass. These termites tend to denude the little standing hay that remains in drought years and some ranchers will go to great lengths to control or exterminate them. This is difficult because the hives are deep (4 m or more). In the long run hodotermitid termites may possibly become more appreciated by ranchers for boosting productivity, by conserving and recycling nitrogen and other nutrients and promoting rainwater infiltration.

3.2 TERMITE- AND ANT-EATING ANIMALS (MYRMECOPHAGES)

The exceptional rates of reproduction and growth of South African termites can be attributed to their unusual trophic strategies which include fungus-culturing, consumption of fresh faeces of herbivores and harvesting of fresh foliage. These functions also affect soils indirectly through the animals that eat termites. Besides the familiar aardvark (*Orycteropus afer*) there are other burrowing animals that are unusually dependent on termites, including the aardwolf (*Proteles cristatus*) and the bat-eared fox (*Otocyon megalotis*).

Aardvark

The aardvark (Figure 3.2.1), weighs about 55 kg, is the largest animal specialised to eat insects and is an extremely rapid excavator, differing from myrmecophages on other continents in being related to ungulates. Although most myrmecophages are known for their slow metabolism, the exceptional physiological power of the aardvark explains its ability to promote other burrow-dependent animals which in turn contribute to the turnover of soils. Whereas edentates such as the giant anteater (*Myrmecophaga tridactyla*) and giant armadillo (*Priodontes maximus*) of South America (both of which weigh approximately 35 kg), lack teeth and have slow metabolism, the aardvark achieves rapid and thorough digestion partly through grinding its food with a set of molars. The aardvark is extremely widespread in South Africa, as far south as Cape Agulhas, and is commonest in areas heavily grazed by ungulates. Although the success of the aardvark is partly explained by its exploitation of the exceptionally fast-growing fungus-culturing termites of the tropics and subtropics, it appears that other termites and ants can sustain this animal in the more temperate and arid regions of southern Africa.

In a study of radio-tracked aardvarks in the Karoo, burrowing was estimated to affect half a percent of the land area per year, with burrows being occupied for between 4 and 38 consecutive nights and then abandoned (van Aarde, 2001). Individuals 'make burrows of three main types: those dug when looking for food; larger, temporary sites, scattered through the home range, that may be used for refuge; and permanent refuges in which the young are born. The latter are deep and labyrinthine, up to 13 m long, and usually have more than one entrance. In the Ruwenzori National Park in Uganda, densities of up to 15 burrows ha^{-1} have been recorded (Melton, 1976); while in the Karoo (where burrows generally occur in clusters) 58 burrows were recorded in an area 40 x 200 m

ANIMALS IN SOIL ENVIRONMENTS

Figure 3.2.1 Aardvark: a.; b.; c. burrow and d. termitarium damaged by feeding of aardvark near Kokstad, KwaZulu-Natal.

(A. Taylor unpubl.). Aardvarks like to change the layout of their homes regularly, and can, if necessary, dig new burrows with considerable speed, disappearing below ground within 5-20 minutes' (van Aarde 2001). The effects of aardvarks on soil formation through time are thus likely to be considerable. A conservative estimate of the annual soil movement by aardvarks – given the formation of 15 burrows ha^{-1} – is 33 tons ha^{-1} (Thomas Lehmann, pers comm, 2006).

Aardwolf

Sliwa (1996) has written a comprehensive account from which the following information is gleaned. Although the Aardwolf (Figure 3.2.2) is a competent soil excavator, its greatest effect on soil properties is arguably through nutrient cycling and concentration of nutrients in middens. During its lickings of termite foraging parties off the soil surface, a single animal consumes up to 300 000 *Trinervitermes trinervoides* in a single night. Like the hyaenids to which it is related the aardwolf defecates mostly at middens – areas of soft, bare sand exposed by frequent diggings. Trenches are dug in the middens (with alternating strokes of the front paws) to house the surprisingly large defecations (as well as urinations), which, once deposited, are covered up with sand. Up to 1 kg of faeces is voided in the first defecation of the evening.

Figure 3.2.2 Aardwolf

Notwithstanding the fact that in sandy areas faeces may comprise 40% sand this is a noteworthy output, considering that the animal's body mass is only 8–12 kg. The productivity of *Trinervitermes* termites is remarkable and, so far, unexplained. Aardwolves consume about 105 million termites per year, with a family consuming about half of the standing crop of *Trinervitermes* termites in the approximately 3 000 mounds in their territory. Given that there are approximately 55 000 termites per mound, and 20 middens per territory (of 100–600 ha), the average midden of 1–4 square metres receives the remains of some 4 million termites annually. Although soil studies have to our knowledge not been conducted on aardwolf middens these sites must represent significant concentrations of organic matter and nutrients in the landscape.

The aardwolf may use old aardvark or porcupine burrows as dens but it usually enlarges springhare burrows or excavates its own. During den construction an aardwolf may excavate more than 650 kg of soil. The final product (which may extend for 6.5 m) has an oval-shaped single entrance (30–40 cm in diameter), rapidly narrowing to 20–30 cm inside the tunnel with a chamber (100 x 40 x 25 cm) at the end, usually providing space for one adult and her cubs. The dens are used regularly for six to eight weeks and may be re-occupied six to eighteen months later.

Bat-eared fox

Information for this section has been drawn from Clark (2005) and www.canids.org. Bat-eared foxes (Figure 3.2.3) can be abundant or rare in their preferred habitats, with densities depending on rainfall, food availability, breeding stage and disease. Recorded densities range from a minimum of 0.3–0.5 foxes per km^2 in the Karoo to 0.7–14 per km^2 in the south-western Kalahari. The fox's prime habitats in southern Africa are short grass plains and areas with bare ground, its range almost completely overlapping those of *Hodotermes* and *Microhodotermes* – the termite genera that make up 80–90% of its diet. The amount of digging that bat-eared foxes undertake during foraging appears to fluctuate according to the availability of certain prey species. In the Serengeti (where the foxes are common in open grassland), dung beetles are the main source of food during the rainy season when termite activity is reduced. When both dung beetles and termites are scarce, beetle larvae are often dug up from the ground. The fox's foraging technique is dependent on prey type but often entails walking slowly, nose close to the ground and ears cocked forward. The prey is detected mostly by sound, with sight and smell playing a lesser role. In southern Africa nocturnal foraging during summer gradually changes to an almost exclusive diurnal pattern in winter, mirroring the activity changes of *Hodotermes mossambicus*. Even the fox's reproduction is closely tied to these termites, the litter size being positively correlated with the density of foraging holes of *Hodotermes*.

Although bat-eared foxes prefer to lick *Hodotermes* or snatch dung beetles from the soil surface, when it comes to den construction and maintenance they are prodigious diggers. During high winds and low temperatures the foxes will frequently rest in self-dug dens and when seeking shade in the middle of the day they will make use of small holes or modify existing ones. Breeding dens may be adapted from disused dens of other

Figure 3.2.3 Bat-eared fox

mammals (e.g. springhare, aardvark and warthog) but are frequently excavated from scratch by breeding adults. These dens may have several entrances, and chambers and tunnels up to 3 m long; presumably the labyrinth design provides alternatives for escape from predators. Cubs are sometimes moved between dens and in the Serengeti the foxes protect their cubs in different 'foraging dens' while on foraging sorties. Dens are zones of frequent soil disturbance since the foxes are fastidious housekeepers, carefully maintaining their dens throughout the year and often using them for several generations. The soil disturbance in a single landscape can be considerable, with breeding dens being highly clustered. In one study in the south-western Kalahari six dens were found in half a square kilometer of dry river bed.

3.3 EARTHWORMS

Earthworms (Annelida: Oligochaeta) are generally most prolific under temperate climates. Their capacity for earthmoving is well known, with no less a biologist than Charles Darwin having devoted much of his time to observing and recording their activities. In the introduction to his book *The Formation of Vegetable Mould*, Darwin (1881) wrote:

'In the year 1869, Mr. Fish rejected my conclusions with respect to the part which worms have played in the formation of vegetable mould, merely on account of their assumed incapacity to do so much work. He remarks that 'considering their weakness and their size, the work they are represented to have accomplished is stupendous'. Here we have an instance of that inability to sum up the effects of a continually recurrent cause, which has often retarded the progress of science, as formerly in the case of geology, and more recently in the principle of evolution.'

This statement could apply to the role of animals in general and not only to that of earthworms in affecting soil formation.

Under tropical and xeric climates earthworms are generally eclipsed by termites as the main agents of litter recycling and tunnelling in soils but the distribution ranges of these two organisms are not mutually exclusive. The earthworm population density of soils can be highly variable. Large populations are usually associated with an abundance of organic substrate, especially a supply of surface litter, but will not materialise if physical (water, air and temperature) and chemical conditions (soil pH and availability of elements such as calcium) are not suitable. The effects of earthworms are most obvious in the casts of digested soil which they produce, either on the soil surface or within the soil. The transition between darker topsoil and a lighter coloured subsoil horizon is where casts are most easily observed because of the contrast in colour between casts containing topsoil material and the matrix consisting of subsoil. Casts can occasionally be found at considerable depths, in excess of one metre. The nature of worm casts has been studied as much for the effects they have on plant nutrient availability as for their physical stability and contribution to soil aggregation. In feeding off surface litter, earthworms create passages (continuous macropores) that promote more rapid infiltration of water as well as aeration of the soil. The stable aggregates formed by their casts also promote better aeration and drainage in the long term. (Once created, pore space will not last long if soil aggregates disintegrate easily because finer particles migrate with moving water and produce 'clugging': terminology invented by the late Haim Frenkel for its educative value, referring to narrow necks of pores becoming plugged, following which the whole pore space becomes clogged).

Earthworms are, in many environments, probably at least as important as termites, ants and other small burrowers in creating the biomantle (Johnson et al., 2005) with its characteristic stoneline often marking the lower limit of sorting (see Section 2.7). Darwin himself observed how a 'stoneline' had formed by exhuming worm-rich soil some years after scattering coarse fragments on the soil surface. The burial of gravel and stones with fine earth to a depth often exceeding that of normal cultivation must be one of the more dramatic impacts which worms and their fellow small burrowers such as termites have had on soil tilth. In some regions this beneficial work has been undone, a notable example being through wheat cultivation in the Swartland region of the Western Cape (see for example, profile XVI described by van der Merwe, 1940, his Appendix X).

So much for the textbook aspects of worms and their effects on soils in general. South Africa has the distinction of being home to the largest recorded earthworm in the world: the giant *Microchaetus rappi* has an average length of 1.36 m, an extreme length of 6.7 m and diameter of up to 2 cm (Carwardine, 1995). Giant earthworms are found on other continents but the unique habitat which they have created over a small belt of country between King William's Town and East London in the Eastern Cape must rank as one of the greatest natural marvels to be found in South Africa. Several species of giant earthworm occur in this area, and they are believed to be the main agents responsible for what are commonly known as kommetjies (Figure 3.3.1), a wavy, reticulated pattern of hollows and intervening mounds that has altered the landscape on a scale at least as dramatic as that of the heuweltjies of Namaqualand (see Section 3.1) or the gilgai of some vertic soils (Section 2.3). The type locality for kommetjies is Debe Nek (the Xhosa word *debe* has similar connotations to that of the Afrikaans word kommetjie in referring to the hollows between the mounds, which fill with water after heavy rain). A Google Earth search for Debe Nek allows anyone with an internet connection to explore from outer space and admire kommetjies at leisure. Some examples are given in Figure 3.3.2. These giant earthworms must have existed in the region for a very long time. Not only are there several species (the most notable perhaps being *Proandricus skeadi*, the one named after C. J. Skead whose unpublished field observations are cared for in the Amathole Museum, King William's Town) but there is an endemic giant golden mole (see Section 3.4) that feeds on them and a recently discovered plant species (*Arctotis debensis*; McKenzie et al., 2006) that is virtually restricted in its distribution to the so-called kommetjie grassland and appears to be morphologically adapted to escaping burial by worm casts (the casts of *P. skeadi* can apparently reach a height of 24 cm).

How do kommetjies form? Based on scant descriptions in the literature (e.g. papers cited by McKenzie et al., 2006) and some personal field observations, the following somewhat speculative explanation has been attempted. The soils affected are confined to upland positions (note the absence of kommetjies on foot slopes and toe slopes in Figure 3.3.2a), are relatively shallow, and generally could be described as either lithic-eluvic or plinthic-eluvic (see Sections 2.9 and 2.13). Figure 3.3.1d shows the low chroma soil colour associated with intermittent wetness and the gravelly base of the profile overlying rock. Such soils, under the sub-humid, summer rainfall climate of the area, are prone to periods of excessive wetness interspersed with droughty conditions. Puddles of rain water, containing an increment of organic residues washed in from the surroundings, would constitute a reservoir of wet soil (and worm food) capable of sustaining a large worm with its feeding end in saturated soil and its casting end in aerated soil. The larger the worm the more effective it would be at withdrawing nutritious mud from the puddle and depositing it after digestion on the puddle shore. Bigger worms would be able better to capitalise on wet spells before going into a dormant state during intervening droughts. Depending on topography (the area is sometimes referred to as the kommetjie flats and has a gentle relief) and on the initial uniformity of shallow depth to bedrock or an impervious plinthic horizon, earthworms might need, be able to achieve, and benefit from an enlargement of the puddles so as to trap increasing amounts of rainwater and secure more extended periods of activity and growth. Growth of the hollows would progress so as never to become wider, along their narrowest dimension, than an earthworm could reach to feed while still breathing (by diffusion through its skin) along sufficient of its length to avoid drowning. The depth of the hollows must in many places have reached a maximum because exposed bedrock is widely in evidence in the hollows. A parallel orientation of hollows and intervening mounds could be thought of as an inevitable

Figure 3.3.1 a. giant earthworm, Fish River reserve, Eastern Cape; b. giant earthworm casts; c. kommetjie topography produced by giant earthworms; d. cross-section through a kommetjie mound; all near Debe Nek, Eastern Cape.

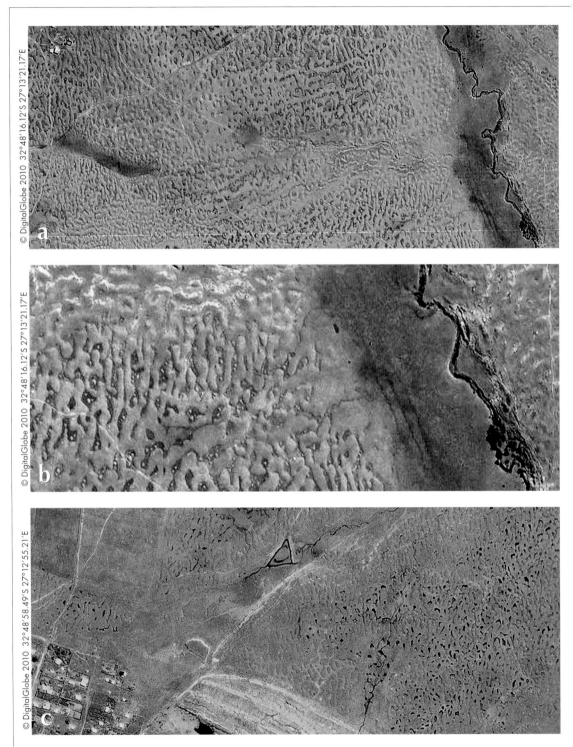

Figure 3.3.2 Satellite images of kommetjie terrain near Debe Nek, Eastern Cape: a, reticulated pattern, with small house at top left providing scale; b. enlargement of lower right section of image in a; c. Kommetjies filled with water after rain (small houses and cultivated lands provide scale at lower left).

consequence of this condition. A fairly substantial degree of mutual orientation is evident in Figure 3.3.2; hence the wavy appearance of kommetjies when seen in the field. The enlargement in Figure 3.3.2b shows small residual islands of soil within some of the hollows. The image in Figure 3.3.2c suggests that considerable volumes of rain water are captured in the kommetjie terrain. The more impervious the bedrock the more likely this capture will have a lasting benefit. The generation of a mound, by casting adjacent to the hollow, augments water capture while enhancing local relief which maximises respiration of the worms while they feed for more extended periods on anaerobic, organic mud. Of course the idea that the giant worms feed like pumps with one end in the wet sediment and the other in dry ground is a bit far fetched and it may simply be that the economy of scale afforded by size of the worms allows them to retire further to well aerated soil after feeding than would be possible for small worms.

So why do we not find kommetjies in places where there are giant earthworms elsewhere? The suburb of Scottsville in Pietermaritzburg has giant earthworms that produce enormous casts but no kommetjies. Some special combination of climate, topography, soil depth, bedrock lithology and spatial extent of this combination must have generated the kommetjie grassland. There are still lots of questions unanswered. Why else does size matter if not only for making kommetjies? Those field notes of C. J. Skead should be scanned and put on the internet.

The giant earthworms produce a prodigious mulch of casts (Figure 3.3.1b) but the quantity of soil moved per year is unknown. Data from other parts of the world suggest that it could be in the order of hundreds of tons/ha. The surface cast production by earthworms in temperate regions ranges from 10–500 t ha^{-1} year^{-1}, with most estimates for pastures and grasslands ranging from 40–50 t ha^{-1} year^{-1}. The latter represents a 3–4 mm annual increment of material with a relatively high clay/silt content on the soil surface (Anderson, 1988). It is thought that casting rates of tropical earthworms have been substantially underestimated through inadequate sampling of geophagous species which have been recorded to consume 5 to 36 times their body weight of soil per day. A study in an Ivory Coast woodland savanna estimated total cast production to be 800–1200 t ha^{-1} year^{-1}, of which only 28–35 t ha^{-1} was deposited on the soil surface. The top 15–20 cm of soil on this site was entirely composed of earthworm casts (Lavelle, 1978).

As already mentioned, the activity of earthworms tends to increase pore volume, water-holding capacity, aggregate stability and infiltration rate. The introduction of a lumbricid earthworm to New Zealand pastures resulted in the doubling of the infiltration rate and an increase in grass rooting depth. Pasture production was stimulated by as much as 72% initially, with a long term increase in yield of 25–30% (Stockdill, 1982; Syers & Springett, 1983). By contrast, a study in Brazilian pastures found that the presence of an exotic earthworm species resulted in a marked decrease in infiltration (Barros et al., 2001). The negative effects on soil structure were ascribed to an imbalance in the macro-invertebrate community due to dominance of the one earthworm species. By restoring the soil faunal balance, through transplanting cores of soils from adjacent intact forests, the soil structure improved and infiltration rates increased. It is evident from this study that manipulating soil structure by altering soil faunal composition requires specialised knowledge of the soil ecosystem.

3.4 GOLDEN MOLES

Earthworms are eaten by other burrowers which in turn amplify the churning effect on soils. Chief among these are the golden moles (Chrysochloridae; Figure 3.4.1). They are uniquely African and include more than 15 species in southern Africa. They mainly eat small animals with some species depending especially on earthworms, in contrast to the vegetarian mole-rats with which they coexist and may share burrows. Most golden moles 'swim' through soil, producing foraging tunnels just below the surface which protrude above, much to the consternation of suburban lawn keepers. Most species also excavate deeper burrows (creating *krotovinas* – the special pedologists' term for infilled animal borrows – in the process) for breeding, refuge, and hygiene and from which they deposit small molehills, typically much smaller than those of mole-rats (Section 3.5), on the surface. They have exceptional

Figure 3.4.1 Juliana's golden mole

orientation ability, linking new tunnels precisely with existing ones.

Eleven species of golden mole are threatened with extinction, due largely to mining and agricultural practices. The giant golden mole (*Chrysospalax trevelyani*, up to 23 cm long) inhabits fire-free patches of forest in the Eastern Cape and appears to depend on the giant earthworm for food. It would be interesting to know by how much its foraging activity aerates the mounds of the kommetjies and thus benefits the growth of the earthworms.

The near-surface foraging habits of golden moles potentially make them one of the most important natural agents of soil tillage. Evidence for this is dramatic in long-term veld burning experiments in the Drakensberg (Mike's Pass, near Cathedral Peak) and on the Ukulinga research farm outside Pietermaritzburg, where plots left unburned for more than thirty years have much softer, more permeable topsoil than adjacent, annually burned plots. Withholding fire allows plant litter to accumulate with a resultant active earthworm population and abundant signs of golden mole activity. It would be instructive to calculate the fuel requirements of the mechanical tillage needed to achieve this softening of hard topsoil with the energy value of unburned plant material supporting the earthworm-golden mole food chain and to find out which system is more economical. Conventional tillage and crop production is inimical to earthworm and mole activity especially where this leads to soil acidification as a result of the excessive use of nitrogen fertilisers and inadequate liming. Earthworms generally disappear when the soil becomes very acidic. The modern trend towards conservation tillage which includes mulching with crop residues could eventually see a resurgence of earthworms and golden moles on South African farms. More judicious use of fire in managing rangelands would also benefit the soil biota, reduce erosion and increase the amount of rain water available for plant use. The promotion of golden mole populations could be of great financial importance to South African farmers.

3.5 RODENTS

Many different lineages of small mammals have become subterranean to the degree of reduction of eyes, extreme specialisation of limbs as digging tools, and foraging underground. In South America, for example, where no true moles (Talpidae) or golden moles (Chrysochloridae) occur, several mammals occupy a similar niche. The Brazilian shrew-mouse *Blarinomys* (Muridae, the same family as the house mouse), is one such fossorial rodent, in which the eyes and ear pinnae are small, giving it a shrew-like appearance. Subterranean mammals flourish conspicuously in South Africa, with mole-rats (Bathyergidae) and golden moles (already discussed) being the two groups that are especially adapted to a subterranean environment.

In the case of the mole-rat (Figure 3.5.1) in Africa five species of these rodents collectively cover most of South Africa (Jarvis, 2001) and the smallest of these is capable of penetrating even compact and clay-rich soils because it tunnels mainly with its teeth. A colony of mole-rats may build a system of foraging tunnels more than a kilometre long. In dry areas there is a flurry of digging after rain with a colony of mole-rats throwing up more than two tons of soil. The Cape dune mole-rat (*Bathyergus suillus*) is among the largest mole-like mammals and is responsible for extreme disturbance of the sandy soils it inhabits in the Western Cape (Figure 3.5.1a). One individual is capable of throwing up as much as half a ton of soil each month (Jarvis, 2001).

Some species of mole-rat have a eusocial organisation in which different individuals adopt different roles, paralleling to a certain degree the eusocial insects such as termites and ants. This division of labour among colony members appears to be linked to a specialised diet of geophytic plants,

Figure 3.5.1 Mole-rats and their environments: a. excavations of a colony near Cathedral Peak in the KwaZulu-Natal Drakensberg; b. cracking red clays in the geophyte-rich Nieuwoudtville district of the Western Cape seem to exclude mole-rats on account of being inhospitable since these self-mulching soils would render the establishment of a permanent burrow system impossible (see Section 2.3 and the vertic, rhodic profile in Chapter 4); c. Namaqua Dune mole-rat (*Bathyergus janetta*); d. Damaraland mole-rat (*Cryptomys damarensis*).

which are exceptionally diverse and abundant in Africa. Although eusociality is best expressed in the naked mole-rat (*Heterocephalus glaber*) of Kenya, it has been shown to a lesser degree in the Damaraland mole-rat (*Cryptomys damarensis*; Figure 3.5.1e) of the Kalahari, which extends into South Africa.

Besides the mole-rat, digging rodents of a wide range of body sizes, including Brant's whistling rat (*Parotomys brantsii*), Cape ground squirrel (*Xerus inauris*), and springhare (*Pedetes capensis*), are common enough to have important impacts on soils in South Africa. Although these rodents forage mainly above ground and have large eyes, they excavate warrens of burrows. So do the gerbils (*Tatera* spp.), which find part of their diet by excavating graminoid tubers. Because gerbils reach extremely high densities in certain places their burrows can rival those of mole-rats in their disturbance of sandy soils.

The Cape porcupine (*Hystrix africaeaustralis*; Figure 3.5.2) is the largest rodent in Africa (up to 20 kg) and one of the most widespread rodents in South Africa. It remains common despite being hunted for its meat and its frequently encountered, shallow excavations are testimony to the considerable impact of this animal. Not only does the porcupine dig burrows up to 18 m long for shelter (as many as 6–8 individuals may share a burrow), but its main foods are roots, tubers and bulbs which it must excavate. Because the least toxic and most palatable corms tend to be the most deeply buried, porcupines may have a disproportionately large effect on soils.

The closely related porcupine, *Hystrix cristata* of Israel, has been shown to propagate various species of plants owing to the collection of water and organic matter in the shallow holes it makes in the soil, and a similar process has recently come to light

Figure 3.5.2 Cape porcupine

(e.g. with *Babiana spp.*) near Nieuwoudtville in the Western Cape, an area with unusual abundance and diversity of geophytes.

3.6 HONEY BADGER

Although many carnivore species dig for prey to some extent, the honey badger or Ratel (*Mellivora capensis*; Figure 3.6.1) is unusual in two ways. Firstly, it is exceptionally energetic as a digger, being prepared to move much earth for what seems like a small reward. Secondly, it actually digs deeply enough in some cases to acquire the status of a fossorial forager, i.e. despite the fact that it is both relatively massive (about 10 kg) and stoutly built (as opposed to renowned tunnel penetrators such as weasels), the honey badger digs so energetically that it actually disappears underground and can catch deeply burrowing prey. It is remarkably energetic in digging out large dung beetles. Certain dung beetles dig deeply to bury a dung ball in which the larva develops. The honey badger digs all the way down to get a single dung ball at a time, just to eat the developing dung beetle inside. On the treeless grasslands of the Serengeti, which is prime habitat for the honey badger, it is common to see the excavations and the empty remains of the exhumed dung-balls.

Dung beetles are not, however, the main subterranean fare for the honey badger. A large proportion of its diet is obtained through digging (Begg *et al.*, 2003) and excavations vary from a shallow scrape for the barking gecko to several large holes with mounds of earth 40 cm high for scorpions, snakes and rodents. Digging time varies accordingly, with the barking gecko taking about 1 minute, skinks 2 minutes, gerbils and mice 3 to 8 minutes and large snakes such as mole snake and Cape cobra up to 10 minutes. Even the springhare is caught through the adept underground maneuverings of the honey badger. This rodent is usually chased above ground but finally cornered and killed in its burrows. Gerbils have extensive burrow systems and numerous escape holes, and the honey badger requires a dif-

Figure 3.6.1 Honey badger mother and cub digging

ferent approach for these rodents. Remaining on the surface, the honey badger disturbs the gerbils in one part of the burrow system by 'paddling' with its hind legs and thrashing with its tail, causing them to use escape holes where its outstretched forepaws are waiting motionless in ambush. The larvae of solitary bees are also a favourite food item and the foraging for these invertebrates can disturb the soil surface over considerable areas. In one study in the Kalahari four male honey badgers were observed digging out a total of some 2 800 of these larvae over a total of 53 hours, on 81 different occasions. The larvae are found in groups of one to four in small chambers from 13–100 cm below the soil surface, with the highest numbers found between 50 and 70 cm depth. Consequently the excavations often penetrate deep into soil B horizons. Indeed, the honey badger often disappears underground when digging out the larvae, and sleeps in the excavations after a bout of foraging. Such excavations may be concentrated on particular soil types. In the Kalahari study, 11–54 holes were found within areas of 30–700 m², primarily in compacted soil in dune troughs where *Rhigozum trichotomum* shrubs predominated. Approximately half of the excavated sites were used as honey badger latrines, which indicates that there is a fertilising as well as ploughing effect when the honey badger exploits the larvae of solitary bees.

3.7 INSECTS AND SPIDERS

Ants

Ants (Hymenoptera: Formicidae) cannot rival termites as excavators and builders of earthen structures but several types of ants in South Africa disturb soils to a significant extent. For example the harvester ant (*Messor capensis*) buries large quantities of seeds underground which attract excavation by mammals and birds that raid these seeds or eat ant larvae and eggs. The army ant (*Dorylus helvolus*) is also partly subterranean and provides yet another example of gigantism in South African burrowers, being the longest ant (wingless queens up to 5.1 cm; Carwardine, 1995). This blind species of ant is most conspicuous in the tropics but extends as far as the southwestern Cape. It eats termites (including *Macrotermes*) and tubers, and is a further example of how termites and geophytes support other burrowing organisms.

Ants, like termites, may construct vertical shafts that penetrate to depths of several metres to find suitable material for building mounds or to provide access to groundwater (Lee and Foster, 1991). Such workings will contribute to the formation of stonelines (see box in Section 2.7), as only fine material will be voided at the surface.

Dung beetles

South Africa supports among the largest dung beetles (Coleoptera: Scarabaeidae), which extend even into the fynbos biome at its southern tip. *Heliocopris*, for example, has a body mass of 10 g. Most species of large dung beetles bury their food rapidly by one method or another before eating it, thus conserving and recycling nutrients (Figure 3.7.1). An extreme example is the deep burial of dung by ball-rolling beetles which may be as much as 1 m deep, and are excavated by the honey badger (see p168).

Dung beetles parallel termites in being detritivores. The quantity and quality of faeces available in South Africa appears to have favoured both groups, producing particularly large and active representatives of these insects here. The faeces are produced by native and domestic ungulates of extreme diversity and abundance, including megaherbivores such as the African elephant (*Loxodonta africana*). Australia provides an instructive contrast, because none of its very diverse fauna of dung beetles is longer than 1 cm, possibly due to a lack of megaherbivores and their dung. The largest species of kangaroo (*Megaleia rufa*) is comparable in size to the springbok (*Antidorcas marsupialis*), and produces fibrous faeces unattractive to dung beetles.

Spiders and crickets

The illustration in Figure 3.7.2 serves as one further example of the role of small burrowers in providing macropores for water infiltration and aeration of soil. In some environments the importance of this role cannot be over-emphasised particularly where plant cover is sparse and soil properties are conducive to crust formation when it rains. Stones woven into the web at the surface of the spider tunnel in Figure 3.7.2 presumably reduce the flow of water into the hole during rainy weather by reinforcing the web that protrudes above ground level. Five litres of water were poured rapidly down this 2.5 cm diameter hole without filling it.

 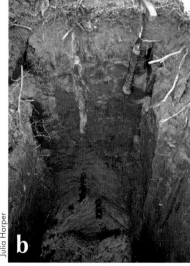

Figure 3.7.1 Dung beetles: a. at work in fresh elephant dung; b. burial of sable antelope dung (northwest Zambia) at about 1 m depth in a humic sandy soil (note backfilling of burrow with light coloured sand excavated before dung burial.

Figure 3.7.2 Web-lined spider hole in an aridisol (between Clanwilliam and Calvinia, Namaqualand). The object above the hole is the base of a walking stick.

Figure 3.8.1 Pied starling

3.8 BIRDS

Although South Africa lacks any counterparts of the burrowing owls of the Americas, it supports dense populations of helmeted guineafowl (Numididae: *Numida meleagris*) and several other species of fowl-like birds (Phasianidae), which routinely make shallow excavations while foraging. The red-winged francolin (*Francolinus levaillantii*), for example, has a staple diet of tubers which it excavates with its robust beak, and is therefore part of the same foraging guild as mole-rats and porcupine.

Although its impact on soils is limited, the ant-eating chat (*Myrmecocichla formicivora*) is the avian counterpart for the yellow mongoose, because its staple in winter is the grass-cutting termite (*Hodotermes mossambicus*). This chat roosts and breeds in the long (up to 1.5 m) burrows it excavates for itself, using the beak as pick and the feet as shovels. These burrows are often made in the roofs of burrows of the aardvark. Among other well-known burrowing birds are the ground woodpecker (*Geocolaptes olivaceus*), giant kingfisher (*Megaceryle maxima*), pied starling (*Spreo bicolor*) (Figure 3.8.1), rollers and brown-throated martin (*Riparia paludicola*). Burrows are primarily for breeding purposes and are usually excavated in the vertical walls of dongas or river banks.

3.9 LARGE HERBIVORES

Pigs (Suidae) disturb topsoil while foraging, using their snouts as trowels. Three species of wild pig, namely karoo warthog (*Phacochoerus aethiopicus aethiopicus*), bushpig (*Potamochoerus spp.*), and common warthog (*Phacochoerus africanus*), formerly covered most of South Africa. Warthogs rely heavily on aardvark burrows for refuge. The karoo warthog, now extinct, was the most arid-adapted suiform. By contrast, the bushpig (Figure 3.9.1) is limited to relatively moist climates and dense vegetation, but obtains its food by digging for tubers and invertebrates. The common warthog differs from the bushpig in that it extends into semi-arid climates and obtains much of its food above ground. Despite its preference for green grass, the common warthog does excavate grass rhizomes in the dry season, and is abundant enough to till large areas in this way. The springhare shares a foraging guild with the common warthog because it also excavates grass rhizomes, but the rodent extends into climates too dry for any pig.

The introduction of the common warthog to subtropical thicket in the Eastern Cape may lead to degradation of soils, possibly because of the different diet and foraging method of the Karoo warthog, the species originally native to this habitat. The diet and associated pedological effects of the Karoo warthog will remain unknown unless it can

Figure 3.9.1 Bushpig (*Potamocheros porcus*), Kruger National Park

be inferred from studies of its endangered relative (*P. a. delamerei*) on the Horn of Africa, before that subspecies also becomes extinct. It is possible that the Karoo warthog ate mainly Karoo shrubs (e.g. *Galenia africana*) and geophytes.

Large herbivorous mammals sometimes eat earth directly, which is thought to aid nutritional supplementation and self-medication. Geophagy is known from numerous locations in South Africa including the Mountain Zebra National Park, Addo Elephant National Park, Kgalagadi Transfrontier National Park, Pilanesberg National Park, Loskop Dam Nature Reserve (Figure 3.9.2), and the Drakensberg, but these records are minor compared to the many geophagic sites known in the national parks of western, eastern, and central Africa, which are less intensely managed than those in South Africa. In particular, the African elephant has been recorded consuming the earth mounds of *Macrotermes* in central Africa, but not in South Africa. It is possible that the provision of many artificial boreholes (e.g. in the Kruger National Park) may have allowed ungulate species to supplement their diets nutritionally by means of drinking groundwater instead of geophagy. The best-documented geophagic sites in South Africa are in the southwest Kalahari, where repeated visits by the local bovids have produced conspicuous excavations in calcretised banks of drainage lines. Geophagy is known to be particularly frequent in the case of the gemsbok (*Oryx gazella*), which is unusual among ungulates worldwide because it combines a staple diet of grass with an extreme tolerance of aridity. Possibly by virtue of direct supplementation from scattered nutrient-rich sites such as mineral licks and pans, the gemsbok appears to tolerate the most nutrient-poor sands of the Kalahari more than other drought-adapted herbivores such as the springbok and eland (*Taurotragus oryx*).

Pans, characteristic of the Kalahari and Highveld, are unusual on a world scale because they are usually not saline enough to account for their formation and maintenance as depressed, treeless, and partly bare landforms. It is thought that small pans have been produced partly by means of the trampling of ungulates, not only because of deflation by wind but also because ungulates can, over many years, remove great quantities of earth on their hooves and

Figure 3.9.2 Earth lick in the Loskop Dam Nature Reserve

in their guts. This situation can, for example, arise where patches of dolerite bedrock, which is nutrient-rich but impermeable to water, force groundwater to the surface in the form of springs which attract wild ungulates. Since ungulates continue to practice geophagy on and immediately around such pans, it is possible that the main reason why ungulates visit pans is nutritional supplementation, and therefore it is possible that geophagy has directly and indirectly been responsible for the maintenance of this distinctive southern African landform. The potential for geophagy to modify the habitat at this scale may be overlooked because of the rapid extermination of wild ungulates on the Highveld, followed by the introduction of lick blocks which provided nutrient supplementation suited to the confinement of livestock on fenced farms. In addition, the springs formerly associated with many pans appear to have dried up owing to depression of water tables by the introduction of boreholes and the planting of trees. Further studies are needed of possible geophagy by wild ungulates reintroduced to farms that include pans.

Analyses in South Africa and elsewhere have confirmed that sodium is one of the main nutrients targeted at 'salt licks.' However, potable groundwaters in the Kalahari are remarkably rich in iodine and selenium and further work is required to test whether the availability of these elements, particularly around pans, has helped to shape the distribution of herbivores in South Africa. Furthermore, geophagy can potentially provide important clues for the pastoral industry (including game ranching) in terms of providing appropriate lick blocks to optimise the health and productivity of ungulates. For example, manganese-deficiency, although unusual in mammals, has been recorded for ungulates in the Karoo, and several cases are known on the highveld of domestic calves (*Bos taurus*) practising geophagy on exceptionally manganese-rich soils to the point of death by manganese toxicity. It remains unknown whether this excessive geophagy was driven by deficiency of manganese or another element. For example, ruminants are known to be copper- and cobalt-deficient in several parts of South Africa and geophagic materials should be analysed for these elements. Interestingly some of the vertic soils in Chapter 4 have unusually high cobalt concentrations.

3.10 ELEPHANTS, TREES AND TERMITES

When large trees are blown down by wind, their roots rip the soil and the horizons are disturbed. This is known as floral pedoturbation. An unusual variation of this, which must once have been common throughout Africa, involves elephants pushing trees over when foraging (Figure 3.10.1). At Etosha in northern Namibia the effect of foraging elephants on soils is spectacular. In the process of pushing over trees to facilitate browsing, elephants effectively plough the soil, breaking up the calcrete horizon in which the trees are rooted and exhuming fragments of calcrete (Figure 3.10.2a). The latter remain on the soil surface after the recovery of vegetation until buried with fine earth by termite activity over a geological time scale (Figure 3.10.2b). Even more remarkable is the brecciated appearance of the calcrete which consist of fragments of calcrete cemented within a calcrete matrix (Figure 3.10.2c). This suggests that rupture and burial, followed by incorporation into a new calcrete hardpan, have happened more than once. The elephants have been at it for a long time. In Section 2.13 it was suggested that the morphology of some soils (e.g. Figure 2.13.7b) might retain some imprint of elephant activity even in regions where elephants were exterminated a century or more ago.

CONCLUDING REMARKS

With our final example we have come full circle, from termites to elephants and back to termites again. There is a host of other animals that depend on soil and affect its properties. No mention was made, for example, of reptiles and amphibians. Not having given space to rain frogs, for example, we should at least mention the desert rain frog (*Breviceps macrops*) whose special relationship with soils has been referred to elsewhere (Francis et al., 2007).

As fascinating as the biomantle may be to the scientist, understanding soil-animal relationships can also have enormous practical significance. In the rehabilitation of land destroyed by strip mining, which affects large parts of South Africa (see Section 2.14) the most difficult problem is to avoid or reduce soil compaction. Restoration is doomed to failure if attention is devoted to revegetation without also finding ways to encourage the repopulation of the soil with a self-sustaining community of animals. In agriculture the modern imperative to pursue conservation farming requires that similar attention be given to achieving a healthy population of organisms capable of effecting bio-tillage. However, in imagining that we can harness soil biota to achieve the sustainable rehabilitation of disturbed land, we need to remember Darwin's words about the additive effects of a 'continually recurrent cause'. The benefits of bioturbation may take decades or centuries to materialise.

Some future research directions are suggested in the following questions:

- For any particular species of soil-dependent animal, what types of preferred soil affect its distribution?

Figure 3.10.1 Elephant pushing over a tree to get at the foliage

Figure 3.10.2 Competing forces of exhumation and burial at the Etosha Game Reserve in Namibia: a. calcrete rubble on the soil surface with some fragments still entangled in the roots of a tree felled by elephant; b. termite spire (probably *Macrotermes michaelseni*) towers above calcrete fragments on the soil surface; c. surface of fractured calcrete with fragments of older calcrete cemented in a calcrete matrix, suggesting polycyclic calcrete formation. The outcome of these opposing processes determines whether coarse fragments are either uniformly distributed in the biomantle or segregated at depth to form a stoneline.

- What kinds of soil properties are affected by animal activity and over what time scale?
- In areas which have largely been depopulated of burrowing animals, what kind of ecological succession is necessary for a return to the kind of complex ecosystem that existed prior to disturbance and depopulation?
- In rehabilitating disturbed land, is it necessary to restock an area with a diverse range of species or do one or two key species have a sufficiently large effect on soil properties that other species are incidental?
- Does the availability of nutrients (both macro- and micro-nutrients) to plant roots have an indirect influence on the rate of bioturbation in different soil types by affecting the energetics of the fauna of the ecosystem?

Edaphology is the study of soil as a medium for plant growth and the agricultural application of soil science has led to a well-developed understanding of soil-plant relationships. There is a growing interest in soil biology as a field of study and the soil science agenda has become as much environmental as agricultural. Edaphology is likely in future to emphasise the ecological application of soil science, one that considers reciprocal relationships between soil properties and both plants and animals. Hopefully some, by reading this chapter, will have been encouraged to make that a reality.

Keith Melvin-Phillips

SOILS
OF
SOUTH AFRICA

Chapter 4

Profile descriptions and analytical data

Martin Fey

Hottentots Holland Nature Reserve, Western Cape

INTRODUCTION

Soil description

First dig a pit and savour the geosmin.
Mark next the slope and aspect, vale or hill.
Horizons in the profile take some skill,
Surface features first; know vegetation.
View pit face with love, imagination.
Feel, observe, describe each special feature.
Systematic'ly move your adventure,
Up and down, admitting information:
Moisture; Munsell hue; texture in the main;
Consistence, this must relate to moisture;
Structure: grade, shape (beauty in creation);
Not forgetting roots; pores and nodules, plain
Or complex; path of burrowing creature;
Last, inspect horizons for transition.

During the 1970s South Africa had the good fortune to have a government department, then known as Agricultural Technical Services, which included a number of specialised institutes one of which was the Soil and Irrigation Research Institute. There was nothing particularly special about this institute (it had its normal share of dead wood and bureaucracy) besides the fact that among its employees were some soil scientists who, with great vision and enthusiasm, initiated what was to become known as the Land Type Survey of South Africa. This was an ambitious project which continued over nearly three decades – classifying, mapping, describing, sampling and analysing soils in relation to natural mapping units termed land types. Each land type consists of a defined combination of soil types, terrain pattern and climate. Land types have been mapped at a scale of 1:250 000 and accompanying memoirs (mostly published) provide quantitative information on the soils of each land type as well as descriptions and analytical data for modal soil profiles that were selected to represent the most abundant soils within each land type. The information was eventually digitised and can be obtained in electronic format from the Agricultural Research Council's Institute for Soil, Climate and Water (ARC-ISCW).

This chapter consists of a selection of thirty-seven modal soil profiles that illustrate each of the soil groups. Each profile is presented as a double page spread, with the description and site information on the left and a table of laboratory data (chemical, physical and mineralogical) on the right. Some of the original data were omitted (for example mineralogical information has been presented for the clay fraction only) because of space considerations. The tabulation design was changed and standardised so that each property can be viewed in the same place on each page. Since the full range of laboratory analyses had not been determined on all soils, there are many blank spaces. If more than three horizons in a profile were analysed, only the first three are shown. The descriptions and site details on the left hand page are transcribed unaltered, as far as possible, from the originals except where easily correctible errors were found. Some of the descriptions leave much to be desired but they generally complement the laboratory data quite adequately.

To all this information gleaned from Land Type memoirs has been added a section entitled 'Points of interest and interpretation'. These notes did not appear in the original memoirs. They are an attempt to add value to the descriptions and data by highlighting some of the more important, interesting, or unusual features of each soil and, where appropriate, comparing them with other soils in the collection and sharing occasional thoughts about genesis and environmental significance. These supplementary remarks should not be viewed as a serious attempt at scholarship; they include some speculative statements and no doubt reveal the bias of their author on occasion. But they should be of value to students and others (less experienced consultants, for example) who often look at a lab report and are not sure what all the numbers mean.

The 37 profiles are arranged in order of soil groups (name in bold); the first qualifier term after the group name identifies the soil form; remaining qualifiers besides final topsoil texture identify the soil family. Traditional form names and family codes appear in parentheses. Italicised qualifiers refer to the surface horizon and all other qualifiers refer to subsurface horizons or materials. The order of the profiles is given by in the following list:

P970:	**Organic**, *sapric,* lithic, *medium sandy loam* (Champagne Ch 2100)	
P587:	**Humic**, xanthirhodic, *thin,* haplic, *clay* (Kranskop Kp 1100)	
P969:	**Humic**, xanthic, *thick,* luvic, *medium sandy loam* (Magwa Ma 2200)	
P1472:	**Humic**, rhodic, *thick,* haplic, *sandy clay* (Inanda Ia 2100)	
P7151:	**Vertic**, aeromorphic, *rhodic, acalcic, clay* (Arcadia Ar 2100)	
P84:	**Vertic**, aeromorphic, *melanic, acalcic, clay* (Arcadia Ar 1100)	
P737:	**Vertic**, aeromorphic, *melanic, acalcic, clay* (Arcadia Ar 1100)	
P644:	**Melanic**, pedocutanic, rhodic, micropedal, calcic, *clay loam* (Bonheim Bo 2120)	
P646:	**Melanic**, pedocutanic, umbric, micropedal, calcic, *clay* (Bonheim Bo 1120)	
P1111:	**Melanic**, lithocutanic, orthosaprolitic, calcic, *clay loam* (Mayo My 1200)	
P794:	**Silicic**, rhodic, haplic, *coarse sand* (Garies Gr 1000)	
P824:	**Silicic**, rhodic, haplic, *coarse sand* (Garies Gr 1000)	
P1180:	**Silicic**, neocalcic, *achromic,* rhodic, haplic, *medium sandy loam* (Trawal Tr 2210)	
P28:	**Calcic**, soft-neocalcic, *chromic,* rhodic, luvic, aeromorphic, *loamy coarse sand* (Addo Ad 1221)	
P783:	**Calcic**, hard-neocalcic, *chromic,* rhodic, haplic, *loamy fine sand* (Prieska Pr 1210)	
P812:	**Calcic**, hard-neocalcic, *chromic,* rhodic, haplic, *coarse sand* (Prieska Pr 1210)	
P826:	**Duplex**, pedocutanic-lithic, *chromic,* rhodic, micropedal, calcic, *sandy clay loam* (Swartland Sw 1212)	
P886:	**Duplex**, pedocutanic-cumulic-hydromorphic, *chromic,* macropedal, calcic, *sandy clay loam* (Sepane Se 1220)	
P1112:	**Duplex**, eluvic-prismacutanic, albic, asombric, *fine sand* (Estcourt Es 1100)	
P2568:	**Podzolic**, cumulic-aeromorphic, friable, *medium sand* (Pinegrove Pg 1000)	
P2569:	**Podzolic**, albic-cumulic-hydromorphic, adensic, friable, *medium sand* (Lamotte Lt 1100)	
P2573:	**Podzolic**, placic, *loamy fine sand* (Jonkersberg Jb 1000)	
P2575:	**Podzolic**, lithic, orthosaprolitic, aeromorphic, *fine sandy loam* (Groenkop Gk 1100)	
P583:	**Plinthic**, soft-xanthic, dystrophic, haplic, *sandy clay loam* (Avalon Av 1100)	
P584:	**Plinthic**, soft-eluvic, albic, *coarse sandy loam* (Longlands Lo 1000)	
P1067:	**Plinthic**, hard-xanthic, mesotrophic, haplic, *coarse sandy loam* (Glencoe Gc 2100)	
P150:	**Oxidic**, rhodic, mesotrophic, haplic, *medium sandy clay* (Hutton Hu 2100)	
P967:	**Oxidic**, pedorhodic, eutrophic, haplic, micropedal, *clay* (Shortlands Sd 2110)	
P1073:	**Oxidic**, xanthic, dystrophic, haplic, *medium sandy clay loam* (Clovelly Cv 1100)	
P1135:	**Oxidic**, xanthirhodic, dystrophic, luvic, *medium sandy loam* (Griffin Gf 1200)	
P578:	**Gleyic**, eluvic, albic, *loamy medium sand* (Kroonstad Kd 1000)	
P1137:	**Gleyic**, orthic, calcic, *medium sandy loam* (Katspruit Ka 2000)	
P638:	**Cumulic**, eluvic, *achromic,* albic, haplic, *medium sand* (Fernwood Fw 1110)	
P778:	**Cumulic**, neocalcic, *chromic,* rhodic, luvic, *coarse sandy loam* (Augrabies Ag 1220)	
P1477:	**Cumulic**, neocutanic, *chromic,* arhodic, luvic, *fine sandy loam* (Oakleaf Oa 1120)	
P1121:	**Lithic**, orthic, *chromic,* acalcic, *fine sandy clay loam* (Mispah Ms 1100)	
P1146:	**Lithic**, glossic, *chromic,* orthosaprolitic, aeromorphic, acalcic, *fine sandy clay loam* (Glenrosa Gs 1111)	

Methods used for describing profiles, sampling and obtaining the laboratory data are summarised in an appendix following the tables.

P970: Organic, *sapric*, lithic, *medium sandy loam*
(Champagne Ch 2100)

Profile No:	P970	**Aspect:**	W
Land type map:	2430 Pilgrim's Rest	**Water table:**	None
Latitude & Longitude:	24°53'00'/30°51'00'	**Occurrence of flooding:**	Nil
Land type No:	Ac89	**Stoniness:**	None
Climate zone No:	970S	**Erosion:**	None apparent
Soil form:	Champagne	**Vegetation/land use:**	Closed grassveld
Soil series:	Champagne	**Parent and underlying material:**	Quartzite
Altitude:	1 390 m		
Terrain unit:	4		
Slope:	02%	**Described by:**	B.C. Geers
Slope shape:	Convex	**Date described:**	Aug-77

Horizon	Depth (mm)	Description	Diagnostic horizon
	0-900	Moist; black 2.2Y2 5/0; medium sandy loam; massive; slightly hard; many pores; many roots; abrupt, smooth transition	Organic
R	900+	Hard rock	Hard rock

Points of interest and interpretation

This soil could be broadly described as an Afromontane, sandy peat. It consists of a relatively thick (0.9 m) black, carbon-rich (18%) organic horizon lying directly on hard quartzite in a lower slope position in the high country of the Drakensberg escarpment near Pilgrim's Rest. Cool temperatures, high rainfall, colluvial aggradation, water seepage from upslope and impermeability of the underlying quartzite coupled with exceptional water retentivity – despite the sandy loam texture of the mineral component – would all have contributed to the abnormal wetness (inferred) and consequent accumulation of organic matter. The quartzite lithology and high rainfall account for the very low levels of exchangeable base cations. The soil is acidic and both exchangeable Al and titratable acidity are very high, as is pH dependent cation exchange capacity indicating a strong buffer capacity. The acidic reaction and low base status conceivably further contribute to the preservation of humic substances. The soil has negligible free sesquioxides (CBD-extractable) and no minerals other than quartz in the clay fraction. Plant nutrient levels, especially basic cations, P and Zn, are very low.

Use of this soil for crop production would be constrained by intermittent wetness, underlying hard rock which limits deep rooting, and a high requirement for lime and fertiliser. Conservation is the most attractive option especially since rare plants are often found in unusual edaphic niches such as this. If the site had to be disturbed for construction the organic material could be valuable if mined as a horticultural potting medium on account of its high water holding capacity and permeability.

P970: Organic, *sapric*, lithic, *medium sandy loam*
(Champagne Ch 2100)

Lab no.	Depth (cm)	Horizon	Coarse fragments (%)	Fine earth (%)				
				CSa	MSa	FSa	Silt	Clay
C4653	0-90	O		8	16	24	23	15

Lab no.	$CaCO_3$ equiv. (%)	Organic C		Fe			Al			Mn	Si $CaCl_2$ (mg kg^{-1})
		Total	Pyr.	CBD	Ox	Pyr.	CBD	Ox	Pyr	CBD	
					(%)						
C4653		18.3		0.1			0.3				

Lab no.	pH			CEC	Exchangeable cations					Acidity		Al Exch
	H_2O	$CaCl_2$	NaF		Ca	Mg	Na	K	Total	Titr	Exch	
					(cmol$_c$ kg^{-1})							
C4653	5.7	4.6		32.6	0.2	0.0	0.1	0.0	0.3	23.7		2.2

Lab no.	Saturation extract							Micronutrients				
	Resist. (ohms)	EC (dS m^{-1})	Satn. water (%)	Ca	Mg	Na	K	Zn	Mn	Cu	B	Co
				(mmol$_c$ L^{-1})				(mg kg^{-1})				
C4653	3 600							0.39	67.6	0.36	0.25	0.36

Lab no.	Bulk density (kg m^{-3})	HC (mm h^{-1})	Air-water permeability ratio	Atterberg		Water retentivity at kPa:			
				LL	PI	-33	-80	-500	-1500
						(%)			
C4653			3			80.6	69.3	56.6	55.9

Lab no.	P Sorp. (%)	P Extr. (mg kg^{-1})	Minerals <2 µm	Mineral legend (abundance classes 1-5)	
C4653	98.5	3.8	Qz5	Qz	**Quartz**

P587: Humic, xanthirhodic, *thin*, haplic, *clay*
(Kranskop Kp 1100)

Profile No:	P587	**Aspect:**	E
Land type map:	2630 Mbabane	**Water table:**	None
Latitude & Longitude:	26°45'50'/30°25'10'	**Occurrence of flooding:**	
Land type No:	Bb35	**Stoniness:**	Class 1
Climate zone No:	110S	**Erosion:**	Gully, class 4
Soil form:	Griffin (previous)	**Vegetation/land use:**	Plantation
Soil series:	Farmhill		
Altitude:	1 400 m	**Parent and underlying material:**	Ecca shale
Terrain unit:	3		
Slope:	3%	**Described by:**	R.W. Fitzpatrick
Slope shape:	Convex	**Date described:**	Sept-72

Horizon	Depth (mm)	Description	Diagnostic horizon
A1	0–350	Dry; moist 90% dark brown 7.5YR3/2; clay; weak subangular blocky; slightly hard; clear smooth transition	Humic
B21	350–700	Dry; moist 90% brown to dark brown 7.5YR4/4; clay; apedal; slightly hard; gradual smooth transition	Yellow-brown apedal
B22	700–1 500+	Moist; 90% red to dark red 2.5YR4/6; few fine faint yellow streaks; clay; apedal; friable; saprolite fragments	Red apedal

Points of interest and interpretation

This profile is representative of humic soils with a thin humic A horizon and a yellow-red subsoil horizon sequence indicative of intermittent reducing conditions associated with organic matter in the upper B horizon. Red hematite is stripped preferentially leaving a yellow residue of more stable goethite (see genesis in Section 2.2; the higher Fe content in the red B2 is consistent with this interpretation). The locality in the eastern highveld of Mpumalanga is characterised by cool temperatures and more than 1 000 mm of (mainly) summer rainfall. The soil would have been covered originally by deciduous grassland. The upper solum is strongly acidic, with potentially toxic levels of exchangeable Al and extremely low levels of exchangeable bases, is poor in P and other nutrients, has a very high P fixing capacity (consistent with kaolinitic, oxidic mineralogy) and is likely to show very strong buffering when limed as a result of the high content of humus. The degree of weathering is high but not exceptionally so (no gibbsite was detected). The aluminous, chloritised vermiculite component emanates from the weathering of mica in the shale under strongly acidic conditions. It is possible that the extremely low base reserves and acidity are partly the result of current land use under commercial forestry. In general these soils are excellent for dryland crop production provided they receive suitable chemical amelioration. Nitrogen mineralisation from humus is often sufficient to satisfy crop requirements although this varies seasonally and is not easy to predict quantitatively. Nitrate storage on anion exchange surfaces in the subsoil is also likely. Natural ecosystems on these soils are mainly sourveld and, occasionally, semi-deciduous forest.

P587: Humic, xanthirhodic, *thin*, haplic, *clay*
(Kranskop Kp 1100)

Lab no.	Depth (cm)	Horizon	Coarse fragments (%)	Fine earth (%)				
				CSa	MSa	FSa	Silt	Clay
C1661	0–5	A1	0	5	4	9	19	55
C1662	35–70	B21	0	6	4	11	20	56
C1664	70–150+	B22	36	3	6	10	18	61

Lab no.	CaCO$_3$ equiv. (%)	Organic C		Fe			Al			Mn	Si CaCl$_2$ (mg kg^{-1})
		Total	Pyr.	CBD	Ox	Pyr.	CBD	Ox	Pyr	CBD	
		(%)									
C1661		2.8		7.4			2.0				
C1662		1.6		7.4			1.5				
C1664		0.3		9.1			1.5				

Lab no.	pH			CEC	Exchangeable cations					Acidity		Al Exch
	H$_2$O	CaCl$_2$	NaF		Ca	Mg	Na	K	Total	Titr	Exch	
					(cmol$_c$ kg^{-1})							
C1661	5.0	4.1		14.7	0.0	0.1	0.1	0.3	0.5			2.5
C1662	5.0	4.3		7.7	0.0	0.0	0.1	0.1	0.2			1.1
C1664	6.7	4.6		5.8	0.4	0.0	0.1	0.1	0.6			0.0

Lab no.	Saturation extract							Micronutrients				
	Resist. (ohms)	EC (dS m^{-1})	Satn. water (%)	Ca	Mg	Na	K	Zn	Mn	Cu	B	Co
				(mmol$_c$ L^{-1})				(mg kg^{-1})				
C1661	3 300							0.49	15.9	4.65	0.75	0.0
C1662	7 100							0.38	12.3	3.14	0.67	0.0
C1664	9 999							0.22	24.7	1.63	0.25	0.0

Lab no.	Bulk density (kg m^{-3})	HC (mm h^{-1})	Air-water permeability ratio	Atterberg		Water retentivity at kPa:			
				LL	PI	-33	-80	-500	-1500
						(%)			
C1661			7			35.5	31.9	25.6	23.6
C1662			5			31.0	28.1	23.5	23.5
C1664			2			37.2	32.8	26.8	26.8

Lab no.	P Sorp. (%)	P Extr. (mg kg^{-1})	Minerals <2 μm	Mineral legend			
C1661	100	0.5	Kt4, Ch(2:1)1, Qz1, Vm1	Qz	**Quartz**	Vm	**Vermiculite**
C1662	99.3	0.5	Kt5, Qz2, Ch(2:1)1, Vm1	Kt	**Kaolinite**		
C1664	99.3	0.6	Kt5, Ch(2:1)2, Vm2, Qz1	Ch	**Chlorite**		

P969: Humic, xanthic, *thick,* luvic, *medium sandy loam*
(Magwa Ma 2200)

Profile No:	P969	**Aspect:**	S
Land type map:	2430 Pilgrim's Rest	**Water table:**	None
Latitude & Longitude:	24°56'00'/30°51'00'	**Occurrence of flooding:**	Nil
Land type No:	Ib193	**Stoniness:**	None
Climate zone No:	988S	**Erosion:**	None apparent
Soil form:	Magwa	**Vegetation/land use:**	Open grassveld
Soil series:	Milford		
Altitude:	1 440 m	**Parent and underlying material:**	Quartzite
Terrain unit:	4		
Slope:	02%	**Described by:**	B.C. Geers
Slope shape:	Convex	**Date described:**	Aug-77

Horizon	Depth (mm)	Description	Diagnostic horizon
A1	0–500	Moist, moist black 2.2Y2 5/0; medium sandy loam; massive; slightly firm; many pores; common roots; gradual smooth transition	Humic
B1	500–730	Moist, moist very dark grayish-brown 10YR3/2; loamy medium sand; massive; firm; many pores; common roots; gradual smooth transition	
B21	730–1 200	Wet; moist yellowish-brown 10YR5/6; fine sandy clay loam; massive; firm; many pores; few roots	Yellow-brown apedal

Points of interest and interpretation

The environment in which this humic soil has formed is similar to that of the organic soil (P970) except that depth is greater and drainage is better, leading to lower humus content and a yellow-brown, luvic B horizon containing kaolinite, aluminous chlorite and mica (contrasting with a dominant quartz component in the clay fraction of the A horizon). Iron content is relatively low but quite high relative to clay content in the lower B. Acidity, exchangeable aluminium and buffer capacity (CEC) are substantial especially in the topsoil. The buffer capacity and high water retentivity are associated with exceptionally high organic matter content, with little contribution coming from the mineral fraction. Despite the high water holding capacity the air-water permeability ratio indicates a well aggregated structure and the soil can be expected to drain freely and be well aerated. Possibly because of the high organic matter content the extractable P is higher than normally encountered on such soils and phosphate retention capacity is also moderate. Plant nutritional problems are likely to occur mainly because of the extremely low reserve of exchangeable bases while some trace element deficiencies may also be expected. Because of the low clay content it would be a priority to use this soil in such a way as to ensure the conservation of organic matter.

P969: Humic, xanthic, *thick*, luvic, *medium sandy loam*
(Magwa Ma 2200)

Lab no.	Depth (cm)	Horizon	Coarse fragments (%)	Fine earth (%)				
				CSa	MSa	FSa	Silt	Clay
C4650	0–500	A1		9	29	19	18	9
C4651	500–730	B1		12	26	47	13	5
C4652	730–1 200	B21		1	19	50	12	22

Lab no.	CaCO$_3$ equiv. (%)	Organic C		Fe			Al			Mn	Si CaCl$_2$ (mg kg^{-1})
		Total	Pyr.	CBD	Ox	Pyr.	CBD	Ox	Pyr	CBD	
		(%)									
C4650		9.9		0.3			0.1				
C4651		1.3		0.4			0.2				
C4652		0.4		2.6			0.9				

Lab no.	pH			CEC	Exchangeable cations					Acidity		Al Exch
	H$_2$O	CaCl$_2$	NaF		Ca	Mg	Na	K	Total	Titr	Exch	
					(cmol$_c$ kg^{-1})							
C4650	5.8	4.5		26.8	0.1	0.0	0.1	0.1	0.3	19.2		2.6
C4651	5.6	4.6		11.0	0.0	0.0	0.1	0.0	0.1	11.1		0.8
C4652	5.6	4.7		4.9	0.0	0.0	0.1	0.0	0.1	4.5		0.3

Lab no.	Saturation extract							Micronutrients			
	Resist. (ohms)	EC (dS m^{-1})	Satn. water (%)	Ca	Mg	Na	K	Zn	Mn	Cu	B	Co
				(mmol$_c$ L^{-1})				(mg kg^{-1})			
C4650	1 700							0.57	29.6	0.39	0.28	0.24
C4651	2 400							0.30	9.5	1.20	0.09	0.00
C4652	1 800							0.12	14.7	0.99	0.02	0.00

Lab no.	Bulk density (kg m^{-3})	HC (mm h^{-1})	Air-water permeability ratio	Atterberg		Water retentivity at kPa:			
				LL	PI	-33	-80	-500	-1500
						(%)			
C4650			3			48.2	39.2	29.3	27.3
C4651			6			11.7	9.3	6.6	6.7
C4652			3			17.6	15.0	11.3	10.4

Lab no.	P Sorp. (%)	P Extr. (mg kg^{-1})	Minerals <2 μm	Mineral legend					
C4650	84.4	9.3	Qz5, Vm2, Un1	Qz	**Quartz**		Vm	**Vermiculite**		
C4651	60.6	1.6	Qz5, Vm4. Mi2	Un	**Unidentified**		Mi	**Mica**		
C4652	45.0	2.8	Kt5, Ch4, Mi3	Kt	**Kaolinite**		Ch	**Chlorite**		

P1472: Humic, rhodic, *thick*, haplic, *sandy clay*
(Inanda la 2100)

Profile No:	P1472	**Aspect:**	E/O
Land type map:	2330 Tzaneen	**Water table:**	None
Latitude & Longitude:	23°42'12'/30°11'36'	**Occurrence of flooding:**	Nil
Land type No:	Ab95	**Stoniness:**	None
Climate zone No:	1020S	**Erosion:**	None apparent
Soil form:	Inanda	**Vegetation/Land use:**	Plantation (forestry)
Soil series:	Inanda		
Altitude:	1 100 m	**Parent and underlying material:**	Granite
Terrain unit:	3		
Slope:	10%	**Described by:**	D.P. Turner
Slope shape:	Straight	**Date described:**	Jan-74

Horizon	Depth (mm)	Description	Diagnostic horizon
A1	0–550	Moist; moist dark reddish-brown 5YR2. 5/2; coarse sandy clay; massive; friable; many pores; few angular fine gravel; many roots; gradual smooth transition	Humic
B2	550–200	Moist; moist dark reddish-brown 2.5YR3/4; coarse sandy clay; massive; friable; many pores; few organic cutans; few angular fine gravel; common roots	Red apedal

Points of interest and interpretation

This third example of a humic soil has intermediate organic matter, clay and sesquioxide content but levels of acidity and base saturation that are similar to those of the previous examples. The presence of gibbsite as a dominant mineral alongside kaolinite in the clay fraction suggests that it is the most strongly weathered of the three examples, and the relatively narrow difference between pH values measured in water and $CaCl_2$ suggest that the B horizon, while still having a net negative charge (CEC), nevertheless is likely to possess enough positive charge to make subsoil anion retention (e.g. nitrate) a potentially important property with respect to plant nutrient supply (although relative P sorption is significantly lower than in profile P587). The concentration of trace elements (Zn, Cu and B) is especially low.

(Note: When the base cation concentration is indicated as 0.0 it should be interpreted as being < 0.1 $cmol_c$ kg^{-1} i.e. below the lower limit of detection by the method used for analysis).

P1472: Humic, rhodic, *thick*, haplic, *sandy clay*
(Inanda la 2100)

Lab no.	Depth (cm)	Horizon	Coarse fragments (%)	Fine earth (%)				
				CSa	MSa	FSa	Silt	Clay
C3208	0–55	A1		14	15	15	11	38
C3209	55–120	B2		20	11	15	12	37

Lab no.	CaCO$_3$ equiv. (%)	Organic C		Fe			Al			Mn	Si CaCl$_2$ (mg kg^{-1})
		Total	Pyr.	CBD	Ox	Pyr.	CBD	Ox	Pyr	CBD	
		(%)										
C3208		3.1		3.4			0.6				
C3209		1.0		4.2			0.7				

Lab no.	pH			CEC	Exchangeable cations					Acidity		Al Exch
	H$_2$O	CaCl$_2$	NaF		Ca	Mg	Na	K	Total	Titr	Exch	
					(cmol$_c$ kg^{-1})								
C3208	5.1	4.3		17.5	0.0	0.0	0.0	0.2	0.2	21.5		1.8
C3209	4.8	4.2		11.5	0.1	0.0	0.1	0.2	0.4	14.4		1.6

Lab no.	Saturation extract								Micronutrients			
	Resist. (ohms)	EC (dS m^{-1})	Satn. water (%)	Ca	Mg	Na	K	Zn	Mn	Cu	B	Co
				(mmol$_c$ L^{-1})				(mg kg^{-1})			
C3208	7 300							0.43	45.6	3.42	0.14	
C3209	5 000							0.37	17.1	3.45	0.02	

Lab no.	Bulk density (kg m^{-3})	HC (mm h^{-1})	Air-water permeability ratio	Atterberg		Water retentivity at kPa:			
				LL	PI	-33	-80	-500	-1500
						(%)			
C3208			4			23.3	20.8	17.5	16.2
C3209			4			20.0	18.1	16.9	15.4

Lab no.	P Sorp. (%)	P Extr. (mg kg^{-1})	Minerals <2 µm	Mineral legend			
C3208	76.9	7.2	Kt5, Gb2, Vm2	Gb	**Gibbsite**	Vm	**Vermiculite**
C3209	62.7	2.2	Kt5, Gb4, Mi2	Kt	**Kaolinite**	Mi	**Mica**

P7151: Vertic, aeromorphic, *rhodic, acalcic,* clay
(Arcadia Ar 2100)

Profile No:	7151	**Aspect:**	N
Land type map:	3119 AC	**Water table:**	None
Latitude & Longitude:	31°23'48'/19°9'24'	**Occurrence of flooding:**	
Land type No:	Ea365	**Stoniness:**	
Climate zone No:	2716W	**Erosion:**	
Soil form:	Arcadia	**Vegetation/land use:**	Shrubveld, open dwarf
Soil series:	Eenzaam		
Altitude:	275 m	**Parent and underlying material:**	Dolerite
Terrain unit:	Footslope		
Slope:	1%	**Described by:**	F. Ellis
Slope shape:	Concave	**Date described:**	May-82

Horizon	Depth (mm)	Description	Diagnostic horizon
A1	0–300	Dry; reddish-brown 2.5YR3/4; clay; strong fine angular blocky; sticky, plastic; medium cracks; few clay cutans; water absorption: 10 seconds; common roots; clear, smooth transition.	Vertic
A2	300–600	Dry; dark reddish-brown 2.5YR3/4; clay; apedal medium angular blocky; medium cracks; common clay cutans; water absorption 10 seconds; few roots; abrupt, smooth transition	Vertic
C	600–700	Dry; brown to dark brown 10YR4/3; coarse sandy clay loam; many fine, faint red and yellow unknown mottles; very many mixed-shape coarse gravel fragments 6–25 mm; common mixed shaped stones 25–75 mm	Saprolite

Points of interest and interpretation

Red vertisols are rare and this profile represents an interesting body derived from dolerite in the Nieuwoudtville district of the Western Cape (see Section 2.3). Noteworthy are the plasticity index (PI) and water retentivity values and the high base status and CEC. Smectite dominates the clay mineralogy while there is a small amount of kaolinite as well as feldspar and quartz, with some calcite also present but only in the saprolite. Base-rich parent material with high clay forming potential and a semi-arid climate are the main agents dictating a pathway of smectite formation. Most vertic soils form in the summer rainfall region and the xeric soil moisture regime is another unusual feature here which may contribute to both the red colour and the low humus content. Extreme stickiness when wet, shiny ped faces, slickensides at depth, wide cracks, and strong self-mulching – all hallmarks of vertic soils – are much in evidence at the sampling locality. Drainage and aeration are probably superior to those of dark vertic soils. Land use would best be confined to winter cereals and not root crops. Perennials may suffer root pruning in dry spells. Building foundations are at risk. The opportunity for tillage is limited and high drawbar pull can be expected. The red vertic soils at this locality are of ecological interest, supporting a diversity of geophytes (flowering bulbs) for which the district is famous. Porcupines flourish; aardvarks probably find the soil tiresome; while mole-rats appear to be absent probably because it is impossible to establish a stable network of foraging burrows. Absence of mole-rats may be part of the explanation for the spectacular flourishing of geophytes.

P7151: Vertic, aeromorphic, *rhodic, acalcic, clay*
(Arcadia Ar 2100)

Lab no.	Depth (cm)	Horizon	Coarse fragments (%)	Fine earth (%)				
				CSa	MSa	FSa	Silt	Clay
C6046	0–30	A1		1.4	4.2	12.4	23.5	54.7
C6047	30–60	A2		2.9	3.3	13.6	16.3	63.2
C6048	60–70	C		24.6	16.6	20.6	12	26.8

Lab no.	$CaCO_3$ equiv. (%)	Organic C		Fe			Al			Mn	Si $CaCl_2$ (mg kg^{-1})
		Total	Pyr.	CBD	Ox	Pyr.	CBD	Ox	Pyr	CBD	
					(%)						
C6046		0.30		1.89			0.11				
C6047		0.20		1.65			0.09				
C6048		0.20		0.48			0.02				

Lab no.	pH			CEC	Exchangeable cations					Acidity		Al Exch
	H_2O	$CaCl_2$	NaF		Ca	Mg	Na	K	Total	Titr	Exch	
							($cmol_c\ kg^{-1}$)					
C6046	7.9			33.5	21.1	13.8	2.6	0.7	38.2		0.02	0.01
C6047	8.0			37.7	22.3	13.4	4.0	0.5	40.2		0.03	0.02
C6048	9.1			13.8	14.0	5.4	2.4	0.2	22.0		0.04	0.02

Lab no.	Saturation extract							Micronutrients				
	Resist. (ohms)	EC (dS m^{-1})	Satn. water (%)	Ca	Mg	Na	K	Zn	Mn	Cu	B	Co
				($mmol_c\ L^{-1}$)				(mg kg^{-1})				
C6046	340	0.47	79.1					0.69	805	3.46	1.8	15.7
C6047	230	0.86	68.9					0.21	799	4.10	2.2	
C6048	300	4.80	46.0					0.16	11	1.10	1.9	0.39

Lab no.	Bulk density (kg m^{-3})	HC (mm h^{-1})	Air-water permeability ratio	Atterberg		Water retentivity at kPa:			
				LL	PI	-33	-80	-500	-1500
						(%)			
C6046				35		43.4	41.7	30.4	23.7
C6047				51		47.2	44.4	32.6	26.6
C6048			132	27	20.7	20.3	14.2	13.4	

Lab no.	P Sorp. (%)	P Extr. (mg kg^{-1})	Minerals <2 μm		Mineral legend			
C6046	24	2.80	St5, Qz2, Fs2, Kt1	Qz	**Quartz**	Kt	**Kaolinite**	
C6047	11	0.90	St5, Qz3, Fs2	Fs	**Feldspar**	Ct	**Calcite**	
C6048	34	0.40	St5, Qz2, Kt2, Ct2	St	**Smectite**			

P84: Vertic, aeromorphic, *melanic, acalcic, clay*
(Arcadia Ar 1100)

Profile No:	P84	**Aspect:**	Level
Land type map:	2528 Pretoria	**Water table:**	None
Latitude & Longitude:	25°00'36'/28°24'00'	**Occurrence of flooding:**	Nil
Land type No:	Ea1	**Stoniness:**	None
Climate zone No:	585S	**Erosion:**	Rill, Class 1
Soil form:	Arcadia	**Vegetation/Land use:**	Sparse dwarf shrubveld
Soil series:	Arcadia		
Altitude:	1 075 m	**Parent and underlying material:**	Basalt
Terrain unit:	4		
Slope:	01%	**Described by:**	J.L. Schoeman
Slope shape:	Concave	**Date described:**	Nov-76

Horizon	Depth (mm)	Description	Diagnostic horizon
A11	0–450	Moist; moist black 7.5YR2. 5/0; silty clay; strong fine angular blocky; very firm; common roots; gradual wavy transition	Vertic
A12	450–800	Moist; moist very dark grey 7.5YR3/0; clay; strong fine angular blocky; very firm; few roots; gradual wavy transition	Vertic
Cca	800–1 000	Moist; moist very dark grey 7.5YR3/0; clay; strong fine angular blocky; very firm; few rounded coarse hard lime nodules; few irregular medium gravel; few roots; gradual wavy transition	

Points of interest and interpretation

From the profile description it is debatable whether this should be classified as the hydromorphic (Rensburg) form instead of aeromorphic, since the A12 and Cca horizons are very dark grey suggesting a tendency toward gleying masked by the presence of organic matter. This is a classic dark, vertic soil (melanic family) rich in smectitic clay with an alkaline reaction, high CEC, full base saturation although non-saline and non-sodic, and exceptionally high air-water permeability ratio and water retentivity. Possibly because it was observed in the moist state there are no morphological indications in the description of features such as cracks, self-mulching and slickensides and in the earlier stages of the land type survey Atterberg tests were not being routinely performed. Nevertheless the clay mineralogy in particular confirms the vertic character. Of particular interest from a chemical point of view are the low carbon content in the black A11 (typical of dark vertic soils) and the very high manganese and associated cobalt extraction values, replicating what was found in the surface horizon of the red vertic profile (P7151). Cobalt is well known to be scavenged by Mn oxides but whether the Mn in this soil is dissolved from an oxide mineral or extracted from cation exchange sites is not known. Certainly the high pH may be one of the prerequisites for Mn persisting in such high concentrations. The Co values represent something of a geochemical anomaly compared with those of most other soils. The physical and engineering limitations of using such soils soil tend to outweigh their chemical fertility, as discussed above. The dark colours, greyish subsoil and low CBD-extractable iron values, despite the Fe-rich parent material, are considered to be indicative of reduction and incorporation of Fe in the smectite structure.

P84: Vertic, aeromorphic, *melanic, acalcic, clay*
(Arcadia Ar 1100)

Lab no.	Depth (cm)	Horizon	Coarse fragments (%)	Fine earth (%)				
				CSa	MSa	FSa	Silt	Clay
C4035	0–450	A11		1	3	19	25	48
C4036	450–800	A12		2	3	16	23	52
C4037	800–1 000	Cca		2	2	13	23	54

Lab no.	$CaCO_3$ equiv. (%)	Organic C		Fe			Al		Mn	Si $CaCl_2$ (mg kg^{-1})
		Total	Pyr.	CBD	Ox	Pyr.	CBD	Ox	Pyr	CBD	
					(%)						
C4035		0.6		0.3			0.1				
C4036		0.6		0.3			0.1				
C4037		0.6		0.1			0.0				

Lab no.	pH			CEC	Exchangeable cations					Acidity		Al
	H_2O	$CaCl_2$	NaF		Ca	Mg	Na	K	Total	Titr	Exch	Exch
					(cmol$_c$ kg^{-1})							
C4035	8.0	6.9		54.6	44.8	10	0.8	0.4	56.1	1.8	0.0	0.0
C4036	8.2	7.2		61.0	43.7	12	1.5	0.3	57.7	0.9	0.1	0.0
C4037	8.3	7.5		61.0	43.5	14	2.0	0.4	60.3	0.0	0.1	0.0

Lab no.	Saturation extract							Micronutrients			
	Resist. (ohms)	EC (dS m^{-1})	Satn. water (%)	Ca	Mg	Na	K	Zn	Mn	Cu	B	Co
				(mmol$_c$ L^{-1})				(mg kg^{-1})				
C4035	230	0.55		8	49	1	0	0.73	410	6.59	0.64	17.99
C4036	240	0.55		8	25	2	0	0.41	395	6.44	0.71	17.56
C4037	270	0.46		6	35	2	0	0.49	395	6.54	0.71	15.98

Lab no.	Bulk density (kg m^{-3})	HC (mm h^{-1})	Air-water permeability ratio	Atterberg		Water retentivity at kPa:			
				LL	PI	-33	-80	-500	-1500
						(%)			
C4035			999			59.5	52.4	46.4	34.2
C4036			999			59.7	52.3		
C4037			999			64.2	57.7	52.2	45.5

Lab no.	P Sorp. (%)	P Extr. (mg kg^{-1})	Minerals <2 µm	Mineral legend			
C4035	43	0.3	St5, Qz3, Ch(2:1), Kt1	Qz	**Quartz**	Kt	**Kaolinite**
C4036	41	0.3	St5, Qz3, Ch(2:1)1, Kt1	Ch	**Chlorite**	St	**Smectite**
C4037	32	0.3	St5, Qz3, Kt2, Ch(2:1)1				

P737: Vertic, aeromorphic, *melanic*, *acalcic*, *clay*
(Arcadia Ar 1100)

Profile No:	P737	**Aspect:**	S
Land type map:	2526 Rustenburg	**Water table:**	None
Latitude & Longitude:	25°05′24′/26°53′24′	**Occurrence of flooding:**	Nil
Land type No:	Ea69	**Stoniness:**	None
Climate zone No:	17S	**Erosion:**	None apparent
Soil form:	Arcadia	**Vegetation/Land use:**	Open treeveld
Soil series:	Arcadia		
Altitude:	1 080 m	**Parent and underlying material:**	Norite
Terrain unit:	4		
Slope:	2%	**Described by:**	P.R. Swanepoel
Slope shape:	Straight	**Date described:**	Jan-81

Horizon	Depth (mm)	Description	Diagnostic horizon
A11	0–300	Moist; moist black 10YR2/1; clay; moderate medium sub-angular blocky; firm; many slickensides; few coarse gravel	Vertic
A12	300–1 200+	Moist; moist black 5Y2.5/1; clay; moderate coarse angular blocky; very firm; many slickensides; few coarse gravel	Vertic

Points of interest and interpretation

Unlike in the previous example, slickensides are prominent in both horizons. Clay content is moderate although smectite is still dominant. Noteworthy is the presence of some talc in the lower A horizon. This vertic soil is quite strongly alkaline, slightly saline, and sodic in the lower A horizon (with ESP of about 25%). Other properties are similar to those of the two examples above except that the air water permeability is more moderate in the upper A horizon (although still much higher than that of the rhodic example) and trace element values are more normal with Mn < 100 and Co < 2 mg kg^{-1}. P sorption is low (this is usually the case in smectitic soils) but extractable P is very low. Crop production on these soils may require irrigation although cracks after the dry season mean that the soil can 'tank up' with spring rain and sustain crops such as wheat and cotton without irrigation. Gypsum on this soil would improve hydraulic conductivity of irrigation water and help to keep Na and soluble salts at low levels through leaching. Some of the most palatable, nutritious (sweet) grazing occurs on these soils as well as those of the melanic group (see p193).

P737: Vertic, aeromorphic, *melanic, acalcic,* clay
(Arcadia Ar 1100)

Lab no.	Depth (cm)	Horizon	Coarse fragments (%)	Fine earth (%)				
				CSa	MSa	FSa	Silt	Clay
C5785	0–30	A11		12	10	23	11	41
C5786	30–120	A12		13	9	20	13	43

Lab no.	CaCO$_3$ equiv. (%)	Organic C		Fe			Al			Mn	Si CaCl$_2$ (mg kg^{-1})
		Total	Pyr.	CBD	Ox	Pyr.	CBD	Ox	Pyr	CBD	
					(%)						
C5785		0.5		1.0			0.1				
C5786		0.3		1.0			0.1				

Lab no.	pH			CEC	Exchangeable cations					Acidity		Al Exch
	H$_2$O	CaCl$_2$	NaF		Ca	Mg	Na	K	Total	Titr	Exch	
					(cmol$_c$ kg^{-1})							
C5785	9.0	7.6		25.4	11.6	14.0	1.4	0.1	27.1	0.0	0.0	0.0
C5786	9.5	8.0		25.0	3.1	15.3	6.6	0.1	25.1	0.0	0.0	0.0

Lab no.	Saturation extract							Micronutrients				
	Resist. (ohms)	EC (dS m^{-1})	Satn. water (%)	Ca	Mg	Na	K	Zn	Mn	Cu	B	Co
				(mmol$_c$ L^{-1})				(mg kg^{-1})				
C5785	400	0.86		6	8	5	0	0.26	49.7	2.04	1.03	1.79
C5786	310	1.65		19	7	12	0	0.30	72.5	1.98	0.96	1.92

Lab no.	Bulk density (kg m^{-3})	HC (mm h^{-1})	Air-water permeability ratio	Atterberg		Water retentivity at kPa:			
				LL	PI	-33	-80	-500	-1500
						(%)			
C5785			207			31.7	28.8	23.0	22.7
C5786			999			45.2	39.0	27.3	27.2

Lab no.	P Sorp. (%)	P Extr. (mg kg^{-1})	Minerals <2 μm	Mineral legend			
C5785	41	0.7	St5, Kt2, Qz2, Tc2	Qz	**Quartz**	St	**Smectite**
C5786	27	0.7	St5, Kt2, Qz2, Tc2, Mi1	Kt	**Kaolinite**	Mi	**Mica**
				Tc	**Talc**		

P644: Melanic, pedocutanic, rhodic, micropedal, calcic, *clay loam*
(Bonheim Bo 2120)

Profile No:	P644	**Aspect:**	S
Land type map:	2632 Mkuze	**Water table:**	Absent
Latitude & Longitude:	27°46'12'/32°07'06'	**Occurrence of flooding:**	
Land type No:	Ea54	**Stoniness:**	Class 0
Climate zone No:	237S	**Erosion:**	Class 1 gully
Soil form:	Bonheim	**Vegetation/land use:**	Bushveld and grass
Soil series:	Bushman	**Parent and underlying material:**	Basalt of the Letaba Formation, Lebombo Group
Altitude:	125 m		
Terrain unit:	4		
Slope:	2%	**Described by:**	D.P. Turner
Slope shape:	Straight	**Date described:**	Oct-76

Horizon	Depth (mm)	Description	Diagnostic horizon
A1	0–300	Moist; moist 95% dark brown 7.5YR3/2 rubbed; clay loam; moderate to strong very fine subangular blocky; slightly firm; few dark brown faint clay skins and not-shiny, not grooved pressure faces; clear smooth boundary	Melanic
B21	300–650	Moist; moist 80% dark-reddish brown 5YR3/4 rubbed; common medium to coarse distinct brown and white (lime) circular mottles; clay loam; moderate fine to medium subangular blocky; slightly firm; common reddish-brown distinct clay skins and not-shiny, not grooved pressure faces; few coarse soft lime nodules; moderate to strongly effervescent; gradual smooth boundary	Pedocutanic
B22	650–1000	Moist; moist 75% dark reddish-brown 5YR3/4 rubbed; many medium to coarse distinct dark red and white (lime) circular mottles; clay loam; moderate fine to medium sub-angular blocky; slightly firm; common reddish-brown distinct clay skins on ped faces; few coarse soft lime nodules; moderate to strongly effervescent; gradual smooth boundary	Pedocutanic

Points of interest and interpretation

This pedon has properties illustrating the close relationship between many melanic and vertic soils, with evidence of pressure faces (although not slickensides) being readily explained by dominant smectite in the clay fraction and very high air-water permeability ratio, water retentivity and CEC. The lower B horizon has free lime nodules and is alkaline, sodic and saline. If plasticity index had been determined this could well have been classified as vertic instead of melanic. Reddish subsoil colours reflect the moderate CBD-Fe values and may be due to the alleviating effects of moderate clay content and mixed mineralogy allowing better drainage and aeration than in the black (but not red) vertic profiles presented earlier. Nevertheless serious degradation could take place under irrigation without careful management in relation to water quality, using amendments such as gypsum when appropriate.

P644: Melanic, pedocutanic, rhodic, micropedal, calcic, *clay loam*
(Bonheim Bo 2120)

Lab no.	Depth (cm)	Horizon	Coarse fragments (%)	Fine earth (%)				
				CSa	MSa	FSa	Silt	Clay
C4282	0–30	A1	2	3	14	31	19	31
C4283	30–65	B21	35	3	13	31	25	30
C4284	65–100	B22	14	1	10	40	20	32

Lab no.	CaCO$_3$ equiv. (%)	Organic C		Fe			Al			Mn	Si CaCl$_2$ (mg kg^{-1})
		Total	Pyr.	CBD	Ox	Pyr.	CBD	Ox	Pyr	CBD	
				(%)							
C4282		1.6		1.6			0.14				
C4283		0.4		1.5			0.10				
C4284		0.2		1.7			0.08				

Lab no.	pH			CEC	Exchangeable cations					Acidity		Al
	H$_2$O	CaCl$_2$	NaF		Ca	Mg	Na	K	Total	Titr	Exch	Exch
					(cmol$_c$ kg^{-1})							
C4282	8.7	7.4		34.1	17.7	13.4	1.9	0.4	33.4	0.0	0.1	0
C4283	9.2	8.2		28.9	8.5	15.7	7.8	0.2	32.2	0.0	0.0	0
C4284	9.1	8.3		29.2	4.6	16	12.3	0.2	33.1	0.0	0.1	0

Lab no.	Saturation extract							Micronutrients			
	Resist. (ohms)	EC (dS m^{-1})	Satn. water (%)	Ca	Mg	Na	K	Zn	Mn	Cu	B	Co
				(mmol$_c$ L^{-1})				(mg kg^{-1})				
C4282	340	1.06		3.3	4.9	6.6	0	0.33	53.7	6.30	1.87	1.35
C4283	200	2.76		1.5	7.5	22.6	0	0.30	5.1	3.93	2.66	0.03
C4284	120	4.14		6.2	9.3	35.6	0	0.21	12.9	3.99	3.15	0.18

Lab no.	Bulk density (kg m^{-3})	HC (mm h^{-1})	Air-water permeability ratio	Atterberg		Water retentivity at kPa:			
				LL	PI	-33	-80	-500	-1 500
						(%)			
C4282			999			40.9	33.4	27.8	25.5
C4283			999			42.6	33.6	31.5	25.8
C4284			999			48.7	36.9	31.4	28.0

Lab no.	P Sorp. (%)	P Extr. (mg kg^{-1})	Minerals <2 μm	Mineral legend			
C4282	26.04	0.7	St5, Qz5, Mi3, Kt3, Tc3, Fs2	Qz	**Quartz**	St	**Smectite**
C4283	23.35	0.4	St5, Qz3, Kt3, Tc2, Mi2, Fs2	Fs	**Feldspar**	Mi	**Mica**
C4284	15.55	0.2	St5, Qz3, Kt2, Tc2, Mi2, Fs2	Tc	**Talc**	Kt	**Kaolinite**

P646: Melanic, pedocutanic, umbric, micropedal, calcic, *clay*
(Bonheim Bo 1120)

Profile No:	P646	**Aspect:**	E/O
Land type map:	2632 Mkuze	**Water table:**	Absent
Latitude & Longitude:	27°57'36'/32°15'06'	**Occurrence of flooding:**	
Land type No:	Ea56	**Stoniness:**	Class 1 stony
Climate zone No:	237S	**Erosion:**	Class 1 gully
Soil form:	Bonheim	**Vegetation/land use:**	Bushveld and grass
Soil series:	Bonheim	**Parent and underlying material:**	Basalt of the Letaba Formation, Lebombo group
Altitude:	120 m		
Terrain unit:	3		
Slope:	3%	**Described by:**	D.P. Turner
Slope shape:	Straight	**Date described:**	Oct-76

Horizon	Depth (mm)	Description	Diagnostic horizon
A1	0–350	Dry; moist 90% very dark greyish-brown 10YR3/2 rubbed; clay; moderate medium subangular blocky; hard; few black faint clayskins on ped faces; few fine soft durinodes and very few fine indurated iron-manganese nodules; slightly effervescent; gradual smooth boundary	Melanic
B21	350–650	Dry; moist 85% dark yellowish-brown 10YR4/4 rubbed; few medium distinct red circular mottles; clay; moderate fine subangular blocky; hard; few yellow faint clayskins on ped faces; few fine soft durinodes and few fine and medium indurated iron-manganese nodules; moderate to strongly effervescent; gradual smooth boundary	Pedocutanic
B22	650–1 000	Dry; moist 70% dark yellowish-brown 10YR4/4 rubbed; common medium distinct red, yellow and black circular mottles; silt clay loam; weak to moderate fine subangular blocky; hard; few yellow and red faint clayskins on ped faces; few fine to medium soft durinodes and common fine indurated iron-manganese nodules; moderate to strongly effervescent; clear smooth boundary	Pedocutanic

Points of interest and interpretation

At first glance this looks like a repetition of the previous profile but in terms of mineralogy (kaolinite dominance over smectite) and CEC (lower despite the higher clay content) it is more characteristically melanic as opposed to vertic. Although somewhat less alkaline than the previous example (also derived from Lebombo basalt) it is equally saline and sodic in the lower B horizon. Despite the dark yellowish brown (umbric) as opposed to reddish brown (rhodic) subsoil colours the CBD-Fe content is substantially higher. Interestingly this is another example of a dark clay soil with high cobalt values (in this case copper extractability is also elevated). Physical properties confirm the expectation that this soil will be easier to manage from the point of view of tillage and irrigation than the previous melanic example and its vertic cousins described earlier, although there is still the same degradation risk that was described previously.

P646: Melanic, pedocutanic, umbric, micropedal, calcic, *clay*
(Bonheim Bo 1120)

Lab no.	Depth (cm)	Horizon	Coarse fragments (%)	Fine earth (%)				
				CSa	MSa	FSa	Silt	Clay
C4247	0–35	A1	9	7	5	16	17	46
C4248	35–65	B21	32	8	4	12	15	59
C4249	65–100	B22	26	2	6	22	30	37

Lab no.	CaCO$_3$ equiv. (%)	Organic C (%)		Fe (%)			Al (%)			Mn (%)	Si CaCl$_2$ (mg kg^{-1})
		Total	Pyr.	CBD	Ox	Pyr.	CBD	Ox	Pyr	CBD	
C4247		1.9		6.10			0.63				
C4248		1.2		6.10			0.60				
C4249		0.1		6.80			0.49				

Lab no.	pH			CEC	Exchangeable cations					Acidity		Al Exch
	H$_2$O	CaCl$_2$	NaF		Ca	Mg	Na	K	Total	Titr	Exch	
					(cmol$_c$ kg^{-1})							
C4247	6.8	7.6		18.1	3.6	4.7	1.7	0.1	10.1	5.3	0.1	0.0
C4248	5.2	6.4		19.8	4.8	7.0	4.0	0.1	15.9	2.1	0.0	0.0
C4249	8.3	7.4		21.9	5.2	9.9	9.3	0.1	24.5	0.0	0.1	0.0

Lab no.	Saturation extract							Micronutrients			
	Resist. (ohms)	EC (dS m^{-1})	Satn. water (%)	Ca	Mg	Na	K	Zn	Mn	Cu	B	Co
				(mmol$_c$ L^{-1})				(mg kg^{-1})				
C4247	900							0.18	28.5	18.2	0.95	14.1
C4248	280	1.52		6.7	6.7	11.3	0	0.21	37.2	12.6	1.40	17.1
C4249	80	6.97		5.2	13	52.3	0	0.12	60.0	8.6	5.28	18.7

Lab no.	Bulk density (kg m^{-3})	HC (mm h^{-1})	Air-water permeability ratio	Atterberg		Water retentivity at kPa:			
				LL	PI	-33	-80	-500	-1500
						(%)			
C4247			33			28.6	25.2	20.5	18.9
C4248			999			36.1	33.7	26.5	24.3
C4249			999			41.3	37.2	30.3	27.9

Lab no.	P Sorp. (%)	P Extr. (mg kg^{-1})	Minerals <2 μm	Mineral legend			
C4247	35.96	1.3	Kt5, Qz3, St2, Fs2	Qz	**Quartz**	St	**Smectite**
C4248	46.56	0.2	Kt5, Qz3, Mi2, St1	Kt	**Kaolinite**	Mi	**Mica**
C4249	37.79	0.0	Kt5, St2, Mi2, Qz2				

P1111: Melanic, lithocutanic, orthosaprolitic, calcic, *clay loam*
(Mayo My 1200)

Profile No:	P1111	**Aspect:**	S
Land type map:	2730 Vryheid	**Water table:**	None
Latitude & Longitude:	27°22'06'/31°41'30'	**Occurrence of flooding:**	Nil
Land type No:	Db144	**Stoniness:**	None
Climate zone No:	372S	**Erosion:**	None apparent
Soil form:	Mayo	**Vegetation/land use:**	Bush land
Soil series:	Pafuri		
Altitude:	275 m	**Parent and underlying material:**	Dolerite
Terrain unit:	3		
Slope:	6%	**Described by:**	D.P. Turner
Slope shape:	Convex	**Date described:**	Oct-77

Horizon	Depth (mm)	Description	Diagnostic horizon
A1	0–300	Dry; moist dark brown 7.5YR3/2; clay loam; moderate fine subangular blocky; slightly hard; hardened free lime; slightly effervescent; few clay cutans; few rounded fine hard lime nodules; clear wavy transition	Melanic
B21	300–900	Dry; moist brown to dark brown 7.5YR4/4 silty loam; weak medium subangular blocky; soft; hardened free lime; strongly effervescent; few clay cutans; many rounded coarse soft lime nodules; few irregular stones, clear wavy transition	Lithocutanic
B3	900–1 100	Dry; coarse sand; weak medium subangular blocky; slightly hard; hardened free lime, strongly effervescent, few clay cutans; many rounded coarse hard lime nodules; few irregular stones	Lithocutanic

Points of interest and interpretation

Although clay mineralogical analysis indicates smectite dominance in the clay fraction, the physical quality of this melanic topsoil is better than in the previous two examples. Lower alkalinity, a higher exchangeable Ca fraction and negligible sodicity, higher organic carbon and more moderate clay contents are probable explanations for this physical improvement (air-water permeability ratios are substantially lower). The most conspicuous feature of this soil is its relatively shallow depth above weathered dolerite on a steep convex slope. Its calcareousness throughout the profile reflects a semi-arid climate. Land use in such situations is normally confined to livestock ranching or wildlife conservation so that management priorities shift towards grazing and burning strategies that ensure the right balance between grass and woody species as well as sufficient cover to limit soil erosion. Such soils are highly prized for the nutritious grazing which they sustain.

P1111: Melanic, lithocutanic, orthosaprolitic, calcic, *clay loam*
(Mayo My 1200)

Lab no.	Depth (cm)	Horizon	Coarse fragments (%)	Fine earth (%)				
				CSa	MSa	FSa	Silt	Clay
C4882	0–30	A1		14	17	34	20	41
C4883	30–90	B21		9	11	30	35	13
C4884	90–110	B3		19	26	41	9	4

Lab no.	$CaCO_3$ equiv. (%)	Organic C		Fe			Al			Mn	Si $CaCl_2$ (mg kg^{-1})
		Total	Pyr.	CBD	Ox	Pyr.	CBD	Ox	Pyr	CBD	
		(%)									
C4882		2.2		2.3			0.18				
C4883		1.4		0.9			0.06				
C4884		0.3		0.3			0.02				

Lab no.	pH			CEC	Exchangeable cations					Acidity		Al
	H_2O	$CaCl_2$	NaF		Ca	Mg	Na	K	Total	Titr	Exch	Exch
				(cmol$_c$ kg^{-1})								
C4882	7.9	7.4		22.6	21.2	4.4	0.3	0.2	26.1	0.0	0.0	0.0
C4883	7.7	8.2		21.7	22.9	4.8	0.5	0.2	30.4	0.0	0.0	0.0
C4884	8.3	7.7		12.1	13.8	4.4	0.3	0.0	18.5	0.0	0.0	0.0

Lab no.	Saturation extract							Micronutrients				
	Resist. (ohms)	EC (dS m^{-1})	Satn. water (%)	Ca	Mg	Na	K	Zn	Mn	Cu	B	Co
				(mmol$_c$ L^{-1})				(mg kg^{-1})				
C4882	600							0.27	68.9	1.71	1.45	0.54
C4883	520							1.02	31.1	3.00	0.00	1.41
C4884	860							0.27	26.7	0.12	0.27	0.18

Lab no.	Bulk density (kg m^{-3})	HC (mm h^{-1})	Air-water permeability ratio	Atterberg		Water retentivity at kPa:			
				LL	PI	-33	-80	-500	-1500
						(%)			
C4882			11			25.0	22.5	20.8	20.3
C4883			19			28.9	25.2	20.8	20.2
C4884			5			11.8	11.1	8.5	8.5

Lab no.	P Sorp. (%)	P Extr. (mg kg^{-1})	Minerals <2 µm
C4882	21.76	1.2	St5, Qz2, Mi1, Fs1
C4883	50.81	1.3	St5, Ct2, Qz1
C4884	38.76	0.6	St5, Kt1, Mi1, Qz1

Mineral legend

Qz	**Quartz**	St	**Smectite**
Fs	**Feldspar**	Mi	**Mica**
Ct	**Calcite**	Kt	**Kaolinite**

P794: Silicic, rhodic, haplic, *coarse sand*
(Garies Gr 1000)

Profile No:	P794	**Aspect:**	W
Land type map:	2916 Springbok	**Water table:**	None
Latitude & Longitude:	29°13'15'/21°27'30'	**Occurrence of flooding:**	Nil
Land type No:	Ae68	**Stoniness:**	None
Climate zone No:	146W	**Erosion:**	Wind, class 3, partially stabilised
Soil form:	Hutton (previous)	**Vegetation/Land use:**	Closed dwarf shrubveld
Soil series:	Moriah		
Altitude:	400 m	**Parent and underlying material:**	Unconsolidated, sandy
Terrain unit:	4		
Slope:	1%	**Described by:**	F. Ellis
Slope shape:	Straight	**Date described:**	Aug-77

Horizon	Depth (mm)	Description	Diagnostic horizon
A	0–150	Dry; moist reddish-brown 5YR4/4; coarse sand; single grain; slightly hard; few medium gravel; clear smooth transition	Orthic
B2	150–450	Dry; moist yellowish-red 5YR4/6; coarse sand; single grain; slightly hard; few cutans (type unknown); few medium gravel; abrupt smooth transition	Red-Apedal
Cdb	450–550	Dry; moist yellowish-red 5YR4/6; loamy coarse sand; weak fine platy; very hard; massive lime and silica cementation; cannot be broken in hands, easily broken with hammer, continuous; hardened free lime, slightly effervescent; few clay cutans; few medium gravel; lamellae; 10% clay in layer with coarsest texture	Dorbank

Points of interest and interpretation

The texture of soils with a dorbank horizon is typically sandy, as illustrated in this and the following two examples. Noteworthy in this profile are the generally alkaline pH, and the sodicity and salinity of the dorbank horizon. Extractable manganese concentration increases sharply in the dorbank, while throughout the profile concentrations of other trace elements are low, especially those of copper. Extractable P concentration is unusually elevated for a virgin sandy soil while exchangeable Ca is anomalously low given the indications of secondary carbonate in the profile description. Besides quartz, mica and kaolinite dominate the clay fraction and there is no indication of sepiolite which is often found in these soils. Silicic soils are characteristic of some of the most arid environments in South Africa and they have ecological significance in regions such as the succulent Karoo. The dorbank horizon represents an important physical barrier to downward water movement and root penetration. Where irrigation schemes have been developed the dorbank is broken by ripping to enhance effective depth and soil drainage for control of salinity.

P794: Silicic, rhodic, haplic, *coarse sand*
(Garies Gr 1000)

Lab no.	Depth (cm)	Horizon	Coarse fragments (%)	CSa	MSa	FSa	Silt	Clay
				\multicolumn{5}{c}{Fine earth (%)}				
C4914	0–15	A	0	39	29	21	5	3
C4915	15–45	B2	0	39	25	25	6	3
C4916	45–55	Cdb	0	44	22	21	11	3

Lab no.	CaCO₃ equiv. (%)	Organic C Total	Organic C Pyr.	Fe CBD	Fe Ox	Fe Pyr.	Al CBD	Al Ox	Al Pyr	Mn CBD	Si CaCl₂ (mg kg⁻¹)
				\multicolumn{7}{c}{(%)}						
C4914		0.20		0.20			0.00				
C4915		0.00		0.30			0.00				
C4916		0.10		0.30			0.00				

Lab no.	pH H₂O	pH CaCl₂	pH NaF	CEC	Ca	Mg	Na	K	Total	Titr	Exch	Al Exch
					\multicolumn{5}{c}{Exchangeable cations (cmol_c kg⁻¹)}	\multicolumn{2}{c}{Acidity}	
C4914	8.3	6.9		1.7	3.5	0.1	0.1	0.2	3.9	0.0		0.0
C4915	9.7	7.1		1.7	0.8	0.2	0.6	0.3	1.9	0.0		0.0
C4916	8.6	6.7		4.2	0.6	1.2	1.1	0.1	3.0	0.7		0.0

Lab no.	Resist. (ohms)	EC (dS m⁻¹)	Satn. water (%)	Ca	Mg	Na	K	Zn	Mn	Cu	B	Co
				\multicolumn{4}{c}{Saturation extract (mmol_c L⁻¹)}	\multicolumn{5}{c}{Micronutrients (mg kg⁻¹)}
C4914	2 700							3.6	74.0	0.36	0.88	0.66
C4915	1 000							1.5	42.1	0.45	0.62	1.59
C4916	180	4.4		4	4	43	0	1.4	545.0	0.69	1.08	1.12

Lab no.	Bulk density (kg m⁻³)	HC (mm h⁻¹)	Air-water permeability ratio	LL	PI	-33	-80	-500	-1500
				\multicolumn{2}{c}{Atterberg}	\multicolumn{4}{c}{Water retentivity at kPa: (%)}
C4914			3			4.0	3.0	1.9	1.4
C4915			2			3.7	2.9	1.7	1.1
C4916			29			7.7	6.6	5.3	3.8

Lab no.	P Sorp. (%)	P Extr. (mg kg⁻¹)	Minerals <2 μm		Mineral legend		
C4914	10.6	31.8	Mi5, Kt3, Qz3, Is2	Qz	**Quartz**	Is	**Interstratified**
C4915	6.8	10.3	Mi5, Qz4, Kt3, Fs2, Is1	Fs	**Feldspar**	Mi	**Mica**
C4916	12.4	12.3	Mi5, Kt3, Qz3, Is2	Kt	**Kaolinite**		

P824: Silicic, rhodic, haplic, *coarse sand*
(Garies Gr 1000)

Profile No:	P824	**Aspect:**	O/E
Land type map:	3018 Loeriesfontein	**Water table:**	None
Latitude & Longitude:	30°49'00'/18°06'48'	**Occurrence of flooding:**	Nil
Land type No:	Ag87	**Stoniness:**	Roundish, 250 mm diameter, 0.3 m apart
Climate zone No:	187W	**Erosion:**	Gully, class 2, partially stabilised
Soil form:	Hutton (previous)	**Vegetation/land use:**	Sparse grassveld
Soil series:	Moriah	**Parent and underlying material:**	Undifferentiated sediments
Altitude:	230 m		
Terrain unit:	3		
Slope:	10%	**Described by:**	L.M. Barkhuysen
Slope shape:	Convex	**Date described:**	Sept-77

Horizon	Depth (mm)	Description	Diagnostic horizon
A1	0–150	Dry; moist yellowish-red 5YR4/8; coarse sand; single grain; loose; few angular fine gravel; gradual smooth transition	Orthic
B2	150–900	Dry; moist yellowish-red 5YR5/8; coarse sand; single grain; soft; few angular fine gravel; clear smooth transition	Red apedal
Cdb	900–1 000	Dry; moist yellowish-red 5YR4/8; loamy coarse sand; massive; very hard; massive silica cementation, cannot be broken in hands, easily broken with hammer, continuous; few angular fine gravel	Dorbank

Points of interest and interpretation

This soil is not as alkaline as the previous example, nor is it sodic although salinity is moderate. In other respects it is physically, mineralogically and chemically quite similar to the previous profile except that boron levels are high, reflecting a common finding in certain areas of Namaqualand where silicic soils are encountered. Another difference is the apparent lack of secondary carbonates, which is consistent with the lower pH values (< 7 in the subsoil).

P824: Silicic, rhodic, haplic, *coarse sand*
(Garies Gr 1000)

Lab no.	Depth (cm)	Horizon	Coarse fragments (%)	Fine earth (%) CSa	MSa	FSa	Silt	Clay
C4972	0–15	A		35	29	29	3	2
C4973	15–90	B2		28	33	31	3	3
C4974	90–100	Cdb		19	13	46	18	5

Lab no.	CaCO$_3$ equiv. (%)	Organic C Total	Pyr.	Fe CBD	Ox	Pyr.	Al CBD	Ox	Pyr	Mn CBD	Si CaCl$_2$ (mg kg^{-1})
C4972		0.3		0.2			0.0				
C4973		0.0		0.2			0.0				
C4974		0.0		0.4			0.0				

Lab no.	pH H$_2$O	CaCl$_2$	NaF	CEC	Exchangeable cations (cmol$_c$ kg^{-1}) Ca	Mg	Na	K	Total	Acidity Titr	Exch	Al Exch
C4972	8.0	7.3		2.2	2.1	0.4	0.7	0.1	3.3	0.0		0.0
C4973	6.0	5.8		1.3	0.6	0.3	0.4	0.1	1.4	0.0		0.0
C4974	6.4	6.1		3.5	2.4	1.2	0.2	0.3	4.1	0.0		0.0

Lab no.	Resist. (ohms)	EC (dS m^{-1})	Satn. water (%)	Saturation extract Ca	Mg	Na	K	Zn	Micronutrients Mn	Cu	B	Co
				(mmol$_c$ L^{-1})					(mg kg^{-1})			
C4972	540							1.29	21.9	0.69	3.81	0.66
C4973	320	4.10		11	11	28	0	0.45	28.8	0.15	4.26	1.02
C4974	670							0.06	146.9	0.09	0.66	0.45

Lab no.	Bulk density (kg m^{-3})	HC (mm h^{-1})	Air-water permeability ratio	Atterberg LL	PI	Water retentivity at kPa: -33	-80	-500	-1500
						(%)			
C4972			9			3.9	3.5	2.6	2.5
C4973			2			3.1	2.7	2.1	2.0
C4974			3			5.8	5.4	4.7	4.8

Lab no.	P Sorp. (%)	P Extr. (mg kg^{-1})	Minerals <2 µm	Mineral legend			
C4972	4.8	9.6	Kt5, Mi4, Qz3, Fs2, St1	Qz	**Quartz**	Is	**Interstratified**
C4973	5.6	2.0	Qz5, Mi3, Kt3, Fs2	Fs	**Feldspar**	Mi	**Mica**
C4974	4.8	2.2	Kt5, Mi4, Qz2	Kt	**Kaolinite**	St	**Smectite**

P1180: Silicic, neocalcic, *achromic*, rhodic, haplic, *medium sandy loam*
(Trawal Tr 2210)

Profile No:	P1180	**Aspect:**	W
Land type map:	3017 Garies	**Water table:**	None
Latitude & Longitude:	30°44'43'/17°52'26'	**Occurrence of flooding:**	Nil
Land type No:	Ag95	**Stoniness:**	None
Climate zone No:	187W	**Erosion:**	Gully, class 2
Soil form:	Hutton (previous)	**Vegetation/land use:**	Open shrubveld
Soil series:	Quaggafontein	**Parent and underlying material:**	Gneiss
Altitude:	150 m		
Terrain unit:	3		
Slope:	10%	**Described by:**	L.M. Barkhuysen
Slope shape:	Concave	**Date described:**	Sept-77

Horizon	Depth (mm)	Description	Diagnostic horizon
A1	0–200	Dry; moist yellowish-red 5YR4/6; medium sandy loam; single grain; soft; non-hardened free lime, moderately effervescent; gradual smooth transition	Orthic
B21	200–600	Dry; moist yellowish-red 5YR5/6; loamy coarse sand; single grain; soft: non-hardened free lime, strongly effervescent; clear smooth transition	Neocarbonate
C1db	600–800+	Dry; moist reddish-yellow 5YR6/6; massive; hard; massive lime and silica cementation, can be broken in hands, discontinuous; hardened free lime, strongly effervescent; common clay cutans	Dorbank

Points of interest and interpretation

A common feature of silicic soils is their extremely low organic carbon levels but this profile appears to contain substantially more C than the previous two examples. An even more conspicuous difference is the extreme salinity in the lower profile, notwithstanding the fact that exchangeable Na levels do not exceed about 15% of total cations. It should be noted, however, that exchangeable Ca is likely to have been overestimated by the method used, probably as a result of carbonate dissolution during extraction. Interestingly the pH measured in $CaCl_2$ differs little from that measured in water, especially in the subsoil. This is probably a result of the high salinity level, meaning that 0.01M $CaCl_2$ has little impact on ionic strength. Extractable P levels are exceptionally high suggesting that besides calcite there may be some apatite present. The clay content is notably higher than in the previous silicic soils especially in the surface horizon. There appears to be a positive correlation between clay content and extractable P concentration. The calcareousness is observed throughout the profile and it could be speculated that termite activity (in the form of heuweltjies which are very common in the region) has influenced the formation of this soil, resulting in clay, organic matter and nutrient enrichment as well as surplus silica and calcium carbonate (see box in Chapter 3).

P1180: Silicic, neocalcic, *achromic*, rhodic, haplic, *medium sandy loam*
(Trawal Tr 2210)

Lab no.	Depth (cm)	Horizon	Coarse fragments (%)	Fine earth (%) CSa	MSa	FSa	Silt	Clay
C4982	0–20	A1		15	21	43	10	7
C4983	20–60	B21		19	23	43	10	4
C4984	60–80+	C1		32	19	35	12	2

Lab no.	$CaCO_3$ equiv. (%)	Organic C Total	Pyr.	Fe CBD	Ox	Pyr. (%)	Al CBD	Ox	Pyr	Mn CBD	Si $CaCl_2$ (mg kg^{-1})
C4982		0.9		0.3		0.0					
C4983		0.3		0.2		0.0					
C4984		0.3		0.2		0.0					

Lab no.	pH H_2O	$CaCl_2$	NaF	CEC	Exchangeable cations Ca	Mg	Na	K	Total (cmol$_c$ kg^{-1})	Acidity Titr	Exch	Al Exch
C4982	8.6	7.9		7.0	9.1	2.0	0.2	0.7	12.0	0.0		0.0
C4983	8.8	8.3		6.4	7.5	2.1	3.5	0.7	13.8	0.0		0.0
C4984	8.5	8.3		7.3	19.1	2.2	4.3	0.5	26.1	0.0		0.0

Lab no.	Resist. (ohms)	EC (dS m^{-1})	Satn. water (%)	Saturation extract Ca	Mg	Na	K (mmol$_c$ L^{-1})	Micronutrients Zn	Mn	Cu	B (mg kg^{-1})	Co
C4982	1 900							1.23	18.6	1.08	1.1	0.78
C4983	140	8.4		15	7	67	4	0.48	18.8	0.27	2.3	0.12
C4984	44	23.2		53	42	147	3	0.21	18.9	0.21	2.2	0.21

Lab no.	Bulk density (kg m^{-3})	HC (mm h^{-1})	Air-water permeability ratio	Atterberg LL	PI	Water retentivity at kPa: -33	-80	-500 (%)	-1500
C4982			3			9.2	7.6	5.3	4.2
C4983			1			11.5	9.2	7.2	5.6
C4984			1			14.1	12.4	10.6	9.3

Lab no.	P Sorp. (%)	P Extr. (mg kg^{-1})	Minerals <2 μm	Mineral legend			
C4982	6.2	144	Qz5, Mi4, Kt4	Qz	**Quartz**	Ct	**Calcite**
C4983	11.3	84	Qz5, Mi3, Kt3	Kt	**Kaolinite**	Mi	**Mica**
C4984	58.1	35	Qz5, Ct4, Mi4, Kt2				

P28: Calcic, soft-neocalcic, *chromic*, rhodic, luvic, aeromorphic, *loamy coarse sand*
(Addo Ad 1221)

Profile No:	P28	**Aspect:**	N
Land type map:	2820 Upington	**Water table:**	Absent
Latitude & Longitude:	28°55'12'/21°04'36'	**Occurrence of flooding:**	
Land type No:	Ag5	**Stoniness:**	Class 0
Climate zone No:	553S	**Erosion:**	Sheet, class 1
Soil form:	Hutton	**Vegetation/land use:**	Karroid and grass
Soil series:	Shigalo	**Parent and underlying material:**	Gneiss
Altitude:	884 m		
Terrain unit:	5		
Slope:	1%	**Described by:**	J.A.V. Dietrichsen
Slope shape:	Concave	**Date described:**	Aug-73

Horizon	Depth (mm)	Description	Diagnostic horizon
A1	0–300	Dry; moist 100% dark red 2.5YR3/6; loamy coarse sand; apedal; loose; very few small quartz fragments; gradual smooth boundary	Orthic
B2	300–900	Dry; moist 100% dark red 2.5YR3/6; coarse sandy loam; apedal; loose; few small quartz fragments; slightly effervescent; gradual smooth boundary	Neocarbonate
Cca	900N	Dry; moist 100% red 2.5YR4/6; coarse sandy clay loam; apedal; slightly hard; cemented by lime; many coarse indurated lime nodules; violently effervescent	Soft carbonate horizon

Points of interest and interpretation

This strongly reddened calcic loamy sand is well developed with a soft carbonate horizon that borders on hardpan carbonate, underlying a neocarbonate B (and judging from pH values the A horizon is probably also calcareous although this was not recorded in the field). Although subsoil EC data are not available the low exchangeable Na values relative to CEC as well as the resistance data indicate that the soil is non-saline. The clay fraction is predominantly micaceous-smectitic, with some kaolinite and, below 90 cm, some pyrophyllite as well. Organic carbon content is extremely low and extractable P values are high throughout the profile.

P28: Calcic, soft-neocalcic, *chromic*, rhodic, luvic, aeromorphic, *loamy coarse sand*
(Addo Ad 1221)

Lab no.	Depth (cm)	Horizon	Coarse fragments (%)	Fine earth (%)				
				CSa	MSa	FSa	Silt	Clay
C3411	0–30	A1	3	25	25	36	2	10
C3412	30–90	B2	15	24	22	35	3	16
C3413	90+	C		35	15	23	3	23

Lab no.	CaCO$_3$ equiv. (%)	Organic C		Fe			Al			Mn	Si CaCl$_2$ (mg kg^{-1})
		Total	Pyr.	CBD	Ox	Pyr.	CBD	Ox	Pyr	CBD	
						(%)					
C3411		0.10		0.80							
C3412		0.10		0.90							
C3413		0.10		0.90							

Lab no.	pH			CEC	Exchangeable cations					Acidity		Al Exch
	H$_2$O	CaCl$_2$	NaF		Ca	Mg	Na	K	Total	Titr	Exch	
					(cmol$_c$ kg^{-1})							
C3411	8.5	7.5		5.1	4.0	1.2	0.2	0.5	5.9	0.0	0.0	0.0
C3412	8.0	7.3		6.9	4.7	2.0	0.1	0.2	7.0	0.8	0.0	0.0
C3413	8.3	7.6		5.2	4.8	1.8	0.2	0.2	7.0	0.0	0.0	0.0

Lab no.	Resist. (ohms)	Saturation extract						Micronutrients				
		EC (dS m^{-1})	Satn. water (%)	Ca	Mg	Na	K	Zn	Mn	Cu	B	Co
				(mmol$_c$ L^{-1})				(mg kg^{-1})				
C3411	1600	0.5		5	0	0	0	0.68	140.3	1.76	0.17	2.33
C3412	1200							0.34	143.8	1.35	0.21	3.04
C3413	1400											

Lab no.	Bulk density (kg m^{-3})	HC (mm h^{-1})	Air-water permeability ratio	Atterberg		Water retentivity at kPa:			
				LL	PI	-33	-80	-500	-1500
						(%)			
C3411			4			5.1	4.2	4.0	3.4
C3412			3			6.8	5.8	5.2	4.8
C3413			2			12.3	11.2	9.8	8.6

Lab no.	P Sorp. (%)	P Extr. (mg kg^{-1})	Minerals <2 μm	Mineral legend			
C3411	22.9	39.3	Mi4, St3, Kt2	Py	**Pyrophyllite**	St	**Smectite**
C3412	29.1	22.1	Mi4, St3, Kt2,	Kt	**Kaolinite**	Mi	**Mica**
C3413	41.7	24.8	Mi3, Py3, St3, Kt2				

P783: Calcic, hard-neocalcic, *chromic*, rhodic, *haplic*, *loamy fine sand*
(Prieska Pr 1210)

Profile No:	P783	**Aspect:**	N
Land type map:	2816 Alexander Bay	**Water table:**	None
Latitude & Longitude:	28°39'48'/17°16'24'	**Occurrence of flooding:**	Nil
Land type No:	Ag38	**Stoniness:**	Both roundish and flat, 500 mm diameter, 0.3 m apart
Climate zone No:	144H	**Erosion:**	Rill, class 1
Soil form:	Hutton	**Vegetation/land use:**	Open dwarf shrubveld
Soil series:	Lowlands	**Parent and underlying material:**	Unconsolidated, sandy
Altitude:	640 m		
Terrain unit:	3		
Slope:	8%	**Described by:**	F. Ellis
Slope shape:	Convex	**Date described:**	Aug-77

Horizon	Depth (mm)	Description	Diagnostic horizon
A	0–50	Dry; moist yellowish-red 5YR5/6; loamy fine sand; single grain; soft; non-hardened free lime, moderately effervescent; few rounded very coarse hard lime nodules; few stones; clear smooth transition	Orthic
B2	50–300	Dry; moist reddish-yellow 5YR6/6; loamy fine sand; single grain; soft; non-hardened free lime, strongly effervescent; common rounded very coarse hard lime nodules; few stones; clear smooth transition	Neocarbonate
C	300+	Calcrete	Hardpan carbonate horizon

Points of interest and interpretation

Sepiolite and palygorskite are common constituents of calcic and silicic soils on the Namaqualand coastal plain (Francis, 2007) and their dominance of the clay fraction in the A and B horizons of this profile suggests that the underlying calcrete, for which no mineralogy is given, could well be a calcareous sepiocrete/palycrete in which case this soil would fall into the silicic group. Copper values are especially low and extractable P values are very high. The maritime location points to a possible apatite component of the sand fraction (phosphatic sands are known to occur in the region). Perhaps the high phosphate helps explain the low copper values. Calculation of metal phosphate mineral saturation indices might be a rewarding avenue of research in such soils (note that the silicic soil examples also exhibited high levels of extractable P).

P783: Calcic, hard-neocalcic, *chromic*, rhodic, haplic, *loamy fine sand*
(Prieska Pr 1210)

Lab no.	Depth (cm)	Horizon	Coarse fragments (%)	Fine earth (%)				
				CSa	MSa	FSa	Silt	Clay
C4723	0–5	A		19	16	51	8	5
C4724	5–30	B2ca		8	13	60	12	6

Lab no.	CaCO$_3$ equiv. (%)	Organic C		Fe			Al			Mn	Si CaCl$_2$ (mg kg^{-1})
		Total	Pyr.	CBD	Ox	Pyr.	CBD	Ox	Pyr	CBD	
C4723		0.4		0.5			0.0				
C4724		0.5		0.5			0.0				

Lab no.	pH			CEC	Exchangeable cations					Acidity		Al Exch
	H$_2$O	CaCl$_2$	NaF		Ca	Mg	Na	K	Total	Titr	Exch	
					(cmol$_c$ kg^{-1})							
C4723	8.8	7.8		5.6	6.3	0.9	0.4	1.0	8.6	0.0		0.0
C4724	8.8	8.0		5.6	9.6	1.2	1.1	0.9	12.8	0.0		0.0

Lab no.	Saturation extract							Micronutrients				
	Resist. (ohms)	EC (dS m^{-1})	Satn. water (%)	Ca	Mg	Na	K	Zn	Mn	Cu	B	Co
				(mmol$_c$ L^{-1})				(mg kg^{-1})				
C4723	1 000							1.26	61.7	0.15	1.23	0.09
C4724	580							1.02	30.0	0.60	2.27	0.09

Lab no.	Bulk density (kg m^{-3})	HC (mm h^{-1})	Air-water permeability ratio	Atterberg		Water retentivity at kPa:			
				LL	PI	-33	-80	-500	-1500
						(%)			
C4723			3			13.2	10.5	6.9	5.3
C4724			2			18.1	14.5	9.8	7.3

Lab no.	P Sorp. (%)	P Extr. (mg kg^{-1})	Minerals <2 μm	Mineral legend			
C4723	31.0	54.4	Sp5, Mi5, Pl4, Qz4, Kt3	Qz	**Quartz**	Kt	**Kaolinite**
C4724	51.4	24.5	Sp5, Mi4, Pl4, Qz3, Kt3	Sp	**Sepiolite**	Mi	**Mica**
				Pl	**Palygorskite**		

P812: Calcic, hard-neocalcic, *chromic*, rhodic, haplic, *coarse sand*
(Prieska Pr 1210)

Profile No:	P812	**Aspect:**	S
Land type map:	2918 Pofadder	**Water table:**	None
Latitude & Longitude:	29°22'03'/18°35'57'	**Occurrence of flooding:**	Nil
Land type No:	Ag62	**Stoniness:**	Roundish, 250 mm diameter, 2.0 m apart
Climate zone No:	183S	**Erosion:**	Wind, class 1, partially stabilised
Soil form:	Hutton	**Vegetation/land use:**	Sparse grassveld
Soil series:	Quaggafontein	**Parent and underlying material:**	Calcrete
Altitude:	860 m		
Terrain unit:	4		
Slope:	02%	**Described by:**	B.H.A. Schloms
Slope shape:	Straight	**Date described:**	Jun-77

Horizon	Depth (mm)	Description	Diagnostic horizon
A	0–50	Moist; moist yellowish red 5YR4/6; coarse sand; single grain; loose; gradual smooth transition	Orthic
B2	50–300	Moist; moist yellowish red 5YR4/8; coarse sand; single grain; loose; hardened free lime, moderately effervescent; few very coarse hard lime nodules; abrupt smooth transition	Neocarbonate
C	300+	Calcrete	Hardpan carbonate horizon

Points of interest and interpretation

This soil is non-saline and only mildly sodic and exhibits chemical and physical properties similar to those of the other two examples of calcic soils. The clay fraction is predominantly micaceous with accessory smectite, kaolinite and quartz. Neither sepiolite nor palygorskite were recorded. Aridity and shallow effective depth are paramount.

P812: Calcic, hard-neocalcic, *chromic*, rhodic, haplic, *coarse sand*
(Prieska Pr 1210)

Lab no.	Depth (cm)	Horizon	Coarse fragments (%)	Fine earth (%)				
				CSa	MSa	FSa	Silt	Clay
C4744	0–5	A		25	37	32	1	4
C4745	5–30	B2		31	37	26	0	6

Lab no.	CaCO$_3$ equiv. (%)	Organic C		Fe			Al			Mn	Si CaCl$_2$ (mg kg^{-1})
		Total	Pyr.	CBD	Ox	Pyr.	CBD	Ox	Pyr	CBD	
					(%)						
C4744		0.1		0.4			0.0				
C4745		0.0		0.4			0.0				

Lab no.	pH			CEC	Exchangeable cations					Acidity		Al Exch
	H$_2$O	CaCl$_2$	NaF		Ca	Mg	Na	K	Total	Titr	Exch	
					(cmol$_c$ kg^{-1})							
C4744	8.2	7.2		2.2	1.6	0.3	0.1	0.3	2.3	0.0		0.0
C4745	8.4	7.7		2.2	3.0	0.6	0.3	0.3	4.2	0.0		0.0

Lab no.	Saturation extract							Micronutrients			
	Resist. (ohms)	EC (dS m^{-1})	Satn. water (%)	Ca	Mg	Na	K	Zn	Mn	Cu	B	Co
				(mmol$_c$ L^{-1})				(mg kg^{-1})			
C4744	2 900							0.45	27.3	0.39	0.05	0.39
C4745	800							0.21	86.1	0.72	0.12	0.96

Lab no.	Bulk density (kg m^{-3})	HC (mm h^{-1})	Air-water permeability ratio	Atterberg		Water retentivity at kPa:			
				LL	PI	-33	-80	-500	-1 500
						(%)			
C4744			3			3.1	2.7	2.2	1.9
C4745			2			3.2	3.1	2.4	2.2

Lab no.	P Sorp. (%)	P Extr. (mg kg^{-1})	Minerals <2 μm	Mineral legend			
C4744	8.2	23.7	Mi5, Kt2, Qz2, St1	Qz	**Quartz**	St	**Smectite**
C4745	8.2	14.3	Mi5, St2, Kt2, Qz1, Fs1	Fs	**Feldspar**	Mi	**Mica**
				Kt	**Kaolinite**		

P826: Duplex, pedocutanic-lithic, *chromic*, rhodic, micropedal, calcic, *sandy clay loam*
(Swartland Sw 1212)

Profile No:	P826	**Aspect:**	S
Land type map:	3018 Loeriesfontein	**Water table:**	None
Latitude & Longitude:	30°35'24'/18°21'48'	**Occurrence of flooding:**	Nil
Land type No:	Ib274	**Stoniness:**	Roundish, 250 mm diameter, 0.1 m apart
Climate zone No:	196W	**Erosion:**	Gully, class 2, not stabilised
Soil form:	Swartland	**Vegetation/land use:**	Open dwarf shrubveld
Soil series:	Broekspruit	**Parent and underlying material:**	Granite
Altitude:	760 m		
Terrain unit:	3		
Slope:	35%	**Described by:**	L.M. Barkhuysen
Slope shape:	Convex	**Date described:**	Sept-77

Horizon	Depth (mm)	Description	Diagnostic horizon
A	0–100	Dry; moist yellowish-red 5YR4/6; coarse sandy clay loam; massive; slightly hard; few rounded fine gravel, clear smooth transition	Orthic
B	100–300	Dry; moist yellowish-red 5YR4/6; coarse sandy clay; moderate medium subangular blocky; hard; non-hardened free lime, moderately effervescent; common clay cutans; common rounded fine gravel	Pedocutanic
R	300+	Saprolite	Saprolite

Points of interest and interpretation

Besides the duplex morphology the shallow depth to saprolite is the key feature of this soil which has more in common with soils of the lithic group than of any other. The soil has mild salinity and negligible sodicity, a high pH, and secondary carbonate in the B horizon. The most interesting data are those for trace elements, with Mn concentrations suggesting acute deficiency whereas B concentrations are well above the toxicity threshold of 4-5 mg/kg, particularly in the B horizon. The origin and extent of these trace element anomalies would be worth investigating. Erosivity, aridity and shallow effective depth are the major limitations to this soil's potential.

P826: Duplex, pedocutanic-lithic, *chromic*, rhodic, micropedal, calcic, *sandy clay loam*
(Swartland Sw 1212)

Lab no.	Depth (cm)	Horizon	Coarse fragments (%)	Fine earth (%)				
				CSa	MSa	FSa	Silt	Clay
C4967	0–10	A		16	18	27	14	25
C4968	10–30	B2		16	14	21	11	39

Lab no.	$CaCO_3$ equiv. (%)	Organic C		Fe			Al			Mn	Si $CaCl_2$ (mg kg^{-1})
		Total	Pyr.	CBD	Ox	Pyr.	CBD	Ox	Pyr	CBD	
					(%)						
C4967		0.9		0.5			0.0				
C4968		0.5		0.7			0.0				

Lab no.	pH			CEC	Exchangeable cations					Acidity		Al Exch
	H_2O	$CaCl_2$	NaF		Ca	Mg	Na	K	Total	Titr	Exch	
					(cmol$_c$ kg^{-1})							
C4967	8.4	7.9		11.6	12.7	2.1	0.6	0.3	15.7	0.0		0.0
C4968	8.3	8.0		17.4	14.5	6.8	0.5	0.0	21.8	0.0		0.0

Lab no.	Saturation extract							Micronutrients				
	Resist. (ohms)	EC (dS m^{-1})	Satn. water (%)	Ca	Mg	Na	K	Zn	Mn	Cu	B	Co
				(mmol$_c$ L^{-1})				(mg kg^{-1})				
C4967	440	1.80		12	0	9	0	0.99	0.7	1.47	6.43	0.30
C4968	180	2.75		19	2	23	0	0.60	0.2	1.05	19.11	0.24

Lab no.	Bulk density (kg m^{-3})	HC (mm h^{-1})	Air-water permeability ratio	Atterberg		Water retentivity at kPa:			
				LL	PI	-33	-80	-500	-1 500
						(%)			
C4967			14			15.3	13.9	10.3	9.9
C4968			38			21.7	19.1	14.4	13.9

Lab no.	P Sorp. (%)	P Extr. (mg kg^{-1})	Minerals <2 µm	Mineral legend			
C4967	34.3	16.1	Mi5, Kt4, Qz3, St3	Qz	**Quartz**	St	**Smectite**
C4968	46.3	8.0	Kt5, Mi3, Qz2	Kt	**Kaolinite**	Mi	**Mica**

P886: Duplex, pedocutanic-cumulic-hydromorphic, *chromic*, macropedal, *calcic*, *sandy clay loam*
(Sepane Se 1220)

Profile No:	P886	**Aspect:**	E
Land type map:	2530 Barberton	**Water table:**	None
Latitude & Longitude:	25°59'00'/30°41'11'	**Occurrence of flooding:**	Nil
Land type No:	Fa166	**Stoniness:**	None
Climate zone No:	118S	**Erosion:**	Gully, class 2
Soil form:	Valsrivier	**Vegetation/land use:**	Sparse grassveld
Soil series:	Valsrivier		
Altitude:	950 m	**Parent and underlying material:**	Gneiss
Terrain unit:	4		
Slope:	2%	**Described by:**	R.W. Fitzpatrick
Slope shape:	Straight	**Date described:**	Aug-72

Horizon	Depth (mm)	Description	Diagnostic horizon
A1	0–100	Dry; moist very dark grey 10YR3/1; coarse sandy clay loam; weak coarse subangular blocky; slightly hard; few coarse gravel; few lime nodules; clear smooth transition	Orthic
B21	100–650	Moist; moist dark grey 10YR4/1; medium sandy clay loam; strong coarse angular blocky; very firm; many clay cutans; common coarse gravel; few lime nodules; clear smooth transition	Pedocutanic
B22	650–1 200+	Moist; moist yellowish-brown 10YR5/4; few coarse distinct yellow, olive and brown mottles; coarse sandy clay loam; moderate coarse angular blocky; firm; non-hardened free lime, strongly effervescent; common coarse gravel; common lime nodules	Pedocutanic

Points of interest and interpretation

Contrasting with the previous example of a duplex soil, this profile consists of a deep clay loam, formed in gneissic colluvium, containing nodules of secondary lime in the B horizon and showing evidence of wetness at depth. Smectite and accessory kaolinite are the dominant clay minerals. Neither salinity nor sodicity are prevalent. Zinc levels are markedly deficient and extractable P is also very low. Such soils can be productively used under irrigation but the duplex character means that artificial drainage would have to be addressed. Hard setting and erodibility are two physical conditions that often make duplex soils less amenable to use, and the texture of this soil is likely to intensify such physical problems. Observation in the field is consistent with this expectation.

P886: Duplex, pedocutanic-cumulic-hydromorphic, *chromic*, macropedal, calcic, *sandy clay loam*
(Sepane Se 1220)

Lab no.	Depth (cm)	Horizon	Coarse fragments (%)	Fine earth (%)				
				CSa	MSa	FSa	Silt	Clay
C1388	0–10	A1		19	19	24	4	21
C1389	10–65	B21		13	15	21	12	31
C1390	65–120	B22		23	16	27	7	27

Lab no.	$CaCO_3$ equiv. (%)	Organic C		Fe			Al			Mn	Si $CaCl_2$ (mg kg^{-1})
		Total	Pyr.	CBD	Ox	Pyr.	CBD	Ox	Pyr	CBD	
		(%)									
C1388		0.9		0.4			0.0				
C1389		0.4		0.4			0.1				
C1390		0.1		0.4			0.0				

Lab no.	pH			CEC	Exchangeable cations					Acidity		Al Exch
	H_2O	$CaCl_2$	NaF		Ca	Mg	Na	K	Total	Titr	Exch	
					(cmol$_c$ kg^{-1})							
C1388	6.3	5.4		8.2	4.7	2.6	0.1	0.3	7.7	5.3		0.0
C1389	7.6	6.1		13.6	11.7	4.2	0.2	0.2	16.3	3.3		0.0
C1390	7.7	6.1		15.5	7.6	4.5	0.3	0.1	12.5	3.7		0.0

Lab no.	Saturation extract							Micronutrients			
	Resist. (ohms)	EC (dS m^{-1})	Satn. water (%)	Ca	Mg	Na	K	Zn	Mn	Cu	B	Co
				(mmol$_c$ L^{-1})				(mg kg^{-1})			
C1388	2000							0.08	236.4	2.62	0.34	3.74
C1389	490							0.00	552.3	3.59	0.28	7.17
C1390	900							0.04	116.5	1.86	0.10	

Lab no.	Bulk density (kg m^{-3})	HC (mm h^{-1})	Air-water permeability ratio	Atterberg		Water retentivity at kPa:			
				LL	PI	-33	-80	-500	-1500
						(%)			
C1388			9			15.4	12.3	9.2	9.1
C1389			43			24.1	20.0	15.3	13.5
C1390			17			20.9	17.3	12.7	10.7

Lab no.	P Sorp. (%)	P Extr. (mg kg^{-1})	Minerals <2 μm	Mineral legend			
C1388	12.8	1.1	St5, Kt3, Mi1, Qz1	Qz	**Quartz**	St	**Smectite**
C1389	13.2	0.5	St5, Kt4, Mi2, Qz1	Kt	**Kaolinite**	Mi	**Mica**
C1390	11.5	0.1	St5, Kt4, Mi2				

P1112: Duplex, eluvic-prismacutanic, albic, asombric, *fine sand*
(Estcourt Es 1100)

Profile No:	P1112	**Aspect:**	S
Land type map:	2730 Vryheid	**Water table:**	None
Latitude & Longitude:	27°19'54'/31°39'30'	**Occurrence of flooding:**	Nil
Land type No:	Db144	**Stoniness:**	None
Climate zone No:	372S	**Erosion:**	Rill, class 2
Soil form:	Estcourt	**Vegetation/land use:**	Bushland
Soil series:	Uitvlugt	**Parent and underlying material:**	Feldspathic sandstone
Altitude:	275 m		
Terrain unit:	3		
Slope:	4%	**Described by:**	B.L. Plath
Slope shape:	Straight	**Date described:**	Jan-81

Horizon	Depth (mm)	Description	Diagnostic horizon
A1	0–200	Dry; dry brown 10YR5/3; fine sand; single grain; soft; gradual smooth transition	Orthic
E	200–450	Dry; dry light yellowish-brown 10YR6/4; few fine faint red and yellow mottles; fine sand; soft; abrupt smooth transition	E-horizon
B21	450–900	Dry; moist brown to dark brown 10YR4/3; common medium distinct yellow, brown and red mottles; sandy clay loam; strong coarse columnar; very hard; many clay cutans	Prismacutanic

Points of interest and interpretation

Mica and kaolinite (with accessory smectite in the B horizon) dominate the clay fraction of this sodic soil with extreme duplex character. The E horizon is sandy and relatively thick, while the B horizon has a characteristic abrupt transition with the overlying E and exhibits some gley morphology. Lateral seepage of perched water in the E and ferrolysis are important natural processes suggested by the profile morphology. Columnar (as opposed to prismatic) structure is especially indicative of the clay destruction that occurs as a result of ferrolysis. The soil is mildly acidic, slightly more so at depth which is unusual. An increase in carbon from the E to the B reflects the dark brown (but nevertheless asombric) cutanic character, associated with an illuvial clay-humus complex. Relatively high levels of both exchangeable Mg and Na are often associated with the B horizon of the eluvic-prismacutanic form of duplex soils. As in the previous example, crusting and erodibility are problems expected on the basis of fine sandy texture, low organic matter content and low salinity. Field observation confirms this.

P1112: Duplex, eluvic-prismacutanic, albic, asombric, *fine sand*
(Estcourt Es 1100)

Lab no.	Depth (cm)	Horizon	Coarse fragments (%)	Fine earth (%)				
				CSa	MSa	FSa	Silt	Clay
C5811	0–20	A1		8	30	51	6	5
C5812	20–45	E		8	29	52	5	6
C5813	45–90	B21		7	21	41	4	26

Lab no.	CaCO$_3$ equiv. (%)	Organic C		Fe			Al			Mn	Si CaCl$_2$ (mg kg^{-1})
		Total	Pyr.	CBD	Ox	Pyr.	CBD	Ox	Pyr	CBD	
					(%)						
C5811		0.4		0.3			0.4				
C5812		0.1		0.3			0.03				
C5813		0.3		0.9			0.08				

Lab no.	pH			CEC	Exchangeable cations					Acidity		Al Exch
	H$_2$O	CaCl$_2$	NaF		Ca	Mg	Na	K	Total	Titr	Exch	
					(cmol$_c$ kg^{-1})							
C5811	6.2	5.2		2.6	0.6	0.5	0.1	0.6	1.8	0.6	0.0	0.0
C5812	6.1	4.9		1.6	0.3	0.5	0.2	0.1	1.1	1.0	-	0.0
C5813	6.1	4.6		5.6	0.4	1.6	1.7	0.3	4.0	0.0	0.5	0.2

Lab no.	Saturation extract							Micronutrients				
	Resist. (ohms)	EC (dS m^{-1})	Satn. water (%)	Ca	Mg	Na	K	Zn	Mn	Cu	B	Co
				(mmol$_c$ L^{-1})				(mg kg^{-1})				
C5811	1 700							0.68	37.85	0.16	0.28	1.79
C5812	2 000							0.12	4.81	0.56	0.48	0.40
C5813	700							0.00	0.32	1.67	0.61	0.00

Lab no.	Bulk density (kg m^{-3})	HC (mm h^{-1})	Air-water permeability ratio	Atterberg		Water retentivity at kPa:			
				LL	PI	-33	-80	-500	-1500
						(%)			
C5811			6			4.7	4.1	2.8	2.7
C5812			6			3.7	3.3	2.3	2.2
C5813			999			21.9	20.5	16.4	15.6

Lab no.	P Sorp. (%)	P Extr. (mg kg^{-1})	Minerals <2 µm	Mineral legend			
C5811			Mi5, Kt4, Qz3	Qz	**Quartz**	St	**Smectite**
C5812				Kt	**Kaolinite**	Mi	**Mica**
C5813			Mi5, Kt5, St2, Qz1				

P2568: Podzolic, cumulic-aeromorphic, friable, *medium sand*
(Pinegrove Pg 1000)

Profile No:	P2568	**Aspect:**	W
Land type map:		**Water table:**	
Latitude & Longitude:	22°52'32'/33°59'06'	**Occurrence of flooding:**	
Land type No:		**Stoniness:**	
Climate zone No:		**Erosion:**	
Soil form:	Pinegrove	**Vegetation/land use:**	Pine plantation
Soil series:	Highbury		
Altitude:	240-260 m	**Parent and underlying material:**	Origin, single, aeolian
Terrain unit:			
Slope:	35%	**Described by:**	J.J.N. Lambrechts & A.B. Oosthuizen
Slope shape:	Convex-concave	**Date described:**	April-92

Horizon	Depth (mm)	Description	Diagnostic horizon
A	0–190	Moist; moist dark reddish-brown 5YR3/2; medium sand; few medium faint white bleached mottles; weak medium crumb; loose; water absorption 4 seconds; many roots; gradual, smooth transition	Orthic A
B1	190–400	Moist; moist dark reddish-brown 5YR3/3; medium sand; few medium faint white bleached mottles; weak medium crumb; loose; water absorption 3 seconds; many roots; gradual, smooth transition	Podzol B
B2	400–590	Moist; moist reddish-brown 5YR4/4; medium sand; common medium faint yellow mottles; apedal, massive; friable; very few round fine soft sesquioxide nodules; water absorption 3 seconds; common roots; gradual, smooth transition	Podzol B

Points of interest and interpretation

The first things to look at in the analyses of podzols are the data for carbon, iron and aluminium as a function of depth in the profile. Pyrophosphate-extractable C, and Fe and Al extracted by not only pyrophosphate but also oxalate and CBD, increase as expected from the A to B horizon. The only visual evidence of illuviation of metal-humic complexes is faint bleaching in the A horizon but the chemical data clearly confirm the podzolic character. Interestingly the trace element depth functions do not show an increase from the A to the B in contrast to some of the other examples which appear below. Trace element levels are nevertheless low (especially those of Mn and Cu). Hydroxy-interlayered vermiculite (also known as pedogenic aluminous chlorite) is the second most abundant clay mineral after quartz and this is consistent with the strongly acidic pH and high levels of exchangeable Al that were recorded. The pH in NaF is a useful confirmation of relatively high levels of amorphous Al hydroxide, as indicated by the effectively full extraction of CBD-soluble Al by oxalate.

P2568: Podzolic, cumulic-aeromorphic, friable, *medium sand*
(Pinegrove Pg 1000)

Lab no.	Depth (cm)	Horizon	Coarse fragments (%)	Fine earth (%) CSa	MSa	FSa	Silt	Clay
C9752	0–19	A1	0.0	0.1	59.0	16.9	10.4	7.5
C9753	19–40	B1	0.0	0.0	60.3	14.5	12.4	7.6
C9754	40–59	B2	0.0	0.0	57.0	16.0	15.4	4.8

Lab no.	CaCO$_3$ equiv. (%)	Organic C Total	Pyr.	Fe CBD	Ox	Pyr.	Al CBD	Ox	Pyr	Mn CBD	Si CaCl$_2$ (mg kg^{-1})
C9752	0	2.55	0.28	0.35	0.10	0.23	0.11	0.09	0.14	tr	3.9
C9753	0	0.96	0.64	0.49	0.15	0.33	0.20	0.20	0.34	tr	5.1
C9754	0	0.78	0.42	0.63	0.15	0.33	0.29	0.29	0.44	tr	9.1

Lab no.	pH H$_2$O	CaCl$_2$	NaF	CEC	Exchangeable cations (cmol$_c$ kg^{-1}) Ca	Mg	Na	K	Total	Acidity Titr	Exch	Al Exch
C9752	5.0	3.9	8.2	7.36	2.01	0.98	0.05	0.24	3.28		1.19	0.80
C9753	4.7	4.1	9.9	5.14	0.36	0.31	0.03	0.07	0.77		1.99	1.91
C9754	4.9	4.4	10.7	3.46	0.12	0.16	0.02	0.10	0.40		1.20	1.01

Lab no.	Resist. (ohms)	EC (dS m^{-1})	Satn. water (%)	Saturation extract Ca	Mg	Na	K (mmol$_c$ L^{-1})	Micronutrients Zn	Mn	Cu	B	Co (mg kg^{-1})
C9752		0.37	35.8					1.27	1.22	0.45	0.96	0.19
C9753		0.31	32.4					1.16	0.22	0.30	0.75	0.04
C9754		0.21	32.6					1.06	<0.10	0.15	0.30	0.01

Lab no.	Bulk density (kg m^{-3})	HC (mm h^{-1})	Air-water permeability ratio	Atterberg LL	PI	Water retentivity at kPa: -33	-80	-500	-1500 (%)
C9752	1 360	118.5	2.8			10.4	7.1	7.0	5.4
C9753	1 430	148.2	2.2			6.0	5.2	5.0	4.6
C9754	1 340	71.7	1.5			8.9	7.6	5.8	5.0

Lab no.	P Sorp. (%)	P Extr. (mg kg^{-1})	Minerals <2 µm		Mineral legend		
C9752			Qz5, HIV2, Kt2, Is2	Qz	**Quartz**	Is	**Interstratified**
C9753			Qz5, HIV3, Is3, Kt1	HIV	**Hydroxy-interlayered vermiculite**		
C9754			Qz5, HIV2, Is2, Kt1	Mi	**Mica**	Kt	**Kaolinite**

P2569: Podzolic, albic-cumulic-hydromorphic, adensic, friable, *medium sand*
(Lamotte Lt 1100)

Profile No:	P2569	**Aspect:**	W
Land type map:		**Water table:**	
Latitude & Longitude:	22°52'32'/33°59'06'	**Occurrence of flooding:**	
Land type No:		**Stoniness:**	
Climate zone No:		**Erosion:**	
Soil form:	Lamotte	**Vegetation/land use:**	Pine plantation
Soil series:	Kruisfontein	**Parent and underlying material:**	Origin single, aeolian
Altitude:	240-260 m		
Terrain unit:			
Slope:	35%	**Described by:**	J.J.N. Lambrechts & A.B. Oosthuizen
Slope shape:	Convex-concave	**Date described:**	April-92

Horizon	Depth (mm)	Description	Diagnostic horizon
A	0–220	Moist; moist very dark grey 7.5YR3/0; medium sand; weak fine crumb; loose; water absorption 3 seconds; common roots; clear, smooth transition	Orthic A
E	220–430	Moist; moist greyish-brown 10YR5/2; medium sand; common fine faint white bleached mottles; apedal single grain; loose; water absorption 1 seconds; abrupt, smooth transition	E
B1	430–650	Moist; moist dark reddish-brown 5YR2.5/2; loamy fine sand; common medium distinct yellow and brown oxidised iron oxide mottles; weak medium subangular blocky; slightly firm; few round fine biocasts; water absorption 3 seconds; clear, smooth transition	Podzol B

Points of interest and interpretation

Here we have a more classic podzol with a clear E horizon interposed between an orthic A and podzol B. The pH and exchangeable Al saturation values indicate extreme acidity, which decreases with depth. Pyrophosphate C, Al and Fe as well as CBD- and oxalate-extractable Fe and Al are at a maximum in the B horizon. Indications from oxalate extraction of amorphous Al are again confirmed by a pH (NaF) > 10 in the B horizon. Soluble Si is also at a maximum in the B. Certain theories of podzol formation (Section 2.8) would be supported by such data. Clay mineralogy is similar to that of the previous podzol profile. Trace element concentrations are generally very low (except Zn) and two (Cu and Co) show B horizon maxima, paralleling the pattern shown by Fe and Al. A similar trend is apparent with exchangeable Ca data. The B horizon is quite deep meaning that the already poor nutrient reserve lies largely below the root zone of many crops. High permeability and low water holding capacity have implications for irrigation scheduling while the low buffer capacity (CEC < 8 $cmol_c$ kg^{-1}) means that fertiliser applications need to be small and frequent to avoid both salinity shock and the loss of nutrients through leaching.

P2569: Podzolic, albic-cumulic-hydromorphic, adensic, friable, *medium sand*
(Lamotte Lt 1100)

Lab no.	Depth (cm)	Horizon	Coarse fragments (%)	Fine earth (%)				
				CSa	MSa	FSa	Silt	Clay
C9759	0–22	A	0	0.2	65	13.9	11.1	4.2
C9760	22–43	E	0	0	63.4	15.9	12.8	1.5
C9761	43–65	B1	0	0.1	54.9	15.2	18.3	4.9

Lab no.	CaCO$_3$ equiv. (%)	Organic C			Fe			Al			Mn	Si CaCl$_2$ (mg kg^{-1})
		Total	Pyr.	CBD	Ox	Pyr.	CBD	Ox	Pyr	CBD		
		(%)											
C9759	0	2.40	0.28	0.07	0.03	0.03	0.03	0.04	0.04	tr		3.5
C9760	0	0.24	0.64	0.05	0.01	0.02	0.01	0.02	0.02	tr		2.3
C9761	0	0.96	0.42	0.43	0.27	0.29	0.25	0.26	0.28	tr		6.1

Lab no.	pH			CEC	Exchangeable cations					Acidity		Al Exch
	H$_2$O	CaCl$_2$	NaF		Ca	Mg	Na	K	Total	Titr	Exch	
					(cmol$_c$ kg^{-1})								
C9759	3.8	2.8	7.0	7.98	0.17	0.44	0.06	0.04	0.71		2.21	1.89
C9760	4.0	3.6	8.0	0.78	0.06	0.02	0.01	0.04	0.13		0.66	0.41
C9761	5.1	4.2	10.5	7.55	0.36	0.25	0.07	0.04	0.72		1.79	1.69

Lab no.	Saturation extract								Micronutrients			
	Resist. (ohms)	EC (dS m^{-1})	Satn. water (%)	Ca	Mg	Na	K	Zn	Mn	Cu	B	Co
				(mmol$_c$ L^{-1})				(mg kg^{-1})					
C9759		0.53	37.1					1.78	0.56	0.30	0.42	0.00
C9760		0.24	21.2					1.18	0.11	0.30	<0.1	0.08
C9761		0.42	32.0					1.37	<0.10	0.84	0.28	0.13

Lab no.	Bulk density (kg m^{-3})	HC (mm h^{-1})	Air-water permeability ratio	Atterberg		Water retentivity at kPa:			
				LL	PI	-33	-80	-500	-1 500
						(%)			
C9759	1 210	130.9	2.4			8.2	6.6	4.7	4.0
C9760	1 460	5.2	6.6			2.4	2.2	1.5	1.5
C9761	1 520	14.2	2.5			9.7	7.6	5.4	4.6

Lab no.	P Sorp. (%)	P Extr. (mg kg^{-1})	Minerals <2 µm	Mineral legend		
C9759			Qz5, HIV2, Is1	Qz	**Quartz**	Is	**Interstratified**
C9760			Qz5, HIV1, Kt1, Is1	HIV	**Hydroxy interlayered vermiculite**		
C9761			Qz5, HIV1, Kt1, Is1	Kt	**Kaolinite**		

P2573: Podzolic, placic, *loamy fine sand*
(Jonkersberg Jb 1000)

Profile No:	P2573	**Aspect:**	South-east
Land type map:		**Water table:**	
Latitude & Longitude:	33°53'48'/22°01'18'	**Occurrence of flooding:**	
Land type No:		**Stoniness:**	
Climate zone No:		**Erosion:**	
Soil form:	Jonkersberg	**Vegetation/land use:**	Natural Fynbos (macchia)
Soil series:	geelhoutboom	**Parent and underlying material:**	Binary origin; local colluvium on residually weathered sandstone
Altitude:	720 m		
Terrain unit:	3		
Slope:	20%	**Described by:**	J.J.N. Lambrechts & A.B. Oosthuizen
Slope shape:	Convex	**Date described:**	April-92

Horizon	Depth (mm)	Description	Diagnostic horizon
A1	0–160	Moist; moist very dark grey 2.5Y3/0; loamy fine sand; weak fine crumb; friable; water absorption 5 seconds; many roots; clear, smooth transition	Orthic A
A2	160–300	Moist; moist very dark grey 2.5Y3/0; loamy fine sand; apedal massive; friable; water absorption 2 seconds; common roots; abrupt, wavy transition	Orthic A
2B1	300–570	Moist; moist very dark grey 10YR3/1; fine sandy loam; weak medium subangular blocky; friable; few fine pores; single placic pan (2 mm thick) in upper part of horizon; water absorption 2 seconds; few roots; abrupt, tonguing transition	Podzol B horizon with placic pan
2B2	570–800	Moist; moist brown to dark brown 10YR4/3; fine sandy loam; weak medium subangular blocky; slightly firm; few fine pores; single placic pan (5 mm thick) in upper part of horizon; water absorption 2 seconds; few roots; clear, wavy transition	Podzol B horizon with placic pan
2B3	800–950	Moist; moist yellowish brown 10YR5/4; fine sandy loam; common medium faint brown oxidised iron oxide mottles; weak medium subangular blocky; slightly firm; common fine organic coated pores; multiple placic pan (6 mm thick) throughout horizon; water absorption 2 seconds; abrupt, tonguing transition to moist; moist brownish yellow 10YR6/6; fine sandy clay loam (unconsolidated BC horizon from 950–1 500 mm)	Podzol B horizon with placic pan

Points of interest and interpretation

This is similar to the previous profile but with a more pronounced A-B contrast in pyrophosphate, oxalate and CBD extraction values. Cobalt is deficient, increasing with depth. The B shows especially strong indications of soluble Si and Al possibly implying an amorphous aluminosilicate phase as has been found in podzols elsewhere.

P2573: Podzolic, placic, *loamy fine sand*
(Jonkersberg Jb 1000)

Lab no.	Depth (cm)	Horizon	Coarse fragments (%)	Fine earth (%) CSa	MSa	FSa	Silt	Clay
C9775	0–6	A1	0	1.7	9.5	26.6	23.1	7.2
C9776	19–30	A2	0	2.9	13.0	25.9	22.3	5.5
C9777	30–57	2B1	2	2.4	11.0	19.9	28.9	5.0

Lab no.	$CaCO_3$ equiv. (%)	Organic C Total	Pyr.	CBD	Fe Ox (%)	Pyr.	Al CBD	Ox	Pyr	Mn CBD	Si $CaCl_2$ (mg kg^{-1})
C9775	0	4.55	1.23	0.89	0.11	0.11	0.90	0.11	0.80	tr	2.00
C9776	0	2.70	1.81	0.34	0.27	0.24	0.31	0.32	0.31	tr	2.00
C9777	0	2.55	2.13	1.08	0.80	1.03	1.43	1.38	1.23	tr	14.60

Lab no.	pH H_2O	$CaCl_2$	NaF	CEC	Exchangeable cations Ca	Mg	Na	K	Total (cmol$_c$ kg^{-1})	Acidity Titr	Exch	Al Exch
C9775	4.5	3.6	7.6	10.11	0.96	0.52	0.04	0.09	1.61		3.12	2.95
C9776	4.6	3.9	9.9	10.18	0.23	0.19	0.02	0.03	0.47		2.79	2.65
C9777	5.1	4.4	11.5	11.74	0.10	0.11	0.04	0.04	0.29		2.76	2.10

Lab no.	Saturation extract Resist. (ohms)	EC (dS m^{-1})	Satn. water (%)	Ca	Mg (mmol$_c$ L^{-1})	Na	K	Micronutrients Zn	Mn (mg kg^{-1})	Cu	B	Co
C9775		0.35	56.9					1.61	1.00	0.84	1.41	0.00
C9776		0.31	47.2					1.28	<0.10	0.45	0.76	0.10
C9777		0.19	48.2					1.27	<0.10	0.45	0.25	0.30

Lab no.	Bulk density (kg m^{-3})	HC (mm h^{-1})	Air-water permeability ratio	Atterberg LL	PI	Water retentivity at kPa: -33	-80	-500 (%)	-1 500
C9775	1 030	93.1	1.4			15.4	11.8	11.2	9.6
C9776	1 300	46.5	2.0			13.0	9.3	6.6	6.4
C9777	1 070	55.7	1.5			24.5	17.2	14.8	9.1

Lab no.	P Sorp. (%)	P Extr. (mg kg^{-1})	Minerals <2 μm		Mineral legend		
C9775			Qz5, Kt1, HIV1	Qz	**Quartz**	Kt	**Kaolinite**
C9776			Qz5, HIV1	HIV	**Hydroxy interlayered vermiculite**		
C9777			Qz5, HIV1				

P2575: Podzolic, lithic, orthosaprolitic, aeromorphic, *fine sandy loam*
(Groenkop Gk 1100)

Profile No:	P2575	**Aspect:**	South
Land type map:		**Water table:**	
Latitude & Longitude:	33°59'18'/22°04'54'	**Occurrence of flooding:**	
Land type No:		**Stoniness:**	
Climate zone No:		**Erosion:**	
Soil form:	Groenkop	**Vegetation/land use:**	Natural Fynbos (macchia)
Soil series:	Noordhoek		
Altitude:	284 m	**Parent and underlying material:**	Single origin suspected; weathering silcrete
Terrain unit:	3		
Slope:	10%	**Described by:**	J.J.N. Lambrechts & A.B. Oosthuizen
Slope shape:	Convex to straight	**Date described:**	April-92

Horizon	Depth (mm)	Description	Diagnostic horizon
A	0–220	Moist; moist very dark grey 10YR3/1; fine sandy loam; weak fine crumb; friable; few angular coarse stones; common mixed-shape coarse gravel fragments; water absorption 6 seconds; common roots; clear, smooth transition	Orthic A
B1	220–500	Moist; moist very dark greyish-brown 10YR3/2; loamy coarse sand; weak fine subangular blocky; friable; few angular coarse ferruginised silcrete stones; common mixed-shape coarse gravel fragments; water absorption 4 seconds; many roots; clear, tonguing transition	Podzol B
B2	500–750	Moist; moist very dark grey 10YR3/1; fine sandy loam; moderate fine subangular blocky; slightly firm; common mixed-shape coarse ferruginised silcrete stones and very few round coarse gravel fragments; water absorption 3 seconds; few fine roots; clear transition	Podzol B

Points of interest and interpretation

Here is a modern soil imprinted on paleo-weathering. Of particular interest is the high clay content and the discrepancy between field texture estimation and lab analysis. The clay fraction consists predominantly of kaolinite. The basal silcrete overlies and protects a kaolinitic pallid zone associated with the early Tertiary (African) erosion surface and it is possible that the silcrete formed through cementation of the kaolinite which is no being liberated through weathering. The underestimation of clay content by feel could be due to an aggregating effect of Fe-Al-humate, producing a sandy loam texture. These aggregates would break up during pretreatment with hydrogen peroxide for organic matter removal and the subsequent dispersion process for texture determination (see methods). Interesting too is the high CEC and the magnesium-rich composition of the exchangeable cation suite in the B (but not A) horizon. Some micronutrients show a second maximum in the lower B horizon but all show greatest concentration in the topsoil, in contrast to the other podzol profiles.

P2575: Podzolic, lithic, orthosaprolitic, aeromorphic, *fine sandy loam*
(Groenkop Gk 1100)

Lab no.	Depth (cm)	Horizon	Coarse fragments (%)	Fine earth (%)				
				CSa	MSa	FSa	Silt	Clay
C9787	0–22	A	59	8.6	4.3	13.4	34.1	24.7
C9788	22–50	B1	65	13.4	6.0	9.6	27.7	33.3
C9789	50–75	B2	69	10.0	4.5	6.7	20.7	50.1

Lab no.	$CaCO_3$ equiv. (%)	Organic C		Fe			Al			Mn	Si $CaCl_2$ (mg kg^{-1})
		Total	Pyr.	CBD	Ox	Pyr.	CBD	Ox	Pyr	CBD	
				(%)							
C9787	0	4.5	0.94	2.25	0.18	0.26	0.36	0.20	0.39	tr	17.00
C9788	0	2.4	0.74	2.77	0.39	0.56	0.72	0.39	0.83	tr	19.60
C9789	0	2.4	0.92	2.75	0.28	0.55	0.81	0.46	1.12	tr	23.20

Lab no.	pH			CEC	Exchangeable cations					Acidity	Al Exch
	H_2O	$CaCl_2$	NaF		Ca	Mg	Na	K	Total	Titr Exch	
					(cmol$_c$ kg^{-1})						
C9787	5.3	4.3	8.4	14.94	4.68	3.37	0.24	0.36	8.65	0.67	0.50
C9788	5.4	4.1	9.3	16.57	2.55	4.18	0.42	0.22	7.37	1.48	1.21
C9789	5.6	4.2	9.6	20.86	2.48	6.91	0.67	0.22	10.28	1.44	1.37

Lab no.	Saturation extract							Micronutrients			
	Resist. (ohms)	EC (dS m^{-1})	Satn. water (%)	Ca	Mg	Na	K	Zn	Mn	Cu	B	Co
				(mmol$_c$ L^{-1})				(mg kg^{-1})				
C9787		0.83	66.8					1.68	6.19	1.06	2.22	0.28
C9788		0.38	54.0					1.14	0.44	0.15	1.83	0.27
C9789		0.35	64.7					1.32	0.44	0.60	2.12	0.10

Lab no.	Bulk density (kg m^{-3})	HC (mm h^{-1})	Air-water permeability ratio	Atterberg		Water retentivity at kPa:			
				LL	PI	-33	-80	-500	-1 500
						(%)			
C9787									
C9788									
C9789									

Lab no.	P Sorp. (%)	P Extr. (mg kg^{-1})	Minerals <2 μm	Mineral legend			
C9787			Kt5, Qz3, Mi2, Is1	Qz	**Quartz**	Is	**Interstratified**
C9788			Kt5, Qz3, Mi2, Is1	Kt	**Kaolinite**	Mi	**Mica**
C9789			Kt5, Mi2, Qz2, Is1				

P583: Plinthic, soft-xanthic, dystrophic, haplic, *sandy clay loam*
(Avalon Av 1100)

Profile No:	P583	**Aspect:**	N
Land type map:	2630 Mbabane	**Water table:**	None
Latitude & Longitude:	26°34'20'/30°29'00'	**Occurrence of flooding:**	
Land type No:	Ac36	**Stoniness:**	None
Climate zone No:	110S	**Erosion:**	None
Soil form:	Avalon	**Vegetation/land use:**	Grassveld
Soil series:	Ruston	**Parent and underlying material:**	Mixed colluvium derived from Ecca sandstone and shale
Altitude:	1 550 m		
Terrain unit:	3		
Slope:	8%	**Described by:**	R.W. Fitzpatrick
Slope shape:	Convex	**Date described:**	Jan-73

Horizon	Depth (mm)	Description	Diagnostic horizon
A1	0–300	Moist; dark brown 7.5YR3/2; medium sandy clay loam; weak subangular blocky; friable; clear, smooth transition	Orthic
B21	300–1 100	Moist; dark brown 7.5YR5/8; medium sandy clay loam; apedal; friable; gradual, smooth transition	Yellow-brown apedal
B22	1 100–1 200+	Moist; moist 80% yellowish-brown 10YR5/6; many coarse distinct red and yellow elongated mottles; medium sandy clay loam; apedal; friable	Soft plinthic

Points of interest and interpretation

This haplic (i.e. non-luvic) loamy textured soil has moderate organic matter status and a deep, yellow-brown (xanthic), apedal, well drained, upper solum overlying a red-mottled, soft plinthic B at a depth of about 1 metre. It is very acidic – 50% acid saturation of ECEC in the topsoil – and extremely low in bases, especially in the subsoil (dystrophic). Phosphate status is low and P sorption capacity is moderate to high. Dolomitic lime would be needed to achieve good crop yields and fertiliser containing Zn would also be advisable. A response might be expected to gypsum, which could be applied to improve rooting depth and increase crop access to subsoil water reserves. The degree of weathering is high, with kaolinite, and accessory goethite and gibbsite, dominating the clay fraction. Some anion exchange capacity and nitrate retention might therefore be expected in the subsoil. The vermiculite component is likely to be an alteration product of a micaceous component in the parent sandstone and is probably chloritic, with aluminium interlayering. Groundwater protection from pollution is moderate to high. The soil is highly suited to dryland cropping, subject to appropriate chemical amelioration. An Avalon soil in this Highveld climate is considered by farmers to be a prize asset.

P583: Plinthic, soft-xanthic, dystrophic, haplic, *sandy clay loam*
(Avalon Av 1100)

Lab no.	Depth (cm)	Horizon	Coarse fragments (%)	Fine earth (%)				
				CSa	MSa	FSa	Silt	Clay
C2015	0–30	A1	0	10	28	25	6	27
C2016	30–110	B21	0	6	24	26	9	29
C2017	110–120	B22	0	10	23	26	13	25

Lab no.	CaCO$_3$ equiv. (%)	Organic C		Fe			Al			Mn	Si CaCl$_2$ (mg kg^{-1})
		Total	Pyr.	CBD	Ox	Pyr.	CBD	Ox	Pyr	CBD	
				(%)							
C2015		1.3		4.8			1.0				
C2016		0.4		7.3			1.4				
C2017		0.2		4.7			1.0				

Lab no.	pH			CEC	Exchangeable cations					Acidity		Al
	H$_2$O	CaCl$_2$	NaF		Ca	Mg	Na	K	Total	Titr	Exch	Exch
					(cmol$_c$ kg^{-1})							
C2015	4.9	4.4		7.3	0.4	0.0	0.1	0.2	0.7			0.7
C2016	5.3	4.5		4.0	0.2	0.0	0.2	0.1	0.5			0.2
C2017	6.0	4.7		4.0	0.3	0.0	0.2	0.1	0.6			0.2

Lab no.	Saturation extract							Micronutrients			
	Resist. (ohms)	EC (dS m^{-1})	Satn. water (%)	Ca	Mg	Na	K	Zn	Mn	Cu	B	Co
				(mmol$_c$ L^{-1})				(mg kg^{-1})				
C2015	3 500							0.38	15.7	2.05	0.25	0.00
C2016	9 300							0.19	7.1	1.29	0.11	0.00
C2017	9 999							0.29	11.9	1.05	0.16	0.00

Lab no.	Bulk density (kg m^{-3})	HC (mm h^{-1})	Air-water permeability ratio	Atterberg		Water retentivity at kPa:			
				LL	PI	-33	-80	-500	-1 500
						(%)			
C2015			2			13.3	11.9	11.5	9.7
C2016			4			16.0	16.0	10.8	10.0
C2017			3			16.6	13.9	12.1	11.4

Lab no.	P Sorp. (%)	P Extr. (mg kg^{-1})	Minerals <2 µm	Mineral legend			
C2015	99.4	2.0	Kt5, Gb3, Vm2, Mi1, Qz1	Qz	**Quartz**	Vm	**Vermiculite**
C2016	99.4	0.9	Kt5, Gb2, Vm1	Mi	**Mica**	Kt	**Kaolinite**
C2017	99.4	0.1	Kt5, Gb2, Ch(2:1)1, Vm1	Gb	**Gibbsite**	Ch	**Chlorite**

P584: Plinthic, soft-eluvic, albic, *coarse sandy loam*
(Longlands Lo 1000)

Profile No:	P584	**Aspect:**	S
Land type map:	2630 Mbabane	**Water table:**	Perched, at 500 mm
Latitude & Longitude:	26°33'50'/30°32'30'	**Occurrence of flooding:**	
Land type No:	Ac35	**Stoniness:**	Class 2
Climate zone No:	111S	**Erosion:**	Sheet, class 1
Soil form:	Longlands	**Vegetation/land use:**	Grassveld
Soil series:	Vaalsand	**Parent and underlying material:**	Mixed colluvium derived from sandstone, shale and possibly tillite
Altitude:	1 340 m		
Terrain unit:	4		
Slope:	3%	**Described by:**	R.W. Fitzpatrick
Slope shape:	Straight	**Date described:**	Oct-72

Horizon	Depth (mm)	Description	Diagnostic horizon
A1	0–300	Dry; moist 100% very dark grey 10YR3/1; coarse sandy loam; single grain; soft; few medium indurated iron-manganese nodules; clear smooth boundary	Orthic
E	300–500	Moist; moist 95% yellowish-brown 10YR5/4; few fine faint grey and black mottles; coarse sandy loam; single grain; loose; few medium indurated iron-manganese nodules; clear irregular boundary	E-horizon
B2	500–1 200+	Moist; moist pale brown 10YR5/3; coarse sandy loam; massive; friable; many medium to coarse indurated iron-manganese nodules	Soft plinthic

Points of interest and interpretation

A moderately high degree of weathering (kaolinitic clay mineralogy), depletion of bases and moderate acidity, sandy loam texture, and morphology indicating a fluctuating water table are the more prominent features of this soil. The low CBD-extractable Fe in the B horizon despite morphology indicating Fe enrichment can be attributed to the bulk of the nodules being removed by sieving prior to analysis (64% coarse fragments). Micronutrients (especially Zn and Co) and P are also low. This soil needs lime and broad-spectrum fertilising for crop production but low buffer capacity will lead to rapid acidification if nitrogen is applied too generously while the risk of leaching losses of nutrients is also high. Intermittent wetness may limit productivity in wetter seasons although in drier years the plinthic horizon could function as a reservoir for deep rooted crops. Groundwater vulnerability would be high if this soil became polluted. Lateral discharge through the E and B would result in the toe slope reception area (likely a wetland, unit 5) being affected by a plume of polluted water.

P584: Plinthic, soft-eluvic, albic, *coarse sandy loam*
(Longlands Lo 1000)

Lab no.	Depth (cm)	Horizon	Coarse fragments (%)	Fine earth (%)				
				CSa	MSa	FSa	Silt	Clay
C1678	0–30	A1	14	15	27	30	10	15
C1679	30–50	E	12	15	21	36	9	12
C1680	50–120+	B2	54	17	22	37	7	12

Lab no.	CaCO$_3$ equiv. (%)	Organic C		Fe			Al			Mn	Si CaCl$_2$ (mg kg^{-1})
		Total	Pyr.	CBD	Ox	Pyr.	CBD	Ox	Pyr	CBD	
				(%)							
C1678		0.9		6.9			1.8				
C1679		1.5		0.4			0.1				
C1680		0.1		0.7			0.1				

Lab no.	pH			CEC	Exchangeable cations					Acidity		Al Exch
	H$_2$O	CaCl$_2$	NaF		Ca	Mg	Na	K	Total	Titr	Exch	
					(cmol$_c$ kg^{-1})							
C1678	5.3	4.4		5.9	0.0	0.0	0.1	0.1	0.2			0.7
C1679	5.6	4.5		5.2	0.2	0.0	0.1	0.1	0.5			0.6
C1680	6.2	5.8		1.8	0.4	0.5	0.1	0.1	1.1			0.0

Lab no.	Saturation extract							Micronutrients				
	Resist. (ohms)	EC (dS m^{-1})	Satn. water (%)	Ca	Mg	Na	K	Zn	Mn	Cu	B	Co
				(mmol$_c$ L^{-1})				(mg kg^{-1})				
C1678	3 900							0.15	9.1	2.29	0.44	0.00
C1679	3 400							0.27	6.6	1.61	0.37	0.00
C1680	4 500							0.13	2.8	0.70	0.22	0.00

Lab no.	Bulk density (kg m^{-3})	HC (mm h^{-1})	Air-water permeability ratio	Atterberg		Water retentivity at kPa:			
				LL	PI	-33	-80	-500	-1 500
						(%)			
C1678			2			12.2	10.3	8.1	7.6
C1679			3			11.4	9.7	7.8	7.2
C1680			2			7.9	6.2	4.5	4.1

Lab no.	P Sorp. (%)	P Extr. (mg kg^{-1})	Minerals <2 µm	Mineral legend			
C1678	100	2.1	Kt5, Ch(2:1)2, Mi2, Qz2, Vm2	Qz **Quartz**		Ch **Chlorite**	
C1679	90	5.1	Kt5, Qz2, Vm2, Ch(2:1)1, Mi1	Mi **Mica**		Vm **Vermiculite**	
C1680	86	2.8	Kt5, Qz3, Vm2, Ch(2:1)1	Kt **Kaolinite**			

P1067: Plinthic, hard-xanthic, mesotrophic, haplic, *coarse sandy loam*
(Glencoe Gc 2100)

Profile No:	P1067	**Aspect:**	W
Land type map:	2730 Vryheid	**Water table:**	None
Latitude & Longitude:	27°38'54'/30°16'00'	**Occurrence of flooding:**	Occasional
Land type No:	Bb54	**Stoniness:**	Flat, 80 mm long, 2.2 m apart
Climate zone No:	267S	**Erosion:**	Gully, class 1, not stabilised
Soil form:	Glencoe	**Vegetation/land use:**	Open grassveld
Soil series:	Glencoe		
Altitude:	1 750 m	**Parent and underlying material:**	Quartzitic sandstone
Terrain unit:	3		
Slope:	1%	**Described by:**	D.P. Turner
Slope shape:	Straight	**Date described:**	May-77

Horizon	Depth (mm)	Description	Diagnostic horizon
A1	0–300	Dry; moist brown to dark brown 10YR4/3; few fine faint grey illuvial clay mottles; coarse sandy loam; massive; slightly hard; few rounded sesquioxide concretions, cannot be crushed by hand; clear smooth transition	Orthic
B21	300–450	Dry, moist dark yellowish-brown 10YR4/4; few fine faint yellowish brown illuvial iron mottles; coarse sandy loam; single grain; slightly hard; few rounded fine sesquioxide concretions, cannot be crushed by hand; gradual smooth transition	Yellow-brown apedal
B22	450–600	Dry; moist dark yellowish-brown 10YR4/4; few fine faint yellowish brown illuvial iron mottles; coarse sandy clay loam; massive; slightly hard; few rounded fine sesquioxide concretions, cannot be crushed by hand; clear smooth transition	Yellow-brown apedal
B23	600–800	Dry; massive sesquioxide cementation, breaks only with sharp blow of hammer, rings; continuous	Hard plinthic

Points of interest and interpretation

This soil is physically, chemically and mineralogically similar to the previous one except for the yellow subsoil colour (despite CBD-Fe being about the same) and the hard plinthic B at 60 cm depth. The midslope position (unit 3) probably provides the explanation for the accumulation of enough Fe oxide to pigment the subsoil, signifying slightly better aerated conditions than in the previous soil, which is located on a footslope (unit 4). Conceivably the latter periodically receives more water than it can transmit. Had a hypoxanthic E (in contrast to albic) been recognised as an option the soil might have been classified as hard-eluvic. No dry colours were reported. Available P is very low although bases and micronutrients are generally in better supply than in the previous soil. No significant acidity exists although the acidification risk with large N fertiliser applications is high because of sandy texture. Compared with the previous soil there is an anomalous difference in P sorption capacity but this could be the result of different methods having been used.

P1067: Plinthic, hard-xanthic, mesotrophic, haplic, *coarse sandy loam*
(Glencoe Gc 2100)

Lab no.	Depth (cm)	Horizon	Coarse fragments (%)	Fine earth (%)				
				CSa	MSa	FSa	Silt	Clay
C4503	0–30	A1	0	38	24	20	2	15
C4504	30–45	B21	1	22	43	35	2	15
C4505	45–60	B22	14	22	30	23	2	21

Lab no.	CaCO$_3$ equiv. (%)	Organic C		Fe			Al			Mn	Si CaCl$_2$ (mg kg^{-1})
		Total	Pyr.	CBD	Ox	Pyr.	CBD	Ox	Pyr	CBD	
					(%)						
C4503		0.7		0.8			0.11				
C4504		0.4		0.8			0.16				
C4505		0.6		1.1			0.22				

Lab no.	pH			CEC	Exchangeable cations					Acidity		Al
	H$_2$O	CaCl$_2$	NaF		Ca	Mg	Na	K	Total	Titr	Exch	Exch
					(cmol$_c$ kg^{-1})							
C4503	4.8	4.5		4.5	1.4	0.9	0.0	0.3	2.6	0.0	0.0	0.0
C4504	5.5	4.6		4.3	1.0	1.1	0.0	0.0	2.1	0.0	0.2	0.0
C4505	6.3	5.4		6.1	1.6	1.8	0.1	0.1	3.6	0.0	0.0	0.0

Lab no.	Saturation extract							Micronutrients				
	Resist. (ohms)	EC (dS m^{-1})	Satn. water (%)	Ca	Mg	Na	K	Zn	Mn	Cu	B	Co
				(mmol$_c$ L^{-1})				(mg kg^{-1})				
C4503	2 400							0.18	94.0	1.20	0.27	1.02
C4504	4 800							0.21	223.0	1.20	0.15	0.12
C4505	2 800							0.15	30.6	1.50	0.21	0.54

Lab no.	Bulk density (kg m^{-3})	HC (mm h^{-1})	Air-water permeability ratio	Atterberg		Water retentivity at kPa:			
				LL	PI	-33	-80	-500	-1 500
						(%)			
C4503			2			6.0	6.0	4.8	3.7
C4504			2			7.2	7.1		
C4505			3			9.8	9.2	7.0	6.3

Lab no.	P Sorp. (%)	P Extr. (mg kg^{-1})	Minerals <2 μm	Mineral legend			
C4503	9.76	0.9	Kt5, Qz2, Mi1, Tc1	Qz	**Quartz**	Kt	**Kaolinite**
C4504	13.9	0.8	Kt5, Qz2, Mi1, Tc1	Tc	**Talc**	Mi	**Mica**
C4505	14.5	0.2	Kt5, Qz3, Mi1, Tc1				

P150: Oxidic, rhodic, mesotrophic, haplic, *medium sandy clay*
(Hutton Hu 2100)

Profile No:	P150	**Aspect:**	O/E
Land type map:	2628 East Rand	**Water table:**	None
Latitude & Longitude:	26°31'30'/28°27'00'	**Occurrence of flooding:**	
Land type No:	Ba1	**Stoniness:**	None
Climate zone No:	24S	**Erosion:**	Sheet, class 1
Soil form:	Hutton	**Vegetation/land use:**	Crops (maize)
Soil series:	Doveton		
Altitude:	1 630 m	**Parent and underlying material:**	Local colluvium derived from Witwatersrand lava
Terrain unit:	1		
Slope:	4%	**Described by:**	J.L. Schoeman
Slope shape:	Convex	**Date described:**	Nov-76

Horizon	Depth (mm)	Description	Diagnostic horizon
A1	0–310	Moist; moist 100% dusky red 2.5YR3/2; medium sandy clay; weak medium subangular blocky; very firm; gradual smooth boundary	Orthic
B21	310–920	Moist; moist 100% dusky red 10R3/4; medium sandy clay; weak medium subangular blocky; very firm; gradual smooth boundary	Red apedal
B22	920–1 200+	Moist; moist 100% dusky red 2.5YR3/2; clay loam; weak medium subangular blocky; very firm; very few medium indurated iron-manganese nodules	Red apedal

Points of interest and interpretation

Deep red soils such as this one are the backbone of the productive maize belt on South African highveld. Their only serious disadvantages are a tendency to be too well drained, storing insufficient water at depth to sustain the maize crop through dry spells that sometimes occur in the later part of the growing season, and their greater than normal requirement for phosphate fertiliser to compensate for P sorption by iron oxides. This pedon has all the best features of these deep, apedal clays: it is well graded texturally, has sufficient Fe and organic matter to provide structural stability (augmented by sufficient weathering to attain predominantly kaolinitic clay mineralogy which increases the effectiveness of Fe oxides in stabilising soil aggregates), and is well endowed with basic cations and micronutrients (with the possible exception of Zn). Current lime requirement is zero and moderate buffer capacity ensures that the frequency of liming to correct for acidification by cropping is not excessive. The nutritional quality of natural grass veld can be expected to be good. The soil is easily cultivated, not markedly erodible and readily managed under irrigation, having well balanced drainage and water retention properties. One can imagine that, for the same reasons, it would make an hospitable environment for burrowing animals to which, in turn, it owes much of its porosity.

P150: Oxidic, rhodic, mesotrophic, haplic, *medium sandy clay*
(Hutton Hu 2100)

Lab no.	Depth (cm)	Horizon	Coarse fragments (%)	Fine earth (%)				
				CSa	MSa	FSa	Silt	Clay
C4100	0–31	A1	8	8	18	26	10	36
C4101	31–92	B21	5	9	19	23	9	38
C4102	92–120+	B22	9	7	15	26	14	38

Lab no.	$CaCO_3$ equiv. (%)	Organic C		Fe			Al			Mn	Si $CaCl_2$ (mg kg^{-1})
		Total	Pyr.	CBD	Ox	Pyr.	CBD	Ox	Pyr	CBD	
				(%)							
C4100		1.2		5.0			0.3				
C4101		0.7		5.1			0.3				
C4102		0.4		5.3			0.3				

Lab no.	pH			CEC	Exchangeable cations					Acidity		Al Exch
	H_2O	$CaCl_2$	NaF		Ca	Mg	Na	K	Total	Titr	Exch	
					(cmol$_c$ kg^{-1})							
C4100	6.3	5.5		9.6	3.3	2.3	0.0	0.7	6.3	4.9	0.0	0.0
C4101	6.1	5.4		8.3	2.9	2.0	0.1	0.2	5.2	4.9	0.0	0.0
C4102	6.3	5.7		7.0	2.3	2.6	0.0	0.1	5.0	4.8	0.0	0.0

Lab no.	Saturation extract							Micronutrients			
	Resist. (ohms)	EC (dS m^{-1})	Satn. water (%)	Ca	Mg	Na	K	Zn	Mn	Cu	B	Co
				(mmol$_c$ L^{-1})				(mg kg^{-1})				
C4100	1 100							1.18	235	11.55	0.38	14.3
C4101	1 300							0.61	137	6.34	0.36	10.5
C4102	1 700							0.43	136	9.25	0.17	14.3

Lab no.	Bulk density (kg m^{-3})	HC (mm h^{-1})	Air-water permeability ratio	Atterberg		Water retentivity at kPa:			
				LL	PI	-33	-80	-500	-1 500
						(%)			
C4100			8			20.5	17.6	13.6	13.9
C4101			11			20.3	18.3	14.3	13.6
C4102			8			22.8	20.1	16.0	15.5

Lab no.	P Sorp. (%)	P Extr. (mg kg^{-1})	Minerals <2 μm	Mineral legend			
C4100	95.0	5.3	Kt4, Qz2, Ch(2:1)1, Mi1, St1	Qz	**Quartz**	Mi	**Mica**
C4101	99.1	2.0	Kt5, Ch(2:1)2, Mi2, Qz1, St1	Fs	**Feldspar**	Ch	**Chlorite**
C4102	30.0	1.0	Kt5, Mi2, Ch(2:1)1, Qz1, St1	Kt	**Kaolinite**	St	**Smectite**

P967: Oxidic, pedorhodic, eutrophic, haplic, micropedal, *clay*
(Shortlands Sd 2110)

Profile No:	P967	**Aspect:**	
Land type map:	2430 Pilgrim's Rest	**Water table:**	None
Latitude & Longitude:	24°53'00'/30°31'00'	**Occurrence of flooding:**	Nil
Land type No:	Ae124	**Stoniness:**	None
Climate zone No:	961S	**Erosion:**	None apparent
Soil form:	Shortlands	**Vegetation/land use:**	Agronomic cash crops
Soil series:	Shortlands	**Parent and underlying material:**	Dolomite
Altitude:	1 280 m		
Terrain unit:	4		
Slope:	02%	**Described by:**	D.P. Turner
Slope shape:	Straight	**Date described:**	Sept-73

Horizon	Depth (mm)	Description	Diagnostic horizon
A1	0–150	Dry; dry dusky red 10R3/4; clay; moderate fine subangular blocky; very hard; many pores; few clay cutans; common roots; gradual smooth transition	Orthic
B2	150–1 200	Dry; dry dusky red 10R3/4; clay; strong, fine, angular blocky; very hard; few pores; common clay cutans; few roots.	Red structured

Points of interest and interpretation

The significance of pedality in oxidic soils is discussed in Section 2.10. The red structured (pedorhodic) B horizon is often considered to result from a clay mineral composition which includes a significant component of swelling (smectitic) clay. This soil, although of mixed clay mineralogy, is strongly dominated by kaolinite with a high proportion of iron oxides. A combination of higher clay content and a drier climate (at least intermittently) than that of the previous profile would explain the strong blocky structure. Simple textural considerations are sufficient to explain why permeability is lower and water retention higher than in the previous soil. Relatively high CEC and full base saturation make this soil fertile once P status has been corrected. Typically this pedorhodic form of oxidic soils is at its productive best under irrigation. Dryland production of crops is generally much more risky than on the rhodic (apedal) forms. High buffer capacity for both cations and anions means that groundwater vulnerability to pollution is low. Although the Fe oxides ameliorate soil physical properties the high clay content has adverse implications for tillage and general physical suitability compared with the previous soil.

P967: Oxidic, pedorhodic, eutrophic, haplic, micropedal, *clay*
(Shortlands Sd 2110)

Lab no.	Depth (cm)	Horizon	Coarse fragments (%)	Fine earth (%)				
				CSa	MSa	FSa	Silt	Clay
C3150	0–15	A1		2	3	10	16	68
C3151	15–120	B2		1	2	10	11	75

Lab no.	CaCO$_3$ equiv. (%)	Organic C		Fe			Al			Mn	Si CaCl$_2$ (mg kg^{-1})
		Total	Pyr.	CBD	Ox	Pyr.	CBD	Ox	Pyr	CBD	
					(%)						
C3150		1.4		9.1			0.2				
C3151		1.4		8.7			0.2				

Lab no.	pH			CEC	Exchangeable cations					Acidity		Al Exch
	H$_2$O	CaCl$_2$	NaF		Ca	Mg	Na	K	Total	Titr	Exch	
					(cmol$_c$ kg^{-1})								
C3150	6.7	5.7		15.6	6.4	5.5	0.2	0.9	13.0			0
C3151	6.2	5.7		17.1	6.9	7.1	0.2	0.5	14.7			0

Lab no.	Saturation extract								Micronutrients			
	Resist. (ohms)	EC (dS m^{-1})	Satn. water (%)	Ca	Mg	Na	K	Zn	Mn	Cu	B	Co
				(mmol$_c$ L^{-1})				(mg kg^{-1})				
C3150	1 000							0.20	142.3	1.40	0.21	
C3151	750							0.73	536.1	14.14	1.04	

Lab no.	Bulk density (kg m^{-3})	HC (mm h^{-1})	Air-water permeability ratio	Atterberg		Water retentivity at kPa:			
				LL	PI	-33	-80	-500	-1 500
						(%)			
C3150			12			32.7	27.0	21.9	21.0
C3151			23			32.5	28.2	24.5	23.8

Lab no.	P Sorp. (%)	P Extr. (mg kg^{-1})	Minerals <2 μm	Mineral legend			
C3150	35.6	0.6	Kt5, Mi4, Qz3	Qz	**Quartz**	Is	**Interstratified**
C3151	46.7	0.4	Kt5, Qz3, Mi3, St1, Is1	Kt	**Kaolinite**	Mi	**Mica**
				St	**Smectite**		

P1073: Oxidic, xanthic, dystrophic, haplic, *medium sandy clay loam*
(Clovelly Cv 1100)

Profile No:	P1073	**Aspect:**	N
Land type map:	2730 Vryheid	**Water table:**	None
Latitude & Longitude:	27°36'12'/30°31'48'	**Occurrence of flooding:**	Nil
Land type No:	Ac113	**Stoniness:**	None
Climate zone No:	275S	**Erosion:**	Gully, class 1
Soil form:	Clovelly	**Vegetation/land use:**	Plantation (forestry)
Soil series:	Oatsdale		
Altitude:	1 450 m	**Parent and underlying material:**	Strongly weathered, Feldspathic sandstone
Terrain unit:	3		
Slope:	3%	**Described by:**	D.P. Turner
Slope shape:	Convex	**Date described:**	May-77

Horizon	Depth (mm)	Description	Diagnostic horizon
A1	0–250	Dry; moist brown to dark brown 10YR4/3; few fine faint red mottles; medium sandy clay loam; massive; slightly firm; clear smooth transition	Orthic
B21	250–600	Moist; moist brown 10YR5/3; few fine faint red mottles; medium sandy clay loam; massive; slightly firm; gradual smooth transition	Yellow-brown apedal
B22	600–900	Moist; moist yellowish-brown 10YR5/6; few fine faint red mottles; few fine faint grey mottles; medium sandy clay loam; massive; slightly firm; clear smooth transition	Yellow-brown apedal
C	900–1 200	Moist; moist yellowish-brown 10YR5/6; many medium distinct red and yellow mottles; many medium distinct grey mottles; fine sandy loam; massive; slightly firm	

Points of interest and interpretation

Yellow colours signify an environment in which goethite is stable but hematite is not. While the explanation for this is complex we often make the generalisation that a moister (and cooler) soil climate is indicated by yellow-brown colours than that favouring red colours (see Sections 2.2, 2.9 and 2.10). This soil has features in common with two profiles described earlier: the humic, rhodic P1472 to which it is mineralogically and chemically similar, and the plinthic, soft-xanthic P583, which it resembles in having mottles in the yellow-brown B horizon suggesting a degree of intermittent wetness greater than is normally associated with soils of the oxidic group. Points of interest that were raised in relation to the acidity, degree of leaching and weathering and hydromorphy of those pedons therefore apply equally well here.

P1073: Oxidic, xanthic, dystrophic, haplic, *medium sandy clay loam*
(Clovelly Cv 1100)

Lab no.	Depth (cm)	Horizon	Coarse fragments (%)	Fine earth (%)				
				CSa	MSa	FSa	Silt	Clay
C4487	0–25	A1	0	10	22	34	7	23
C4488	25–60	B21	0	9	20	32	8	27
C4489	60–90	B22	0	9	20	33	7	27

Lab no.	CaCO$_3$ equiv. (%)	Organic C		Fe			Al			Mn	Si CaCl$_2$ (mg kg^{-1})
		Total	Pyr.	CBD	Ox	Pyr.	CBD	Ox	Pyr	CBD	
					(%)						
C4487		0.9		1.4			0.46				
C4488		0.6		1.8			0.63				
C4489		0.4		1.8			0.52				

Lab no.	pH			CEC	Exchangeable cations					Acidity		Al
	H$_2$O	CaCl$_2$	NaF		Ca	Mg	Na	K	Total	Titr	Exch	Exch
					(cmol$_c$ kg^{-1})							
C4487	4.8	3.9		8.1	0.0	0.0	0.0	0.2	0.2	10.9	2.4	1.9
C4488	4.8	3.9		7.7	0.0	0.0	0.0	0.1	0.1	9.4	2.3	2.2
C4489	4.9	3.9		7.2	0.1	0.0	0.0	0.2	0.3	8.7	4.6	3.7

Lab no.	Saturation extract							Micronutrients			
	Resist. (ohms)	EC (dS m^{-1})	Satn. water (%)	Ca	Mg	Na	K	Zn	Mn	Cu	B	Co
				(mmol$_c$ L^{-1})				(mg kg^{-1})				
C4487	14 000						0.33	1.90	0.60	0.24	0.06
C4488	14 000						0.18	1.50	0.39	0.28	0.12
C4489	16 000						0.18	1.50	0.36	0.18	0.12

Lab no.	Bulk density (kg m^{-3})	HC (mm h^{-1})	Air-water permeability ratio	Atterberg		Water retentivity at kPa:			
				LL	PI	-33	-80	-500	-1 500
						(%)			
C4487			3			13.7	12.0	9.0	9.9
C4488			4			15.5	13.7	10.3	10.7
C4489			2			16.0	13.6	10.2	10.4

Lab no.	P Sorp. (%)	P Extr. (mg kg^{-1})	Minerals <2 μm	Mineral legend				
C4487	38.1	1.2	Kt5, Mi3, Gb3, Vm2, Qz2	Qz	**Quartz**		Vm	**Vermiculite**
C4488	37.2	0.6	Kt5, Mi3, Gb2, Vm2, Qz1	Gb	**Gibbsite**		Mi	**Mica**
C4489	35.8	0.2	Kt5, Mi3, Vm2, Gb2, Qz2	Kt	**Kaolinite**			

P1135: Oxidic, xanthirhodic, dystrophic, luvic, *medium sandy loam*
(Griffin Gf 1200)

Profile No:	P1135	**Aspect:**	S
Land type map:	2730 Vryheid	**Water table:**	None
Latitude & Longitude:	27°49'12'/31°01'00'	**Occurrence of flooding:**	Nil
Land type No:	Ac117	**Stoniness:**	None
Climate zone No:	288S	**Erosion:**	None apparent
Soil form:	Griffin	**Vegetation/land use:**	Plantation (forestry)
Soil series:	Griffin	**Parent and underlying material:**	Quartzitic sandstone of the Vryheid Formation, Ecca Group
Altitude:	1 550 m		
Terrain unit:	3		
Slope:	3%	**Described by:**	D.P. Turner
Slope shape:	Convex	**Date described:**	Nov-77

Horizon	Depth (mm)	Description	Diagnostic horizon
A1	0–450	Moist; moist very dark greyish-brown 10YR3/2; medium sandy loam; massive; slightly firm; clear smooth transition	Orthic
B21	450–1 100	Moist; moist yellowish-brown 10YR5/6; sandy clay; massive; slightly firm; gradual smooth transition	Yellow-brown apedal
B22	1 100–1 400	Moist; moist yellowish-red 5YR4/8; sandy clay; massive; slightly firm; gradual smooth transition	Red apedal
B3	1 400–1 800	Moist; moist yellowish-red 5YR5/8; sandy clay loam; single grain; slightly firm; clear smooth transition	Red apedal
C	1 800–2 400	Moist; moist yellow 10YR8/6; loamy coarse sand; single grain; friable; common irregular fine gravel	

Points of interest and interpretation

Again, comparison with another profile (xanthirhodic P587 of the humic group) is useful. In this case the orthic A has a particle size distribution suggesting that the luvic character arises from a lithological discontinuity in the transition to the underlying B horizons, reinforced by the predominantly micaceous clay fraction of the A horizon which also contains no gibbsite and has a neutral pH, as opposed to the B horizons which are free of mica, rich in gibbsite and strongly acidic. Colluvial aggradation of less weathered material to form the A horizon is suggested by these data. Remarks made previously about genesis and management of highly weathered pedons (P587, P583) also apply here, at least in relation to the subsurface horizons.

P1135: Oxidic, xanthirhodic, dystrophic, luvic, *medium sandy loam*
(Griffin Gf 1200)

Lab no.	Depth (cm)	Horizon	Coarse fragments (%)	Fine earth (%)				
				CSa	MSa	FSa	Silt	Clay
C4900	0–45	A1		13	36	32	4	15
C4901	45–110	B21		11	21	24	8	36
C4902	110–140	B22		12	19	24	7	37

Lab no.	CaCO$_3$ equiv. (%)	Organic C		Fe			Al			Mn	Si CaCl$_2$ (mg kg^{-1})
		Total	Pyr.	CBD	Ox	Pyr.	CBD	Ox	Pyr	CBD	
					(%)						
C4900		0.2		0.5			0.06				
C4901		0.7		1.8			0.80				
C4902		0.5		1.9			0.46				

Lab no.	pH			CEC	Exchangeable cations					Acidity		Al
	H$_2$O	CaCl$_2$	NaF		Ca	Mg	Na	K	Total	Titr	Exch	Exch
					(cmol$_c$ kg^{-1})							
C4900	7.2	5.5		5.1	2.3	0.8	0.3	0.1	3.5	1.0	0.0	0.0
C4901	5.8	4.4		7.6	0.0	0.0	0.0	0.0	0.0	4.6	1.7	1.1
C4902	5.1	4.5		6.3	0.0	0.0	0.0	0.0	0.0	5.7	0.5	1.0

Lab no.	Saturation extract							Micronutrients			
	Resist. (ohms)	EC (dS m^{-1})	Satn. water (%)	Ca	Mg	Na	K	Zn	Mn	Cu	B	Co
				(mmol$_c$ L^{-1})				(mg kg^{-1})				
C4900	2 100							0.18	294.9	0.18	0.26	0.18
C4901	10 600							0.03	143.2	0.48	0.88	0.06
C4902	10 000							0.03	41.0	0.24	0.68	0.06

Lab no.	Bulk density (kg m^{-3})	HC (mm h^{-1})	Air-water permeability ratio	Atterberg		Water retentivity at kPa:			
				LL	PI	-33	-80	-500	-1 500
						(%)			
C4900									
C4901									
C4902									

Lab no.	P Sorp. (%)	P Extr. (mg kg^{-1})	Minerals <2 µm	Mineral legend				
C4900	5.3	0.3	Mi5, Kt4, Qz2, Fs1	Qz	**Quartz**		Ch	**Chlorite**
C4901	64.2	0.5	Kt5, Gb4, Ch(2:1)4, Qz2	Mi	**Mica**		Kt	**Kaolinite**
C4902	66.3	0.7	Kt5, Vm4, Gb4, Qz3, Fs1	Gb	**Gibbsite**		Vm	**Vermiculite**

P578: Gleyic, eluvic, albic, *loamy medium sand*
(Kroonstad Kd 1000)

Profile No:	P578	**Aspect:**	W
Land type map:	2630 Mbabane	**Water table:**	None
Latitude & Longitude:	26°28'10'/30°13'50'	**Occurrence of flooding:**	
Land type No:	Bb21	**Stoniness:**	Class 2
Climate zone No:	51S	**Erosion:**	Sheet, class 2
Soil form:	Kroonstad	**Vegetation/land use:**	Grassveld
Soil series:	Umtentweni	**Parent and underlying material:**	Ecca sandstone
Altitude:	1 500 m		
Terrain unit:	5		
Slope:	1%	**Described by:**	R.W. Fitzpatrick
Slope shape:	Straight	**Date described:**	July-72

Horizon	Depth (mm)	Description	Diagnostic horizon
A	0–250	Dry; dry 100% grey 10YR5/1, moist 100% very dark grey 10YR3/1; loamy medium sand; single grain; loose; clear boundary	Orthic
E	250–600	Moist; dry 100% light grey 10YR7/2, moist brown 10YR5/3; medium sand; single grain; loose; abrupt boundary	E-horizon
G1	600–700	Moist; moist 75% brown 10YR5/3; many fine distinct grey elongated mottles; medium sandy clay loam; moderate prismatic to angular blocky; firm; common clayskins on ped faces; very few medium sandstone fragments; gradual transition	G horizon
G2	700–900+	Dry; moist 75% brown 10YR5/3; many fine distinct yellow elongated mottles; medium sandy clay; moderate prismatic to angular blocky; extremely hard; few medium sandstone fragments	G horizon

Points of interest and interpretation

Originally the G horizon was classified as a gleycutanic B but this has fallen away in favour of a G in the current classification. The main difference between this example of a gleyic soil and the one that follows is the presence of a prominent, sandy E horizon overlying, via an abrupt transition, a gleyed sandy clay loam. There is consequently a strong relationship with duplex, eluvic-prismacutanic soils such as the example given earlier (P1112). There is a similarly strong relationship with plinthic, soft-eluvic soils such as P584, the main difference being that in the present case the water table, when present, is perched above and moves laterally across the surface of the G whereas in P584 the restrictive layer lies beneath the soft plinthic B horizon, within which lateral flow is probably as strong as it is within the overlying E horizon. In all three cases ferrolysis is a crucial soil forming process. From a land use perspective such soils are fickle: in wet years they are prone to waterlogging, whereas in growing seasons that are drier than average they may produce bumper maize crops when better drained soils are barren.

P578: Gleyic, eluvic, albic, loamy medium sand
(Kroonstad Kd 1000)

Lab no.	Depth (cm)	Horizon	Coarse fragments (%)	Fine earth (%)				
				CSa	MSa	FSa	Silt	Clay
C1352	0–25	A	0	10	32	42	2	8
C1353	25–60	E	0	12	41	44	0	4
C1354	60–70	B21	10	9	27	28	1	30

Lab no.	$CaCO_3$ equiv. (%)	Organic C		Fe			Al			Mn	Si $CaCl_2$ (mg kg^{-1})
		Total	Pyr.	CBD	Ox	Pyr.	CBD	Ox	Pyr	CBD	
				(%)							
C1352		0.5		0.2			0.0				
C1353		0.1		0.1			0.0				
C1354		0.1		1.2			0.0				

Lab no.	pH			CEC	Exchangeable cations					Acidity		Al
	H_2O	$CaCl_2$	NaF		Ca	Mg	Na	K	Total	Titr	Exch	Exch
					(cmol$_c$ kg^{-1})							
C1352	4.9			1.6	1.0	0.3	0.1	0.2	1.6	4.0		
C1353	5.4			0.4	0.2	0.0	0.0	0.2	0.4	1.2		
C1354	5.5			4.8	0.4	1.7	0.2	0.4	4.8	2.0		

Lab no.	Saturation extract							Micronutrients			
	Resist. (ohms)	EC (dS m^{-1})	Satn. water (%)	Ca	Mg	Na	K	Zn	Mn	Cu	B	Co
				(mmol$_c$ L^{-1})				(mg kg^{-1})				
C1352	2 000							0.44	12.8	0.41	0.18	
C1353	4 400							0.20	0.1	0.44	0.04	
C1354	1 100							0.17	436.2	7.77	1.32	

Lab no.	Bulk density (kg m^{-3})	HC (mm h^{-1})	Air-water permeability ratio	Atterberg		Water retentivity at kPa:			
				LL	PI	-33	-80	-500	-1 500
						(%)			
C1352			2			5.8	4.2	3.1	3.0
C1353			2			2.5	1.7	1.1	0.8
C1354			214			16.4	14.1	10.5	8.9

Lab no.	P Sorp. (%)	P Extr. (mg kg^{-1})	Minerals <2 μm	Mineral legend			
C1352	33.4	4.3	Kt5, Qz5, Ch(2:1), Mi1, Vm1	Qz	**Quartz**	St	**Smectite**
C1353	18.0	1.1	Kt5, Qz4, St2, Ch(2:1)1, Mi1	Ch	**Chlorite**	Mi	**Mica**
C1354	50.0	0.3	Kt5, Qz4, St2, Mi1	Kt	**Kaolinite**	Vm	**Vermiculite**

P1137: Gleyic, orthic, calcic, *medium sandy loam*
(Katspruit Ka 2000)

Profile No:	P1137	**Aspect:**	N
Land type map:	2730 Vryheid	**Water table:**	None
Latitude & Longitude:	27°52′48′/31°02′42′	**Occurrence of flooding:**	Occasional
Land type No:	Bb66	**Stoniness:**	Both roundish and flat, 320 mm diameter, 4.7 m apart
Climate zone No:	292S	**Erosion:**	Gully, class 2, not stabilised
Soil form:	Katspruit	**Vegetation/land use:**	Open grassveld
Soil series:	Killarney	**Parent and underlying material:**	Unconsolidated, clayey
Altitude:	1 040 m		
Terrain unit:	3		
Slope:	2%	**Described by:**	D.P. Turner
Slope shape:	Straight	**Date described:**	Aug-77

Horizon	Depth (mm)	Description	Diagnostic horizon
A1	0–250	Dry; moist very dark greyish-brown 10YR3/2; common medium faint grey mottles; common medium faint olive mottles; medium sandy loam; moderate coarse prismatic; very hard; few clay cutans; clear smooth transition	Orthic
G1	250–650	Moist; moist very dark greyish-brown 2.5Y3/2; common medium prominent grey mottles; common coarse prominent olive mottles; medium sandy clay loam; moderate coarse subangular blocky; hard; non-hardened free lime, moderately effervescent; few clay cutans; clear smooth transition	G horizon
G2	650–750	Moist; moist brownish-yellow 10YR6/6; many coarse prominent red mottles; many coarse prominent grey, yellow and olive mottles; coarse sandy clay loam; weak coarse subangular blocky; very hard; non-hardened free lime, strongly effervescent; few clay cutans; many irregular medium sesquioxide concretions, cannot be crushed by hand; many irregular medium lime nodules, cannot be crushed by hand; (clear smooth transition to similar but redder material from 750–1 000 mm.)	G horizon

Points of interest and interpretation

The structure and colour of the G horizon and the position in the landscape suggest that this profile is not a classic gleyic soil in a wetland situation and that it is transitional in certain respects to wetter variants of soils in the cumulic group (neocutanic B) and the duplex group (pedocutanic B). Calcareousness and a slightly saline, moderately sodic condition are prominent features of this pedon, as is the very low permeability of the G horizon.

P1137: Gleyic, orthic, calcic, *medium sandy loam*
(Katspruit Ka 2000)

Lab no.	Depth (cm)	Horizon	Coarse fragments (%)	Fine earth (%)				
				CSa	MSa	FSa	Silt	Clay
C4850	0–25	A1		12	24	35	8	17
C4851	25–65	G1		11	23	33	11	20
C4852	65–75	G2		13	18	27	10	29

Lab no.	$CaCO_3$ equiv. (%)	Organic C		Fe			Al			Mn	Si $CaCl_2$ (mg kg^{-1})
		Total	Pyr.	CBD	Ox	Pyr.	CBD	Ox	Pyr	CBD	
					(%)						
C4850		0.6		0.9			0.11				
C4851		0.4		0.4			0.09				
C4852		0.4		1.9			0.36				

Lab no.	pH			CEC	Exchangeable cations					Acidity		Al Exch
	H_2O	$CaCl_2$	NaF		Ca	Mg	Na	K	Total	Titr	Exch	
					(cmol$_c$ kg^{-1})								
C4850	7.5	6.3		9.9	2.9	5.7	0.7	0.1	9.4	0.0	0.0	0.0
C4851	8.4	7.4		10.2	2.8	6.6	1.9	0.1	11.4	0.0	0.0	0.0
C4852	8.5	7.5		13.3	4.1	6.7	2.6	0.1	13.5	0.0	0.0	0.0

Lab no.	Resist. (ohms)	EC (dS m^{-1})	Satn. water (%)	Ca	Mg	Na	K	Zn	Mn	Cu	B	Co
				(mmol$_c$ L^{-1})				(mg kg^{-1})			
C4850	800							0.21	6.4	2.88	0.31	3.2
C4851	600							0.12	47.8	2.31	0.19	3.4
C4852	240	2.10		2.1	19.2	23.5	0	0.24	6.0	1.29	0.15	1.5

Lab no.	Bulk density (kg m^{-3})	HC (mm h^{-1})	Air-water permeability ratio	Atterberg		Water retentivity at kPa:			
				LL	PI	-33	-80	-500	-1 500
						(%)			
C4850			23			14.3	12.4	8.5	7.2
C4851			999			21.3	16.9	11.2	9.5
C4852			999			27.4	21.7	17.0	16.5

Lab no.	P Sorp. (%)	P Extr. (mg kg^{-1})	Minerals <2 µm		Mineral legend			
C4850	16.4	0.9	Qz5, Kt3, Tc2, Fs1	Qz	**Quartz**		Kt	**Kaolinite**
C4851	20.8	0.8	Kt5, Mi4, Qz3	Fs	**Feldspar**		Mi	**Mica**
C4852	46.2	1.1	Kt5, St4, Qz3, Mi2	Tc	**Talc**		St	**Smectite**

P638: Cumulic, eluvic, *achromic*, albic, haplic, *medium sand*
(Fernwood Fw 1110)

Profile No:	P638	**Aspect:**	Level
Land type map:	2632 Mkuze	**Water table:**	Absent
Latitude & Longitude:	27°24'00'/32°29'18'	**Occurrence of flooding:**	
Land type No:	Ha21	**Stoniness:**	Class 0
Climate zone No:	229S	**Erosion:**	Class 1 sheet
Soil form:	Fernwood	**Vegetation/land use:**	Bushveld and grass
Soil series:	Fernwood	**Parent and underlying material:**	Blown sand
Altitude:	75 m		
Terrain unit:	3		
Slope:	1%	**Described by:**	D.P. Turner
Slope shape:	Convex	**Date described:**	Dec-76

Horizon	Depth (mm)	Description	Diagnostic horizon
A1	0–300	Dry; moist 100% grey 10YR6/1; medium sand; single grain; loose; diffuse smooth boundary	Orthic
E	300–1 200	Dry; moist 100% grey 10YR6/1; medium sand; single grain; loose	E horizon

Points of interest and interpretation

This profile could have been classified in the arenic (Namib) form of cumulic soils in view of its recent origin from wind blown sands. In general, however, grey regic sands, particularly those on the Zululand coastal plain where this one is situated, are considered to consist of a thick E horizon because of the common association of a water table with such soils during the wet season. The achromic topsoil signifies that this wetness is less protracted than in the umbric families which have an abnormal accumulation of organic matter associated with a wetland environment. This soil has little for which to recommend it, being almost devoid of clay, humus and oxide colloids. (An informal Australian term for such soils is 'gutless grey sand'). Groundwater vulnerability to pollution is at a maximum. Although not specifically referred to in the description, these sands are often strongly hydrophobic in dry weather as a result of the coating of sand grains with products of fungal activity. Differential (fingered) infiltration of rainwater is the result and has important consequences for groundwater recharge. An ecological as well as commercial forestry-related benefit of these deep grey sands is their capacity to function as a shallow sand aquifer and support forest vegetation in an environment that would otherwise be unfavourable because of greater runoff and/or less storage volume in the regolith. Deficiencies of both macronutrients and trace elements are common on such soils. This has both ecological and agricultural significance and may even affect the health of human communities that rely largely on local produce (Ceruti *et al.*, 2003). Very often these deep grey coastal sands may have a podzol B or a G horizon at considerable depth (2 or more metres).

P638: Cumulic, eluvic, *achromic*, albic, haplic, *medium sand*
(Fernwood Fw 1110)

Lab no.	Depth (cm)	Horizon	Coarse fragments (%)	Fine earth (%)				
				CSa	MSa	FSa	Silt	Clay
C4292	0–30	A1	0	4	61	32	1	3
C4293	30–120	E	0	4	56	38	1	3

Lab no.	CaCO₃ equiv. (%)	Organic C		Fe			Al			Mn	Si CaCl₂ (mg kg⁻¹)
		Total	Pyr.	CBD	Ox	Pyr.	CBD	Ox	Pyr	CBD	
		(%)										
C4292		0.2		0			0				
C4293		0.1		0			0				

Lab no.	pH			CEC	Exchangeable cations					Acidity		Al Exch
	H₂O	CaCl₂	NaF		Ca	Mg	Na	K	Total	Titr	Exch	
					(cmol_c kg⁻¹)								
C4292	6.1	5.5		1.7	0.9	0.3	0.0	0.0	1.2	1.8	0.0	0.0
C4293	6.3	5.5		1.7	0.5	0.1	0.1	0.0	0.7	0.6	0.1	0.1

Lab no.	Resist. (ohms)	EC (dS m⁻¹)	Satn. water (%)	Saturation extract				Micronutrients			
				Ca	Mg	Na	K	Zn	Mn	Cu	B	Co
				(mmol_c L⁻¹)				(mg kg⁻¹)			
C4292	4 800							0.18	2.50	0.09	0.16	0.24
C4293	4 400							0.06	0.30	0.03	0.18	0.09

Lab no.	Bulk density (kg m⁻³)	HC (mm h⁻¹)	Air-water permeability ratio	Atterberg		Water retentivity at kPa:			
				LL	PI	-33	-80	-500	-1 500
						(%)			
C4292			22			17.1	15.5	12.2	11.7
C4293			6			5.5	4.9	3.7	2.8

Lab no.	P Sorp. (%)	P Extr. (mg kg⁻¹)	Minerals <2 µm	Mineral legend			
C4292	0.63	1.9	Qz5, Kt5, Mi2	Qz	**Quartz**	Kt	**Kaolinite**
C4293	2.13	0.9	Kt5, Qz3, Mi2	Mi	**Mica**		

P778: Cumulic, neocalcic, *chromic*, rhodic, *luvic*, *coarse sandy loam*
(Augrabies Ag 1220)

Profile No:	P778	**Aspect:**	O/E
Land type map:	2816 Alexander Bay	**Water table:**	None
Latitude & Longitude:	28°56'12'/16°58'36'	**Occurrence of flooding:**	Occasional
Land type No:	Ae69	**Stoniness:**	None
Climate zone No:	141W	**Erosion:**	Wind, class 2, partially stabilised
Soil form:	Oakleaf	**Vegetation/land use:**	Barren
Soil series:	Letaba		
Altitude:	240 m	**Parent and underlying material:**	Undifferentiated sediments
Terrain unit:	5		
Slope:	1%	**Described by:**	F. Ellis
Slope shape:	Straight	**Date described:**	Aug-77

Horizon	Depth (mm)	Description	Diagnostic horizon
A1	0–300	Dry; moist yellowish-red 5YR4/8; coarse sandy loam; massive; slightly hard; non-hardened free lime, moderately effervescent; few clay cutans; few roots; clear smooth transition	Orthic
B2	300–450	Dry; moist yellowish-red 5YR4/6; medium sandy clay loam; weak fine subangular blocky; non-hardened free lime, moderately effervescent; few sesquioxide cutans; few roots; clear smooth transition	Neocutanic
Cca	450–600	Dry; moist yellowish-red 5YR5/6; medium sandy loam; weak fine subangular blocky; non-hardened free lime, moderately effervescent; few sesquioxide cutans; few rounded fine gravel; few roots	

Points of interest and interpretation

Originally classified as a calcareous series of the Oakleaf form, this soil has a B horizon which qualifies as a diagnostic neocarbonate B in the present classification. Based on the description of the Cca horizon it does not have sufficiently developed free lime to qualify as a soft carbonate B which would have placed the soil in the soft-neocalcic (Addo) form of calcic soils. Physically and chemically this soil has features in common with the examples of calcic soils presented earlier (especially profiles P28 and P783). Extractable P values are particularly high. The soil is weakly saline and interestingly the saturation extract is dominated by Mg. Pedogenetically this soil can be regarded as an immature calcic soil developed in unconsolidated sediments. Ecological and land use considerations are essentially the same as those pertaining to the calcic group.

P778: Cumulic, neocalcic, *chromic*, rhodic, luvic, *coarse sandy loam*
(Augrabies Ag 1220)

Lab no.	Depth (cm)	Horizon	Coarse fragments (%)	Fine earth (%)				
				CSa	MSa	FSa	Silt	Clay
C4736	0–30	A		15	24	35	15	12
C4737	30–45	B2		10	25	35	7	25
C4738	45–60	C		10	22	35	10	25

Lab no.	$CaCO_3$ equiv. (%)	Organic C		Fe			Al			Mn	Si $CaCl_2$ (mg kg^{-1})
		Total	Pyr.	CBD	Ox	Pyr.	CBD	Ox	Pyr	CBD	
		(%)									
C4736		0.1		0.8			0.0				
C4737		0.1		0.8			0.0				
C4738		0.1		0.7			0.1				

Lab no.	pH			CEC	Exchangeable cations					Acidity		Al
	H_2O	$CaCl_2$	NaF		Ca	Mg	Na	K	Total	Titr	Exch	Exch
				($cmol_c\ kg^{-1}$)								
C4736	8.8	8.1		8.4	7.4	2.4	0.7	1.4	11.9	0.0		0.0
C4737	8.6	8.0		10.4	7.7	1.3	0.7	0.8	10.5	0.0		0.0
C4738	8.7	8.0		10.1	7.2	1.8	0.6	0.8	10.4	0.0		0.0

Lab no.	Saturation extract							Micronutrients				
	Resist. (ohms)	EC (dS m^{-1})	Satn. water (%)	Ca	Mg	Na	K	Zn	Mn	Cu	B	Co
				($mmol_c\ L^{-1}$)				(mg kg^{-1})				
C4736	580							0.93	139.1	1.65	1.05	1.86
C4737	360	1.58		9	23	6	0	0.39	40.2	2.34	0.42	3.27
C4738	460	1.35		8	13	5	0	0.27	53.5	2.28	0.61	2.70

Lab no.	Bulk density (kg m^{-3})	HC (mm h^{-1})	Air-water permeability ratio	Atterberg		Water retentivity at kPa:			
				LL	PI	-33	-80	-500	-1 500
						(%)			
C4736			7			10.9	9.5	5.0	4.5
C4737			5			10.6	9.9	8.0	7.2
C4738			8			14.3	13.0	10.8	9.8

Lab no.	P Sorp. (%)	P Extr. (mg kg^{-1})	Minerals <2 µm
C4736	18.9	141.8	Mi5, Kt4, Qz4, Is2, Vm2
C4737	18.0	107.4	Mi5, Kt5, Is2, Qz2
C4738	18.6	175.9	Mi5, Kt5, Is2, Qz2

Mineral legend

Qz	**Quartz**	Is	**Interstratified**
Kt	**Kaolinite**	Vm	**Vermiculite**
Mi	**Mica**		

P1477: Cumulic, neocutanic, *chromic*, arhodic, *luvic*, *fine sandy loam*
(Oakleaf Oa 1120)

Profile No:	P1477	**Aspect:**	E/O
Land type map:	2330 Tzaneen	**Water table:**	None
Latitude & Longitude:	23°47'57'/30°14'16'	**Occurrence of flooding:**	Nil
Land type No:	Ab94	**Stoniness:**	None
Climate zone No:	1017S	**Erosion:**	None apparent
Soil form:	Oakleaf	**Vegetation/Land use:**	Abandoned field/disturbed land
Soil series:	Jozini		
Altitude:	670 m	**Parent and underlying material:**	Granite
Terrain unit:	3		
Slope:	8%	**Described by:**	P.G. Ross
Slope shape:	Concave	**Date described:**	April-84

Horizon	Depth (mm)	Description	Diagnostic horizon
A1	0–320	Dry; dry very dark greyish-brown 10YR3/2; fine sandy loam; massive; soft; many pores; few angular fine gravel; common roots; gradual smooth transition	Orthic
B22	320–650	Moist; moist strong brown 7.5YR4/6; fine sandy clay loam; weak medium subangular blocky; friable; many pores; many clay cutans; few angular fine gravel; common roots; gradual wavy transition	Neocutanic
B23re	650–1200	Moist; moist red 2.5YR4/8; sandy clay; single grain; loose; many pores; few angular fine gravel; few roots	

Points of interest and interpretation

Inevitably, when interpreting the profile descriptions of others, one will sometimes encounter evidence suggesting that the soil may have been inappropriately classified. Here is a case in point. The B22 horizon would probably qualify as a red apedal B (despite the fact that its structure is not described). The B21 horizon is 33 cm thick and could be considered as a transitional (AB) horizon, in which case this soil would have been classified in the rhodic (Hutton) form of oxidic soils. This difficulty of determining whether colour variegation due to cutanic character and faunal activity in the upper B horizon is sufficient to qualify the transitional material as a neocutanic B will persist until some thickness criterion is introduced. The definition of the neocutanic B warns about this pitfall. A simple solution would be to disqualify any horizon as neocutanic if it overlies, within 1.5 m of the surface, material that would qualify as a red or yellow-brown apedal B horizon. This soil is relatively leached although only mildly acidic. The B22 horizon has 3.6% CBD-extractable Fe and kaolinite dominates the clay fraction, explaining the low CEC of 6 cmol$_c$ kg^{-1} despite a clay content of 41%. The profile is strongly luvic. The fine sand grade is surprising in view of the granite parent material, but location on a steep midslope position makes it difficult to discount lithological admixture through colluviation.

P1477: Cumulic, neocutanic, *chromic*, arhodic, luvic, *fine sandy loam*
(Oakleaf Oa 1120)

Lab no.	Depth (cm)	Horizon	Coarse fragments (%)	Fine earth (%) CSa	MSa	FSa	Silt	Clay
C6877	0–32	A1		11	18	48	9	14
C6878	32–65	B21		8	16	40	9	28
C6879	65–120	B22		7	12	37	5	41

Lab no.	$CaCO_3$ equiv. (%)	Organic C Total	Pyr.	Fe CBD	Ox	Pyr.	Al CBD	Ox	Pyr	Mn CBD	Si $CaCl_2$ (mg kg^{-1})
C6877		0.6		0.9			0.1				
C6878		0.4		2.1			0.2				
C6879		0.2		3.6			0.4				

Lab no.	pH H_2O	$CaCl_2$	NaF	CEC	Exchangeable cations Ca	Mg	Na	K	Total	Acidity Titr	Exch	Al Exch
C6877	5.4	4.7		4.9	1.5	0.6	0.1	0.2	2.4	2.7		
C6878	5.2	4.9		5.1	1.4	0.6	0.1	0.2	2.3	2.7		
C6879	5.3	5.0		6.0	1.6	1.2	0.1	0.1	3.0	4.6		

(cmol$_c$ kg^{-1})

Lab no.	Resist. (ohms)	EC (dS m^{-1})	Satn. water (%)	Ca	Mg	Na	K	Zn	Mn	Cu	B	Co
C6877	6 700							0.19	72.3	0.67	0.27	1.76
C6878	3 500							0.20	42.9	0.70	0.32	2.70
C6879	6 800							0.11	11.6	0.40	0.67	1.54

Saturation extract (mmol$_c$ L^{-1}) — Micronutrients (mg kg^{-1})

Lab no.	Bulk density (kg m^{-3})	HC (mm h^{-1})	Air-water permeability ratio	Atterberg LL	PI	Water retentivity at kPa: -33	-80	-500	-1 500
C6877			10			10.4	8.8	6.0	5.7
C6878			7			12.1	11.2	9.5	8.7
C6879			6			18.6	16.7	14.1	13.6

(%)

Lab no.	P Sorp. (%)	P Extr. (mg kg^{-1})	Minerals <2 µm	Mineral legend			
C6877	4.1	2.8	Kt5, Mi2	St **Smectite**		Is **Interstratified**	
C6878	13.6	0.7	Kt5, Mi2, Is1	Kt **Kaolinite**		Mi **Mica**	
C6879	26.8	0.4	Kt5, Mi2, St1				

P1121: Lithic, orthic, *chromic, acalcic,* fine sandy clay loam
(Mispah Ms 1100)

Profile No:	P1121	**Aspect:**	E/O
Land type map:	2730 Vryheid	**Water table:**	None
Latitude & Longitude:	27°36'18'/31°23'06'	**Occurrence of flooding:**	Nil
Land type No:	Fa379	**Stoniness:**	Flat, 100 mm long, 65 m apart
Climate zone No:	372S	**Erosion:**	Sheet, class 3
Soil form:	Mispah	**Vegetation/Land use:**	Open grassveld
Soil series:	Mispah	**Parent and underlying material:**	Moderately soft horizontal blue shale, Pietermaritzburg Formation, Ecca Group
Altitude:	850 m		
Terrain unit:	3		
Slope:	15%	**Described by:**	D.P. Turner
Slope shape:	Convex	**Date described:**	Jul-77

Horizon	Depth (mm)	Description	Diagnostic horizon
A1	0–200	Dry; moist dark brown 10YR3/3; fine sandy clay loam; weak medium subangular blocky; slightly hard; clear wavy transition	Orthic
Cso	200–400N	Dry; silty clay loam; massive	Hard rock

Points of interest and interpretation

Horizontally bedded shale qualifies as hard rock even though it has softened through weathering. In this case the weathering is quite advanced, as confirmed by kaolinite dominance of the clay fraction in the orthic A horizon and by strong acidity (acid saturation of effective CEC is about 60%). Soils this shallow are best left under natural vegetation cover, although the degree of weathering of the shale would probably tempt some grape farmers in the Western Cape (if this soil was there) to apply lime and rip it deeply.

P1121: Lithic, orthic, *chromic, acalcic, fine sandy clay loam*
(Mispah Ms 1100)

Lab no.	Depth (cm)	Horizon	Coarse fragments (%)	Fine earth (%)				
				CSa	MSa	FSa	Silt	Clay
C4607	0–20	A1	1	3	12	37	12	32
C4608	20–40	C	87	3	4	29	37	28

Lab no.	CaCO$_3$ equiv. (%)	Organic C		Fe			Al			Mn	Si CaCl$_2$ (mg kg^{-1})
		Total	Pyr.	CBD	Ox	Pyr.	CBD	Ox	Pyr	CBD	
				(%)							
C4607		1.5		2.5			0.39				
C4608		0.7		1.8			0.23				

Lab no.	pH			CEC	Exchangeable cations					Acidity		Al
	H$_2$O	CaCl$_2$	NaF		Ca	Mg	Na	K	Total	Titr	Exch	Exch
					(cmol$_c$ kg^{-1})							
C4607	5.1	4.1		15.9	0.9	0.9	0.3	0.2	2.3	16.9	3.7	2.8
C4608	5.2	4.1		14.7	2.1	1.5	0.3	0.3	4.2	14.6	6.7	5.8

Lab no.	Saturation extract							Micronutrients				
	Resist. (ohms)	EC (dS m^{-1})	Satn. water (%)	Ca	Mg	Na	K	Zn	Mn	Cu	B	Co
				(mmol$_c$ L^{-1})				(mg kg^{-1})				
C4607	3 400							0.18	8.6	1.08	0.29	0.42
C4608	3 000							0.45	10.0	2.46	0.18	0.57

Lab no.	Bulk density (kg m^{-3})	HC (mm h^{-1})	Air-water permeability ratio	Atterberg		Water retentivity at kPa:			
				LL	PI	-33	-80	-500	-1 500
						(%)			
C4607			5			20.3	18.8	13.9	11.8
C4608			4			26.1	24.0	18.2	15.0

Lab no.	P Sorp. (%)	P Extr. (mg kg^{-1})	Minerals <2 µm	Mineral legend			
C4607	42.0	1.3	Kt5, Qz4, Ch(2:1)2	Qz	**Quartz**	Kt	**Kaolinite**
C4608	46.8	0.2		Ch	**Chlorite**		

P1146: Lithic, glossic, *chromic*, orthosaprolitic, aeromorphic, acalcic, *fine sandy clay loam*
(Glenrosa Gs 1111)

Profile No:	P1146	**Aspect:**	W
Land type map:	2730 Vryheid	**Water table:**	None
Latitude & Longitude:	27°58'12'/31°11'06'	**Occurrence of flooding:**	Nil
Land type No:	Fb235	**Stoniness:**	None
Climate zone No:	334S	**Erosion:**	Gully, class 1
Soil form:	Glenrosa	**Vegetation/Land use:**	Closed grassveld
Soil series:	Williamson		
Altitude:	550 m	**Parent and underlying material:**	Tillite
Terrain unit:	3		
Slope:	3%	**Described by:**	D.P. Turner
Slope shape:	Convex	**Date described:**	Aug-77

Horizon	Depth (mm)	Description	Diagnostic horizon
A1	0–200	Dry; moist very dark grayish-brown 10YR3/2; fine sandy clay loam; weak fine blocky; hard; few clay cutans; few stones; clear wavy transition	Orthic
A3	200–300	Dry; gradual wavy transition; stoneline	Lithocutanic
B21	300–700	Dry, moist reddish-brown 5YR4/4; many medium faint red and yellow mottles; many medium faint grey mottles; clay loam; weak fine subangular blocky; very hard, few rounded medium hard sesquioxide concretions	Lithocutanic
B22	700–800	Dry; moist strong brown 7.5YR5/6; many medium distinct red and yellow mottles; clay loam; weak medium sub angular blocky; very hard; few rounded medium hard sesquioxide concretions; common flaggy medium gravel; gradual wavy transition	Lithocutanic
B23	800–900	Dry; moist brown to dark brown 10YR4/3; common medium distinct grey and yellow mottles; clay loam; weak medium sub angular blocky; hard; few rounded medium hard sesquioxide concretions; clear broken transition	Lithocutanic
Cso	900–1 000+	Dry, moist brown to dark brown 10YR4/3; loam; weak medium granular; hard; many flaggy coarse gravel	

Points of interest and interpretation

The old Williamson series is classically derived from Dwyka tillite and perhaps its most interesting feature is the ubiquitous stoneline consisting of glacial erratics – stones of various lithological origin, often rounded and more resistant to weathering than the mudstone matrix within which they were seated. Their accumulation at the interface between the orthic A and the lithocutanic B can be ascribed to faunal working of fine earth material to create a biomantle above the stoneline (see Section 2.13 and the box in Section 2.7). The hard consistence of the B horizon means that this soil has a shallow effective depth. The texture of the A horizon is conducive to crusting and hard setting.

P1146: Lithic, glossic, *chromic*, orthosaprolitic, aeromorphic, acalcic, *fine sandy clay loam*
(Glenrosa Gs 1111)

Lab no.	Depth (cm)	Horizon	Coarse fragments (%)	Fine earth (%)				
				CSa	MSa	FSa	Silt	Clay
C4815	0–20	A1		9	15	43	10	22
C4816	30–70	B21		13	9	28	15	36
C4817	70–80	B22		6	7	32	19	37

Lab no.	CaCO$_3$ equiv. (%)	Organic C		Fe			Al			Mn	Si CaCl$_2$ (mg/kg)
		Total	Pyr.	CBD	Ox	Pyr.	CBD	Ox	Pyr	CBD	
		(%)									
C4815		0.8		1.8			0.24				
C4816		0.6		4.7			0.67				
C4817		0.3		3.8			0.46				

Lab no.	pH			CEC	Exchangeable cations					Acidity		Al Exch
	H$_2$O	CaCl$_2$	NaF		Ca	Mg	Na	K	Total	Titr	Exch	
					(cmol$_c$ kg^{-1})							
C4815	6.2	5.5		8.7	2.2	2.8	0.0	0.3	5.3	2.4	0.0	0.0
C4816	6.8	6.0		12.2	2.8	5.1	0.1	0.2	8.2	5.0	0.0	0.0
C4817	7.2	6.2		13.9	2.5	7.8	0.3	0.3	10.9	2.9	0.0	0.0

Lab no.	Saturation extract							Micronutrients				
	Resist. (ohms)	EC (dS m^{-1})	Satn. water (%)	Ca	Mg	Na	K	Zn	Mn	Cu	B	Co
				(mmol$_c$ L^{-1})				(mg kg^{-1})				
C4815	1 800							0.30	1.9	2.61	0.19	3.39
C4816	2 000							0.21	393.0	2.37	0.24	4.92
C4817	1 000							0.24	245.9	1.98	0.34	1.68

Lab no.	Bulk density (kg m^{-3})	HC (mm h^{-1})	Air-water permeability ratio	Atterberg		Water retentivity at kPa:			
				LL	PI	-33	-80	-500	-1 500
						(%)			
C4815									
C4816									
C4817									

Lab no.	P Sorp. (%)	P Extr. (mg kg^{-1})	Minerals <2 µm	Mineral legend				
C4815	15.8	0.5		Qz	**Quartz**	Kt	**Kaolinite**	
C4816	41.6	0.3	Kt5, Mi2, Qz1	Mi	**Mica**			
C4817	34.9	0.03						

APPENDIX: METHODS OF SOIL DESCRIPTION AND ANALYSIS

PROFILE DESCRIPTION METHOD

The method is based on that in the Soil Survey Manual (Soil Survey Staff, 1951). Abbreviations used are those contained in the Binomial System (MacVicar et al., 1977). The Munsell system was used to determine soil colour. Classes used for stoniness and erosion are as follows:

Stoniness

Stones larger than 250 mm in diameter and rock outcrops are grouped together since they both have an important bearing on soil use. On the one hand they interfere with the use of agricultural machinery and on the other they dilute the soil mass.

Class 0 – no stones or bedrock or too few stones to interfere with tillage
Class 1 – sufficient stones or bedrock to interfere with tillage but not to make inter-tilled crops impracticable
Class 2 – sufficient stones or bedrock to make tillage of inter-tilled crops impracticable, but soil can be worked for hay crops or improved pasture if other soil characteristics are favourable
Class 3 – use of all but very light machinery and hand tools impracticable; forestry and grazing possible
Class 4 – use of all machinery impracticable; forestry and grazing possible
Class 5 – more than 90% of the land surface covered by stones or exposed bedrock

Erosion

Kind:
Sheet erosion – uniform removal of topsoil from an area without the development of conspicuous water channels
Rill erosion – removal of soil through the cutting of numerous small but conspicuous water channels
Gully erosion – removal of soil giving rise to deep channels or gullies
Wind erosion – removal of topsoil by wind
Class:
Class 1 – none apparent or slight
Class 2 – moderate loss of topsoil and/or some slight dissection by run-off channels or gullies
Class 3 – severe loss of topsoil and/or marked dissection by run-off channels or gullies
Class 4 – total loss of topsoil and exposure of subsoil and/or deep intricate dissection by gullies

SAMPLE PREPARATION

The sample preparation procedure underwent several changes and improvements, the biggest change being that from sample crushing and sieving at SIRI (1972–1979) to crushing and sieving by hand under the pedologist's supervision (1979–present). The two procedures followed were, respectively:

(a) Samples collected in the field were air-dried and sent to SIRI in plastic bags contained within canvas sample bags. Samples were dried further if necessary, and a quantity of soil sufficient for the laboratory analysis was machine crushed using a porcelain mortar and steel pestle (Nasco-Aspelin grinder), and sieved through a 0.065 inch (1.68 mm) mesh sieve.
(b) Samples were air-dried and a quantity of soil sufficient for laboratory analysis was crushed by hand using a porcelain mortar and pestle and sieved through a 2 mm sieve. The mass of material >2 mm was noted. The crushed and remaining uncrushed soil was sent to SIRI for analysis and storage.

A portion of the sieved soil (<2 mm) was sent to the Winter Rainfall Region laboratories for micronutrient analysis.

PHYSICAL METHODS

Particle size distribution

Details of the method used are described by Day (1965).

Clay (< 2 µm settling diameter) and silt (2–20 µm) were determined by sedimentation and pipette sampling. The sand fractions, fine (0.02–0.2 mm), medium (0.2–0.5 mm) and coarse (0.5–2 mm) were determined by dry sieving. The results are expressed as a percentage of the mass of oven-dried soil.

Air to water permeability ratio

Method used is as described by Reeve (1965). Sufficient soil which had been crushed to pass a 2 mm mesh sieve was packed into a 50 mm diameter plastic permeameter to form a column about 40 mm high after tamping. The air permeability of the soil column was determined. The water permeability of the same column was measured after de-ionised water had flowed through the sample for 4 hours. A single air to water permeability ratio was measured for each sample. According to Reeve, individual determinations can be expected to be within 10% of the mean 75% of the time and within about 20% of the mean 80% of the time.

Modulus of rupture

Method used is as described by Reeve (1965). Soil which had been crushed to pass through a 2 mm mesh sieve was packed into a rectangular frame mould, saturated for at least 6 hours with de-ionised water and dried at about 45°C in a forced draught oven. The resulting soil briquette was supported at either end and a steadily increasing force was applied to a knife edge across the centre of the briquette until failure occurred. The modulus of rupture was calculated from the force required to break the briquette and the dimensions of the briquette and the knife edge assembly. Two briquettes were prepared for each sample, the results reported being mean values.

Water retentivity

The method described by Richards (1965) was applied to disturbed samples. Single determinations at each of four potentials (-33, -80, -500 and -1 500 kPa) were made using appropriate ceramic plate extractors. Samples were contained in rubber rings 10 mm high and 50 mm in diameter. The samples were saturated with tap water for at least 16 hours and at least 3 days were allowed for equilibration under pressure.

Dispersion ratio

A method quoted by Willen (1965) was modified slightly. De-ionised water (1L) was added to air-dry soil (20 g) contained in a cylinder. After 20 s, the cylinder and its contents were shaken end over end for 40 s (about 2 s per cycle). The concentration of < 20 µm particles was determined by pipette sampling after the appropriate sedimentation period. The dispersion ratio is this mass expressed as a percentage of the mass of silt and clay in the sample.

CHEMICAL METHODS

Total exchangeable acidity and exchangeable aluminium

The procedure outlined by McLean (1965) with modifications set out below was used. A 10 g soil sample was shaken with 70 cm^3 1 mol/L KCl solution in a 100 cm^3 slopy neck plastic bottle on a reciprocating shaker at a minimum of 180 oscillations per minute for 1 hour. The suspension was filtered through Whatman No. 2V filter paper into a 500 cm^3 Erlenmeyer flask and washed with an additional 30 cm^3 1 mol/L KCl solution. Total exchangeable acidity was determined on the KCl extract by titration with 0.1 mol/L NaOH. Excess NaF was added to the titration solution and the OH$^-$ so released were titrated with 0.1 mol/L HCl to determine exchangeable aluminium. In both titrations phenolphthalein was used as indicator. Exchangeable hydrogen was obtained by difference. All results were expressed as cmol$_c$/kg soil.

Exchangeable and soluble cations, cation exchange capacity, titratable acidity and electrical conductivity

The procedure outlined by Peech (1965), with modifications set out below was used. A LiCl solution served as extractant for exchangeable plus soluble cations and at the same time saturated the soil exchange complex with Li. The LiCl extract was also used for determining titratable acidity. Li was then displaced from the Li-saturated soil with Ca(NO$_3$)$_2$ and taken as an index of the CEC. Soluble cations were determined separately in soils containing significant quantities of soluble salts. These were subtracted from the LiCl-extractable cations to obtain the exchangeable cations.

The LiCl extracting solution (0.5 mol/L LiCl buffered at pH 8 with triethanolamine) was used instead of 0.25 mol/L BaCl$_2$ as recommended by Peech (1965). A 25 g sample of air-dry soil was transferred to an 800 cm^3 beaker to which 150 cm^3 LiCl extracting solution was added, stirred with a glass rod and left to react overnight. Extraction was done through a 110 mm diameter Buchner funnel using Whatman No. 40 or 50 filter paper discs. An amount of 350 cm³ of LiCl extracting solution was added continuously in small volumes to prevent the soil from cracking. The leachate, approximately 450 cm³, was collected and quantitatively transferred to a 500 cm³ volumetric flask and made up to volume with the LiCl extracting solution.

The cations Ca and Mg were determined in the leachate using a continuous flow auto-analyser system, Ca by flame emission and Mg colorimetrically with Technicon Industrial Method No. 237/72A. An EEL flame photometer was used for the determination of Na and K in the leachate. Results were expressed in $cmol_c$/kg soil and accepted as a measure of exchangeable cations except in cases where soil contained significant quantities of soluble salts.

Titratable acidity was determined by titrating a 100 cm³ portion of the LiCl soil extract with 0.2 mol/L HCl and subtracting the value so obtained from that obtained on a 100 cm³ buffered 0.5 mol/L LiCl blank solution.

CEC was determined as follows. The Li-saturated soil on the Buchner funnel was washed with 150 cm³ 80% ethanol added in 3–4 portions allowing complete drainage between portions. The soil was transferred with the filter paper to an 800 cm³ beaker and 500 cm³ 0.25 mol/L $Ca(NO_3)_2$ solution was added. The suspension was heated on a water bath at 80-90°C for 30 min and stirred at intervals to completely disintegrate any lumps. The suspension was filtered through a Buchner funnel (110 mm disc, Whatman No. 40) until suction drainage was complete. Lithium was determined in the filtrate using an EEL flame photometer and used as the measure of CEC expressed in $cmol_c$/kg soil at pH 8.

For soils with an electrical resistance less than 460 ohms, the exchangeable cations were estimated by subtracting the cations in the saturation extract (see below) from the total extractable cations. In soils containing lime or gypsum or those with a very high salt content, not all the water soluble salts were dissolved in the saturation extract. In these cases, the figure for total exchangeable cations is higher than the CEC.

Cations in the saturation extract were determined as follows. Based on the method of Longenecker and Lyerly (1964), square sample holders were made from Whatman No. 52 filter paper of 180 mm diameter. A 250 g air-dry soil sample was transferred to each filter paper holder, which was then placed on sand (about 40 mm thick) in a plastic container of de-ionised water in which the water level was controlled to saturate the bottom 10 mm of sand. The sample was allowed to absorb water for 24 hours. Some soils high in sodium or clay do not saturate satisfactorily and the method of preparing a saturated paste in a porcelain dish as described by the USSL Staff (1954) was used. The extraction was done on a Buchner funnel using 90 mm diameter Whatman No. 40 filter paper.

Conductivity of the saturation extract was measured and has been expressed in dS/m (equivalent to mS/cm). A 5 cm³ portion of the extract was diluted to 100 cm³ in a volumetric flask and Na and K determined with an EEL flame photometer. Ca and Mg were determined on a 20 cm³ portion of the extract by EDTA titration as described by Horwitz (1965). The only modifications were that in the preparation of the calcein indicator no charcoal was used and 0.12 g thymolphthalein per 0.2 g calcein was added.

pH measurements

Two pH measurements were made, one on a 1:2.5 soil to water suspension and one on a suspension prepared by adding 75cm³ 0.01 mol/L $CaCl_2$ solution to 15 g soil. In both instances, suspensions were stirred intermittently for 15 min and allowed to stand for at least 1 hour. The electrodes were positioned in the supernatant liquid.

Electrical resistance of the saturated paste

A saturated paste was prepared as described by the USSL Staff (1954). The soil paste was allowed to stand for 4 hours and the resistance determined in a US Bureau of Soils cup using a Thornley and Yntema Type BC12 resistance bridge. Results were given in ohms.

Organic carbon

The Walkley-Black method as described by Allison (1965) was used with the following modifications. The soil was ground to pass a 44 mesh (approx. 0.355 mm) sieve and masses of 0.5 g, 1 g or 2 g were used, depending on the amount of carbon present. Instead of 20 cm^3, 15 cm^3 conc. H_2SO_4 was used. An amount of 196 g $(NH_4)_2Fe(SO_4)_2.6H_2O$ plus 5 cm^3 conc. H_2SO_4 was made up to 1L to replace the ferrous sulfate o-phenanthroline monohydrate solution used to back-titrate the unreacted $K_2Cr_2O_7$ (initially 0.5 mol/L).

Phosphorus status

Phosphorus was extracted using a slightly modified version of the ISFEI method described by Hunter (1975). Instead of a 1:10 soil to solution ratio on a volume bases, a 1:10 ratio on a mass basis was used. Ten minutes of stirring was replaced by 30 min of shaking at 180 oscillations per minute. Phosphorus was determined in the extract by means of the molybdenum blue method using an auto-analyser. The P status is expressed as mg/kg soil.

Phosphorus sorption

Sorption of P from solution was measured in order to identify soil materials likely to fix large quantities of fertiliser P. Air-dried soil (2 g, later 8 g) was shaken with 50 cm^3 (later 200 cm^3) of a 10 mg/L P solution (prepared from KH_2PO_4 to which was added 4.46 g $CaCl_2$ and 10 cm^3 CH_2O per 20L of solution) in an end-over-end shaker at 40 rpm for 24 hours. The soil suspensions were filtered through Whatman No. 42 filter paper and P determined in the clear filtrate by the molybdenum blue method using an auto-analyser. The sorption reaction is temperature-sensitive and was therefore carried out under controlled temperature (20°C ± 2°) conditions.

Sorption was expressed as a percentage of the P added (250 mg/kg soil). In the case of soils giving sorption percentages higher than 98%, the procedure was repeated using a 100 mg/L P solution instead of a 10 mg/L P solution. This gave a better discrimination among soils with high P-sorption characteristics. The results provide an empirical index of the soils' P-sorption ability not necessarily directly related to fertiliser requirements. This procedure using the lower and where necessary the higher P concentration was followed for samples from 25 land type maps (SE 27/20 Witdraai, 2522 Bray, 2524 Mafeking, 2528 Pretoria, 2622 Morokweng, 2624 Vryburg, 2626 Wes-Rand, 2628 East Rand, 2630 Mbabane, 2720 Noenieput, 2722 Kuruman, 2724 Christiana, 2726 Kroonstad, 2820 Upington, 2822 Postmasburg, 2824 Kimberley, 2826 Winburg, 2920 Kenhardt, 2922 Prieska, 2924 Koffiefontein, 2926 Bloemfontein, 2930 Durban, 3030 Port Shepstone, 3320 Ladismith, 3420 Riversdale). Samples from all other areas were treated only with 100 mg/L P solution and centrifuged.

Micronutrients (Cu, Mn, Zn, Co, B)

For the determination of micronutrients in soil samples, the multiple soil extractant used was di-ammonium EDTA (Beyers and Coetzer, 1971). Soil (10 g) was shaken up with 100 cm^3 of 0.02 mol/L $(NH_4)_2$EDTA at pH 4.6 for 60 min, centrifuged, filtered, and 80 cm^3 of the filtrate evaporated to dryness in a silica dish and ashed in a furnace at 500 °C. The ash was moistened with distilled water and a drop of HNO_3, whereafter it was again ashed at 500 °C for 15 min. After cooling, the residue was taken up in 2.5 cm^3 of 1:1 constant boiling point HCl solution containing Li (2%) and Sr (4%).

Up until the analysis of sample C3340 a direct reading spectrometer was used to determine Cu, Zn, Mn, Co and B in the prepared solutions. From sample C3341 to C4496 solutions obtained in the same way were analysed for all these elements, with the exception of B, on an atomic absorption spectrometer. Dilution of the extracts was however necessary because of high acid concentrations as well as the need to adjust the readings to the range of the atomic absorption apparatus. Boron was extracted with the warm water reflux extraction procedure and the B content of the extract determined colorimetrically according to the curcumin method described by Jackson (1958).

In order to increase throughput, precision and accuracy, and thereby reduce costs, a modified procedure for determining Cu, Zn, Co and Mn was introduced from sample C4497 onwards. The $(NH_4)_2$EDTA solution concentration and time of extraction remained unchanged but the soil to extractant ratio was adjusted and the ashing procedure omitted. In the case of Cu, Zn and Co determinations 5 g of soil was shaken up with 15 cm^3 of $(NH_4)_2$EDTA for 60 min in a plastic bottle. For Mn the original soil to solution ratio of 1:10 was maintained. It should be noted that for soils containing free lime the buffer action of EDTA is inadequate to prevent an increase in pH of the extract. After shaking, the soil suspension was centrifuged in the same plastic bottle and thereafter filtered through 90 mm diameter Whatman No. 40 filter paper. Concentrations of elements in the extracts were determined directly with an atomic absorption spectrometer. Only in the case of soils with high Mn contents was it necessary to dilute the extract 10-fold. Standards were prepared and diluted with a solution of 1 000-1 250 mg/L $LaCl_3$.

Before any change in procedure was accepted, duplicate analyses were carried out using the original and modified methods. Results were statistically analysed and no significant differences were found. Micronutrient content is expressed as mg/kg of soil.

MINERALOGICAL METHODS

Pretreatment

The methods described by Jackson (1956) were modified to facilitate the handling of a large number of samples. All samples received the same pretreatment. Sufficient soil to yield between 6 and 12 g clay was weighed into a plastic 500 cm³ centrifuge bottle and treated with 200 cm³ 1 mol/L NaOAc (buffered to pH 5) in a water bath at 70 °C with intermittent stirring to dissolve carbonates. After centrifuging, the NaOAc was poured off and discarded. To remove as much Ca as possible the sample was shaken up with an additional 200 cm³ 1 mol/L NaOAc, centrifuged, and the supernatant discarded. Organic matter was removed by adding 50 cm³ 30% H_2O_2 and stirring. After the initial vigorous reaction had subsided, the removal was brought to completion on a water bath. The procedure was repeated with 20 cm³ H_2O_2 for most soils, but with 50 cm³ for soils rich in organic matter. To the peroxide treated samples 300 cm³ 1 mol/L NaCl was added, shaken by hand, centrifuged and the supernatant decanted.

Extractable Fe and Al

Deferration was carried out by adding 200 cm³ Nacitrate/bicarbonate buffer (pH 8.5) solution (0.3 mol/L Nacitrate and 0.1 mol/L $NaHCO_3$), shaking the sample into suspension and adding about 10 g Na-dithionite and allowing it to react with intermittent stirring on a water bath at 70 °C. After the colour became completely grey (about 30 min), the suspension was centrifuged and the citrate dithionite extract was poured off into pre-weighed plastic bottles. A further washing with 200 cm³ Na-citrate/bicarbonate solution was done and after centrifuging, the supernatant was added to the plastic bottles. Iron, and in certain samples also aluminium, was determined in the extract by atomic absorption and the results recorded as per cent *(m/m)* Fe and Al on a soil (< 2 mm) basis.

Particle size separation and Mg-saturation

The centrifuged sample was washed with 300 cm³ 1 mol/L NaCl and thereafter with 300 cm³ de-ionised water. The sample was then wet-sieved (50 μm sieve), collecting the less than 50 μm fraction in 3.8L glass bottles and discarded the sand fraction. The suspension was flocculated by adding 50-100 g NaCl and the supernatant siphoned off. The less than 50 μm fraction was transferred to plastic centrifuge bottles for separation of the clay (< 2 μm) and silt (for convenience 2-50 μm) fractions as described by Jackson (1956). The 2-50 μm fraction was saturated with Mg (using 0.5 mol/L $Mg(OAc)_2$ and $MgCl_2$) and dried in an oven at 70 °C. The clay fraction was flocculated by adding 100 cm³ 0.5 mol/L $Mg(OAc)_2$ and the supernatant siphoned off. The clay fraction was transferred to 250 cm³ glass centrifuge bottles and centrifuge-washed twice with 100 cm³ 0.5 mol/L $MgCl_2$. The samples were centrifuge-washed with de-ionised water until Cl- free ($AgNO_3$ tested) and freeze-dried.

X-ray diffraction analysis

X-ray diffraction (XRD) analyses were carried out on a Philips unit with PW 1010/25 generator, PW 1050/25 goniometer and AMR 3 – 202 graphite monochromator, using Co Kα radiation, a 1.0° divergence slit and 0.1° receiving slit, and a proportional counter. Standard experimental conditions were 30 kV, 20 mA, a scanning speed of 2° 2 θ/min, paper speed 25.4 mm per min, and variable rate meter settings (usually R4 or R8).

Parallel or 'preferred' orientated deferrated clay specimens were examined by XRD after preparation on unglazed ceramic tiles. The method used was similar to that described by Rich (1969). The 2–50 μm fraction was initially examined in the Mg-saturated form on glass slides (first 1 000 samples) and thereafter in randomly orientated form by pressing samples in aluminium holders against a filter paper surface (Fitzpatrick, 1978). Ferricrete and calcrete samples were finely ground in an agate mortar and examined by XRD in a randomly orientated form as above.

To differentiate between various layer silicates, Mg-saturated samples solvated with glycerol, and K-saturated samples heated at 110 °C and 550 °C were examined (Jackson, 1956). In cases where confirmation of certain minerals was required, the chemical method described by Alexiades and Jackson (1966) together with thermal (differential thermal analysis and thermogravimetric analysis) and infrared analysis were used.

Relative intensities or peak heights of X-ray diffraction peaks were used as estimates of the approximate amounts of minerals present in the sample and are expressed as very strong (peak height 75–100 in relative units), strong (50–75), medium (25–50), weak (5–25) and very weak (0–5). It should be noted that quantitative estimations of different minerals in the separates by comparison of intensities of X-ray diffraction peaks is not reliable because individual minerals differ in mass absorption coefficient, orientation of grains, crystal perfection and chemical composition. Therefore, at most, even when the interpretation of the diffractograms is made with caution, the peak height values can serve only as a semi-quantitative indicator of the amounts of minerals present.

Modifications

From sample C5797, the following procedures were adopted. The <2 μm and 2–50 μm fractions were collected in 250 cm^3 bottles after the sedimentation phase of particle size analysis (see PHYSICAL METHODS, Particle size distribution). The clay was saturated with Mg (using 1 mol/L $MgCl_2$) and repeatedly shaken up with de-ionised water and centrifuged until chloride-free ($AgNO_3$ tested) and freeze-dried. The 2–50 μm fraction was dried overnight at 70 °C.

X-ray diffraction analysis was carried out on both size fractions using the method described above. Extractable Fe and Al were determined separately on 1 g samples as described above.

References

Alexiades, C.A. and Jackson, M.L., 1966. Quantitative clay mineralogical analysis of soils and sediments. *Clays and Clay Minerals*, 14: 35–52.

Allen, B.L. and Fanning, D.S., 1983. Composition and soil genesis. In: *Pedogenesis and Soil Taxonomy*. L.P. Wilding, N.E. Smeck and G.F. Hall (eds.), Elsevier Publishing Co., Amsterdam, pp144–192.

Allison, L.E., 1965. In: *Methods of Soil Analysis*. Part 2. Black, C.A., Evans, D.E., White, J.L., Ensminger, L.E. and Clark, F.E. (eds.) American Society of Agronomy, Madison, Wisconsin.

Anderson, J.M., 1988. Spatiotemporal effects of invertebrates on soil processes. *Biology and Fertility of Soils*, 6: 216–227.

Beater, B.E., 1957. *Soils of the Sugar Belt, Part 1: Natal North Coast. Part 2(1959): Natal South Coast. Part 3(1962): Zululand*. Oxford University Press, Cape Town.

Begg, C.M., Begg, K.S., Du Toit, J.T. and Mills, M.G.L., 2003. Sexual and seasonal variation in the diet and foraging behaviour of a sexually dimorphic carnivore, the honey badger (*Mellivora capensis*). *Journal of Zoology*, 260: 301–316.

Beyers, C.P. de L. and Coetzer, F.J., 1971. Effect of concentration, pH and time on the properties of di-ammonium EDTA as a multiple soil extractant. *Agrochemophysica*, 3: 49–54.

Birkeland, P.W., 1984. *Soils and Geomorphology*. Oxford University Press, New York.

Blume, H.P., 1967. Zum Mechanismus der Marmorierung, und Konkretionsbildung in Stauwasser Boden. *Z. Pfl. Ernähr. Düng. Bodenk*, 119: 124–134.

Borchardt, G.A., 1977. Montmorillonite and other smectite minerals. In: *Minerals in Soil Environments*. J.B. Dixon and S.B. Weed (eds.), Soil Science Society of America, Madison, Wisconsin, pp. 293–330.

Botha, G. and Porat, N., 2007. Soil chronosequence development in dunes on the southeast African coastal plain, Maputaland, South Africa. *Quaternary International*, 162–163: 111–132.

Boyer, P., 1975. Action de certains termites constructeurs sur l'évolution des sols tropicaux II. Etude particuliére de trois termitieres de Bellicositermes et de leur action sur les sols tropicaux. *Annales des Sciences Naturelles Zoologie Paris sér*, 12(17): 273–446.

Brinkman, R., 1970. Ferrolysis, a hydromorphic soil forming process. *Geoderma*, 3: 199–206.

Buchanan, F., 1807. *A Journey from Madras through the Countries of Mysore, Canara and Malabar*. East India Company, London.

Bühmann, C. and Bühmann, D., 1990. Phyllosilicates in hydrothermally altered dolerite sills. *South African Journal of Geology*, 93: 446–453.

Bühmann, C. and Grubb, P.L.C., 1991. A kaolin-smectite interstratification sequence from a red and black complex. *Clay Minerals*, 26: 343–358.

Bühmann, C., 1991. Clay mineralogical aspects of thermally induced parent material discontinuity. *Applied Clay Science*, 6: 1–19.

Bühmann, C., 1992. Smectite-to-illite conversion in a geothermally and lithologically complex Permian sedimentary sequence. *Clays and Clay Minerals*, 40: 53–64.

Bühmann, C., 1994. Parent material and pedogenic processes in South Africa. *Clay Minerals*, 29: 239–246.

Bühmann, C., 1995. Mineralogical and chemical characteristics of crusting and self-mulching vertisols from South Africa. ARC-ISCW Procs. Wise Land Use Symp. Pretoria. Sigma Press. Pretoria, pp. 54–60.

Bühmann, C., de Villiers, J.M. and Fey, M.V., 1988. The mineralogy of four heaving clays. *Applied Clay Science*, 3: 219–136.

Bühmann, C., Fey, M.V. and de Villiers, J.M., 1985. Aspects of the X-ray identification of swelling clay minerals in soils and sediments. *South African Journal of Science*, 81: 505–510.

Bühmann, C., Rapp, I., Laker, M.C. and van der Merwe, G.M.E., 1998. Aggregation: the soil texture approach. *Bulletin of the Egyptian Geographical Society*, 71: 113–135.

Bühmann, C., van der Merwe, G.M.E., Rapp, I. and Laker, M.C., 1999. Water infiltration and crust formation characteristics of South African soils. Proceedings of the 2nd International Land Degradation Conference, Khon Kaen, Thailand.

Buol, S.W., Hole, F.D., McCracken, R.J. and Southard, R.J., 1997. *Soil Genesis and Classification* 4th Edition. Iowa State University Press, Ames.

Carson, M.A. and Kirkby, M.J., 1972. Hillslope *Form and Process*. Cambridge University Press.

Carwardine, M., 1995. *The Guinness Book of Animal Records*. Guinness Publishing, Middlesex, England.

Ceruti, P.O., Fey, M.V. & Pooley, J., 2003 Soil nutrient deficiencies in an area of endemic osteoarthritis (Mseleni Joint Disease) and dwarfism in Maputaland, South Africa. Chapter 24 In: H.C.W. Skinner and A.R. Berger (eds.) *Geology and Health: Closing the Gap*. New York, Oxford University Press. Pp. 151–154.

Ceruti, P., 1999. Crushed rock and clay amelioration of a nutrient deficient, sandy soil of Maputaland. MSc thesis, University of Cape Town.

Chadwick, O.A. and Chorover, J., 2001. The Chemistry of pedogenic thresholds. *Geoderma*, 100: 321–53.

Clark, H.O., 2005. *Otocyon megalotis*. Mammalian species No. 766. pp. 1–5.

Clough, M.E. and Payne, T.W., 1988. The Fieldes and Perrott field test as an aid to identification of podzol B horizons. *South African Journal of Plant and Soil*, 1: 43–45.

Darwin, C. 1881. The formation of vegetable mould through the action of worms with observations on their habits. Reprinted edition, 1945. Faber & Faber, London.

Day, P.R., 1965. Physical and mineralogical properties, including statistics of measurement and sampling. In: *Methods of Soil Analysis*. Part 1. Black, C.A., Evans, D.E., White, J.L.,

De Bruyn, L.A. and Conacher, A.J., 1990. The role of termites and ants in soil modification: a review. *Australian Journal of Soil Research*, 28: 55–93.

DeGraff, J.M. and Aydin, A., 1987. Surface morphology of columnar joints and its significance to mechanics and direction of joint growth. *Geological Society of America Journal*, 99: 605–617.

De Villiers, J.M., 1964a. A case of subsoil mottling developed under non-hydromorphic conditions. *South African Journal of Agricultural Science*, 7: 653–658.

De Villiers, J.M., 1964b. The genesis of some Natal soils. I. Clovelly, Kranskop and Balmoral series. *South African Journal of Agricultural Science*, 7: 417–438.

Donkin, M.J. and Fey, M.V., 1993. Relationships between soil properties and climatic indices in southern Natal. *Geoderma*, 59:197–212.

Dowding, C.E., 2004. Morphology, mineralogy and surface chemistry of manganiferous oxisols near Graskop, Mpumalanga Province. MSc thesis, University of Stellenbosch.

Duchaufour, P., 1982. *Pedogenesis and Classification*. Allen & Unwin, London.

Elkins, N.Z., Sabol, G.V., Ward, J.J. and Whitford, W.G., 1986. The influence of subterranean termites on the hydrological characteristics of a Chihuahuan desert ecosystem. *Oecologia*, 68: 521–528.

Ellis, F., 2002. Contribution of termites to the formation of hardpans in soils of arid and semi-arid regions of South Africa. Paper delivered at the 17th World Congress of Soil Science, Bangkok, Thailand, August 2002.

Emerson, W.W. and Bakker, A.C., 1973. The comparative effects of exchangeable calcium, magnesium and sodium on some physical properties of red-brown earth soils. II. The spontaneous dispersion of aggregates in water. *Australian Journal of Soil Research*, 11: 151–157.

Ensminger, L.E. and Clark, F.E. (eds.) American Society of Agronomy, Madison, Wisconsin.

FAO, 2001. *Lecture Notes on the Major Soils of the World.* P. Driessen, J. Dekkers, O. Spaargaren and F. Nachtergaele (eds.). World Soil Resources Reports 94. FAO, Rome.

Farina, M. P. W., Channon, P. and G. R. Thibaud, 2000. A Comparison of strategies for ameliorating subsoil acidity: II. Long-term soil effects. Soil Sci. Soc. Am. J. 64:652–658.

Farmer, V.C., 1981. Possible roles of mobile hydroxyaluminium orthosilicate complex (proto-imogolite) and other hydroaluminium and hydroiron species in podzolisation. In: *Migrations organominérales dans les sols tempérés. Colloques Internationaux du Cent. Natl. de la Rech. Sci.* 303, pp. 275–279.

Farmer, V.C., Russell, J.D. and Berrow, M.L., 1980. Imogolite and proto-imogolite allophane in spodic horizons: Evidence of a mobile aluminium silicate complex in podzol formation. *European Journal of Soil Science*, 31: 673–684.

Fey, M.V. and Donkin, M.J., 1994. The base status criterion in South African soil classification. *South African Journal of Plant and Soil*, 11: 149–151.

Fey, M.V. and Manson, A.D., 2004. Soil chemistry in South Africa: a quarter century of progress and prospects for the future. *South African Journal of Plant and Soil*, 21: 278–287.

Fey, M.V., 1982. Hypothesis for the pedogenic yellowing of red soil materials. In: E. Verster (ed.), *Proceedings of the 10th Congress of Soil Science Society of South Africa*, Dept. Agric. Tech. Comm. No. 180: 130–136.

Fey, M.V., 1985. The genesis of soils with humic A horizons. *Proceedings of the Combined Conference of Soil, Crop & Weed Science Societies of South Africa*, Stellenbosch.

Fey, M.V., Mills, A.J. and Yaalon, D.H., 2006. The alternative meaning of pedoderm and its use for soil surface characterization. *Geoderma*, 133: 474–477.

Fitzpatrick, R.W. and le Roux, J., 1977. Mineralogy and chemistry of a Transvaal black toposequence. *Journal of Soil Science*, 28: 165–179.

Fitzpatrick, R.W., 1978. Occurrence and properties of iron and titanium oxides in soils along the eastern seaboard of South Africa. Ph.D. thesis. University of Natal, Pietermaritzburg.

Francis, M.L., 2007. Soil formation on the Namaqualand coastal plain. PhD thesis, University of Stellenbosch.

Francis, M.L., Fey, M.V., Prinsloo, H.P., Ellis, F., Mills, A.J. and Medinski, T., 2007. Soils of Namaqualand: compensations for aridity. *Journal of Arid Environments*, 70: 588–603.

Frenkel, H., Levy, G.J. and Fey, M.V., 1992. Clay dispersion and hydraulic conductivity of clay-sand mixtures as affected by the addition of various anions. Clays & Clay Minerals, 40: 515–521.

Gay, F.J. and Calaby, J.H., 1970. Termites of the Australian region. In: *Biology of Termites*. K. Krishna and F.M. Weesner (eds.), Academic Press, New York, Vol. 2, pp. 393–448.

Hanlon, R.D.G., 1981. Some factors influencing microbial growth on soil animal faeces: I. Bacterial and fungal growth on particulate leaf litter. *Pedobiologia*, 21: 257–263.

Hanlon, R.D.G. and Anderson, J.M., 1980. Influence of macroarthropod feeding activities on macroflora in decomposing oak leaves. *Soil Biology and Biochemistry*, 12: 255–261.

Jarvis, J.U.M., 2001. African mole-rats. In: *The new encyclopedia of mammals*. D. Macdonald (ed.) Oxford University Press. pp. 690–693.

Gubevu, J.S., 1997. A study of the pH-dependent charge of a selection of highly weathered humic soils. MSc thesis, University of Cape Town.

Hawker, L.C. and Fitzpatrick, R.W., 1989. Iron oxide mineralogy of placic horizons in podzols near George. Summaries of papers. Combined Cong. SASCP, SAWSS, SSSSA, Wild Coast Sun, Transkei.

Hawker, L.C. and Thompson, J.G., 1988. Weathering sequence and alteration products in the genesis of the Graskop manganese residua, Republic of South Africa. *Clays & Clay Minerals*, 36: 448–454.

Hawker, L.C., 1986. A mineralo-chemical study of podzols and podzolized soils in a slope sequence near George, southern Cape. MSc thesis, University of South Africa.

Hawker, L.C., Van Rooyen, T.H. and Fitzpatrick, R.W., 1992. A slope sequence of Podzols in the southern Cape, South Africa 1. Physical and micromorphological properties. *South African Journal of Plant and Soil*, 9: 94–102.

Horwitz, W. (ed.), 1965. *Official methods of analysis of the Association of Official Agricultural Chemists*. Association of Official Agricultural Chemists, Washington DC.

Hunter, A.H., 1975. In: *Soil management in tropical America. New techniques and equipment for routine plant analytical procedures.* Bornemisza, E. and Alvorado, A. (eds.) North Carolina State University, Raleigh.

IUSS Working Group WRB, 2006. *World Reference Base for Soil Resources* 2nd edition. World Soil Resources Reports 103. FAO, Rome.

Jackson, M.L., 1956. (5th Print 1969). Soil chemical analysis. Advanced course. Published by M.L. Jackson, Madison, Wisconsin.

Jackson, M.L., 1958. *Soil Chemical Analysis*. London: Constable and Co.

Jackson, M.L., 1965. Clay transformations in soil genesis during the Quaternary. *Soil Science*, 99: 15–22.

Johnson, D.L., Domier, J.E.J. and Johnson, D.N., 2005. Reflections on the nature of soil and its biomantle. *Annals of the Association of American Geographers*, 95: 11–31.

Johnson, D.L., 1990. Biomantle evolution and the redistribution of earth materials and artifacts. *Soil Science* 149: 84–102.

Jones, D.E. and Holtz, W.G., 1973. Expansive Soils: The Hidden Disaster. *Civil Engineering* August: 49–51.

King, L.C., 1972. *The Natal Monocline*. University of Natal, Durban.

Lavelle, P., 1978. Les vers de terre de la savane de Lamto (Cote d'Ivoire): peuplements, poulations et fonctions dans l'ecosysteme. Doctoral Thesis, Univ Paris, VI Publ Lab Zool, ENS 12.

Le Roux, P.A.L. & Du Preez C.C., 2006. Nature and distribution of South African plinthic soils: Conditions for occurrence of soft and hard plinthic soils. S. Afr. J. Plant Soil 23, 120–125.

Lee, K.E. and Foster, R.C., 1991. Soil fauna and soil structure. *Australian Journal of Soil Research*, 29: 745–775.

Leistner, A.O., 1967. The Plant Ecology of the Southern Kalahari. *Memoirs of the Botanical Survey of South Africa*, 38:1–172.

Lepage, M., 1974. *Les termites d'une savane sahélienne (Ferlo septentrional, Sénégal): peuplement, populations, consommation, role dans l'écosystème.* Thesis, Université de Dijon, pp. 344.

Longenecker, D.E. and Lyerly, P.J., 1964. Making soil pastes for salinity analysis. A reproducible capillary procedure. *Soil Science*, 97: 268–275.

MacVicar, C.N., de Villiers, J.M., Loxton, R.F., Lambrechts, J.J.N., le Roux, J., Merryweather, F.R., van Rooyen, T.H., Harmse, H.J. Von M. and Verster, E., 1977. *Soil Classification – a Binomial System for South Africa*. Department of Agricultural Technical Services, Pretoria.

MacVicar, C.N., Fitzpatrick, R.W. and Sobczyk, M.E., 1984. Highly weathered soils in the east coast hinterland of Southern Africa with thick, humus-rich A1 horizons. *European Journal of Soil Science*, 35: 103–115.

MacVicar, C.N., Hutson, J.L. and Fitzpatrick, R.W., 1985. Soil formation in the coast eolianites and sands of Natal. *Journal of Soil Science*, 36: 373–387.

Maud, R.R., 1965. Laterite and lateritic soil in coastal Natal, South Africa. *European Journal of Soil Science*, 16: 60–72.

McBride, M.B., 1994. *Environmental Chemistry of Soils.* Oxford University Press, New York.

McKenzie, R.J., Mitchell, S.D. and Barker, N.P. 2006. A new species of *Arctotis* (Compositae, Arctotideae) from kommetjie grassland in Eastern Cape Province, South Africa. *Botanical Journal of the Linnean Society* 151, 581–588.

McKissock, I., Gilkes, R.J. and Walker, E.L., 2002. The reduction of water repellency by added clay is influenced by clay and soil properties. *Applied Clay Science*, 20: 225–241.

McLean, E.O., 1965. Chemical and microbiological properties. In: *Methods of Soil Analysis.* Part 2. Black, C.A., Evans, D.E., White, J.L., Ensminger, L.E. and Clark, F.E. (eds.) American Society of Agronomy, Madison, Wisconsin.

Melton, D.A., 1976. The biology of aardvark (Tubulidentata-Orycteropodidae). *Mammal Review*, 6: 75–88.

Merryweather, F.R., Le Roux, J., Van Rooyen, T.H. and Harmse, H.J., 1977. *Soil Classification. A Binomial System for South Africa.* Department of Agricultural Technical Services, Pretoria.

Midgley, J.J. and Schafer, G., 1992. Correlates of water colour in streams rising in Southern Cape catchments vegetated by fynbos and or forest. *Water SA*, 18(2): 93–100.

Mielenz, R.C. and King, M.E., 1955. Physical-chemical properties and engineering performance of clays, In: J.A. Pask and M.D. Turner (eds.), *Clays and Clay Technology: California Division of Mines Bulletin*, 169: 196–254.

Millot, G., 1970. *Geology of clays.* Springer Verlag, New York.

Mills, A.J., Fey, M.V., Gröngröft, A., Petersen, A. and Medinski, T.V., 2006. Unraveling the effects of soil properties on water infiltration: segmented quantile regression on a large data set from arid south-west Africa. *Australian Journal of Soil Research*, 44: 783–797.

Milne, G., 1935. Some suggested units of classification and mapping particularly for East African soils. *Soil Research*, 4: 183–198.

Mohr, E.C.J. and van Baren, F.A., 1954. *Tropical Soils: a Critical Study of Soil Genesis as Related to Climate, Rock and Vegetation.* Interscience, New York, pp. 498.

Mucina, L. & Rutherford, M.C. (eds) 2006. The vegetation of South Africa, Lesotho and Swaziland. *Strelitzia* 19. South African National Biodiversity Institute, Pretoria.

Munnik, M.C., Van Rooyen, T.H. and Verster, E., 1996. Spatial pattern and variability of soil and hillslope properties in a granitic landscape 3 – Phalaborwa Area, *South African Journal of Plant and Soil*, 13(1): 9–16.

Netterberg, F., 1969. The geology and engineering properties of South African calcretes. PhD thesis, University of the Witwatersrand.

Netterberg, F., 1980. Geology of southern African calcretes: 1. Terminology, description and classification. *Transactions of the Geological Society of South Africa*, 83: 255–283.

Netterberg, F., 1985. Pedocretes. In: Brink, A.B.A. (ed.), *Engineering geology of southern Africa*, 4: 286–307, Building Publications, Silverton.

Nortcliff, S., 1992. *Soils.* Nelson, Walton-on-Thames, Surrey, England.

Northcote, K.H., 1979. *A Factual Key for the Recognition of Australian Soils* 4th edition. Rellim Tech. Pubs, Adelaide, SA, Australia.

Oberholster, R.E., 1969a. Genesis of two different soils on basalt. II. Contribution of transported material. *Agrochemophysica*, 1: 73–78.

Oberholster, R.E., 1969b. Genesis of two different soils on basalt. I. Mineralogical characteristics. *Agrochemophysica*, 1: 53–62.

Paton, T.R., Humphreys, G.S. and Mitchell, P.B., 1995. *Soils. A new global view.* UCL, London.

Peech, M., 1965. In: *Methods of Soil Analysis.* Part 2. Black, C.A., Evans, D.E., White, J.L., Ensminger, L.E. and Clark, F.E. (eds.) American Society of Agronomy, Madison, Wisconsin.

Pomeroy, D.E., 1976. Studies on a population of large termite mounds in Uganda. *Ecological Entomology*, 1: 49–61.

Provincial Government Western Cape, 2003. *Western Cape Olifants/Doring River Irrigation Study. Soils and Irrigation Potential.* Prepared by J.J.N. Lambrechts, F. Ellis and B.H.A. Schloms in Association with ARCUS GIBB (Pty) Ltd. Cape Town. PGWC Report No 259/2004/10.

Reeve, R.C., 1965. In: *Methods of Soil Analysis.* Part 1. Black, C.A., Evans, D.E., White, J.L., Ensminger, L.E. and Clark, F.E. (eds.) American Society of Agronomy, Madison, Wisconsin.

Rich, C.I., 1969. Suction apparatus for mounting clay specimens on ceramic tiles for X-ray diffraction. *Soil Science Society of America Proceedings*, 33: 815.

Richards, L.A., 1965. In: *Methods of Soil Analysis.* Part 1. Black, C.A., Evans, D.E., White, J.L., Ensminger, L.E. and Clark, F.E. (eds.) American Society of Agronomy, Madison, Wisconsin.

Schaetzl, R.J., 2001. Morphologic evidence of lamellae forming directly from thin, clayey bedding planes in a dune. *Geoderma*, 99: 51–63.

Schloms, B.H.A., 2003. Gypsic horizons – A need for formal recognition in the South African soil classification system. Conference of the Soil Science Society of South Africa, Stellenbosch.

Schwertmann, U. and Taylor, R.M., 1977. Iron oxides. In: *Minerals in Soil Environments.* J.B. Dixon and S.B. Weed (eds.), Soil Science Society of America, Madison, pp145–180.

Schwertmann, U., 1991. Solubility and dissolution of iron oxides. *Plant and Soil*, 130(1–2): 1–25.

Scotney, D.M., 1970. Soils and land use in the Howick extension area. PhD Thesis, University of Natal, Pietermaritzburg.

Simonson, R.W., 1959. Outline of a generalised theory of soil genesis. *Soil Science Society of America Proceedings*, 23: 152–156.

Sliwa, A., 1996. A functional analysis of scent marking and mating behaviour in the aardwolf, *Proteles cristatus* (Sparrman, 1783). PhD thesis, University of Pretoria, Pretoria.

Snyman, K., Fey, M.V. and Cass, A., 1984. Physical properties of some highveld Vertisols. *South African Journal of Plant and Soil,* 2: 18–20.

Soil Classification Working Group, 1991. *Soil Classification – a Taxonomic System for South Africa.* Memoirs on the Agricultural Natural Resources of South Africa No. 15. Department of Agricultural Development, Pretoria.

Soil Survey Staff, 1951. *Soil Survey Manual.* Agriculture Handbook No. 18. USDA, Washington DC.

Soil Survey Staff, 1975. *Soil Taxonomy. A basic system for soil classification for making and interpreting soil surveys.* Agriculture Handbook No. 436. USDA, Washington DC.

Soil Survey Staff, 1999. *Soil Taxonomy. A Basic System of Soil Classification for Making and Interpreting Soil Surveys.* Second Edition. United States Department of Agriculture, Natural Resources Conservation Service, Agriculture Handbook No.436, U.S. Government Printing Office, Washington, DC. 20402.

Soil Survey Staff, 2003. *Keys to Soil Taxonomy.* Ninth Edition. United States Department of Agriculture, Natural Resources Conservation Service, U.S. Government Printing Office, Washington, DC. 20402.

Sommer, M., Halm, D., Geisinger, C., Andruschkewitsch, I., Zarei, M. and Stahr, K., 2001. Lateral podzolisation in a sandstone catchment. *Geoderma,* 103: 231–247.

South African Sugar Association Experiment Station, 1999. *Identification & Management of the Soils of the South African Sugar Industry,* Third edition. SASEX, Mount Edgecombe.

Stockdill, S.M.J., 1982. Effects of introduced earthworms on the productivity of New Zealand pastures. *Pedobiologia,* 24: 29–35.

Syers, J.K. and Springett, J.A., 1983. Earthworm ecology in grassland soils. In: *Earthworm ecology.* J.E. Satchell (ed.), Chapman and Hall, London, pp. 67–83.

Turner, J.S., Marais, E., Vinte, M., Mudengi, A. and Park, W.L., 2006. Termites, water and soils. *Agricola* 16: 40–45.

Taylor, K.P., 1972. An investigation of the clay fraction of soils from the Springbok Flats, Transvaal. MSc. thesis, University of Natal.

United States Salinity Laboratory Staff, 1954. *Diagnosis and improvement of saline and alkali soils.* Agriculture Handbook No. 60. USDA, Washington DC.

Van Aarde, R.J., 2001. Aardvark. In: *The New Encyclopedia of Mammals.* D. Macdonald (ed.), Oxford University Press. pp. 452–3.

Van der Eyk, J.J., MacVicar, C.N. and de Villiers, J.M., 1969. *Soils of the Tugela Basin. A Study in Subtropical Africa.* Town and Regional Planning Commission, Natal: Pietermaritzburg. Vol. 15.

Van der Merwe, C.R., 1940. Soil Groups and Sub-Groups of South Africa. Science Bulletin 231, Department of Agriculture and Forestry, Pretoria, 316 pp.

Van der Merwe, G.M.E., Laker, M.C. and Bühmann, C., 2002a. Clay mineral associations in melanic soils of South Africa. *Australian Journal of Soil Research,* 40: 115–126.

Van der Merwe, G.M.E., Laker, M.C. and Bühmann, C., 2002b. Factors that govern the formation of melanic soils in South Africa. *Geoderma,* 107: 165–176.

Van Huyssteen, C.W., 2004. The relationship between the water regime and morphology of soils in the Weatherley catchment, northerly Eastern Cape. PhD thesis, University of the Free State, Bloemfontein.

Van Huyssteen, C.W., Ellis, F. and Lambrechts, J.J.N., 1997. The relationship between subsoil colour and degree of wetness in a suite of soils in the Grabouw district, Western Cape. II. Predicting duration of water saturation of colour defined horizons. *South African Journal of Plant and Soil,* 14: 154–157.

Van Reeuwijk, L.P. and de Villiers, J.M., 1985. The origin of textural lamellae in Quaternary coastal sands of Natal. *South African Journal of Plant and Soil,* 2: 38–44.

Verster, E., de Villiers, J.M. and Scheepers, J.C., 1973. Gilgai in the Rustenburg area. *Agrochemophysica,* 4: 57–62.

Willen, D.W., 1965. Surface soil textural and potential erodibility characteristics of some Southern Sierra Nevada forest sites. *Soil Science Society of America Proceedings,* 29: 213–218.

Glossary

As explained in Chapter 1, an alternative terminology has been developed in this book for naming the forms and families in the current South African classification. In all cases this terminology relates to existing definitions of forms and families and the classification itself remains unaltered. The definitions of soil horizons, materials and family differentiating criteria (MacVicar et al., 1991) are first provided here (in their original form for the reader's convenience), followed by the full list of group, form and family terminology used in this book (arranged in order of the fourteen soil groups) alongside the definitions to which the terms refer. In a few cases the definitions have been slightly modified to convey the intended meaning more accurately. These modifications include bleached or not bleached (achromic/chromic) A horizons, signs of wetness (aeromorphic/hydromorphic), hard or not hard (hyper-/orthosaprolitic) in the lithocutanic horizon of relevant forms in the humic, melanic, podzolic and lithic groups, and the friable/firm distinction in the podzol B horizon. It should also be noted that if the recommendation concerning the Constantia form at the end of Section 2.10 is accepted then the need to define podzolic character beneath a diagnostic yellow-brown apedal B horizon would fall away since the podzolic character would qualify for a diagnostic podzol B.

TOPSOIL HORIZONS

Five surface horizons (organic, humic, vertic, melanic and orthic) have been defined as diagnostic by virtue of the presence of such distinctive features as a high organic carbon content, wetness, expansive properties, dark colour and a high base status – or their absence (in the orthic horizon).

Organic O horizon
(i) has sufficient organic carbon to ensure an average content of at least 10% throughout a vertical distance of 200 mm;
(ii) is saturated with water for long periods in most years unless drained (evidence of wetness is invariably present in the subsoil).

Humic A horizon
(i) contains, in some part, more than 1.8% organic carbon;
(ii) contains less than 4 $cmol_c$ of exchangeable cations (Ca, Mg, K, Na) per kg clay for every one per cent of organic carbon present;
(iii) does not show evidence of stripping of sesquioxide coatings from mineral particles;
(iv) does not directly overlie gleyed material, a G horizon, an E horizon, a placic pan, a podzol B horizon, a soft plinthic B horizon, a hard plinthic B horizon, a prismacutanic B horizon or a red structural B horizon.

Vertic A horizon
(i) has strongly developed structure;
(ii) has at least one of the following:
- clearly visible, regularly occurring slickensides in some part of the horizon or in the transition to an underlying layer,
- a plasticity index greater than 32 (using the SA Standard Casagrande cup to determine liquid limit) or greater than 36 (using the British Standard cone to determine liquid limit).

Melanic A horizon
(i) has dark colours in the dry state such that both value and chroma are 3 or less but with the exclusion of 10YR 3/3; red colours of hues 5YR or redder are not permitted;
(ii) lacks slickensides that are diagnostic of vertic horizons;
(iii) has a plasticity index equal to or less than 32 (using the SA Standard Casagrande cup to determine liquid limit, or equal to or less than 36 (using the British Standard cone to determine liquid limit);
(iv) has 4 or more $cmol_c$ of exchangeable cations (Ca, Mg, K, Na) per kg clay for every one per cent of organic carbon present;
(v) has less organic carbon than that required for a diagnostic organic O horizon.

Orthic A horizon
(i) is a surface horizon that does not qualify as an organic, humic, vertic or melanic topsoil although it may have been darkened by organic matter.

SUBSOIL HORIZONS AND MATERIALS

E horizon
(i) has a Munsell colour value that is at least one unit higher than that of the overlying topsoil horizon unless the latter has been removed;
(ii) has, in the dry state unless otherwise specified, one of the following 'grey' matrix colours:
- if hue is 2.5Y, then values of 5 or more and chromas of 2 or less; or values of 6 or more and chromas of 4 or less;
- if hue is 10YR, then values of 4 and chromas of 2 or less; or values of 5 or more and chromas of 3 or less; or values of 6 or more with a chroma of 4;
- if hue is 7.5YR, then values of 5 or more with a chroma of 2 or less or values of 6 or more with a chroma of 4 or less;
- if hue is 5YR, then values of 5 or more and chromas of 2 or less; or values of 6 or more with a chroma of 3; or values of 6 or more with a chroma of 4 in both the dry and most states;

(iii) may contain discernible mottling or streaking with a higher chroma than that of the matrix, the result of periodic saturation with water;
(iv) is loose or friable in the moist state, is non-plastic and, depending on texture, can be very hard and brittle (fragipan character) when dry;
(v) has very weakly developed structure, usually macroscopically structureless;
(vi) has undergone marked in situ net removal of colloidal matter (iron oxides, silicate clay, organic matter) as evidenced by a comparison of its properties with those of overlying and underlying horizons and, when possible, with the properties of its parent material;
(vii) does not qualify as diagnostic regic sand.

G horizon
(i) underlies a vertic, melanic or orthic A horizon, or an F horizon;
(ii) is saturated with water for long periods unless artificially or naturally drained;
(iii) is dominated, especially on macrovoid and ped surfaces, by grey, low chroma matrix colours, often with blue or green tints, with or without mottling; randomly patterned sesquioxide mottles may be yellowish brown, olive brown, red or black;
(iv) has not undergone marked net removal of colloidal matter (silicate clay, organic matter); indeed, accumulation of colloidal matter has usually taken place in the horizon;
(v) has consistence at least one grade firmer than that of an overlying orthic A or E horizon.
(vi) lacks the saprolite character needed to qualify as a lithocutanic B horizon;
(vii) lacks the plinthic character needed to qualify as a diagnostic soft plinthic B horizon;
(viii) has any type or degree of development of structure except moderate to strong permanent prismatic or columnar with uniformly coloured dark ped faces.

Red apedal B horizon
(i) has one or more of the following 'red' colours in both the moist and dry states unless otherwise specified:
- if hue is 5YR, then values of 3 to 5 and chromas of 4 or more; or values of 3 to 4 and a chroma of 3; values of 6 to 7 and chromas of 4 or more in the dry state if values are 5 or less in the moist state; or 5YR 5/3 in the dry state only;
- if hue is 2.5YR, then values of 3 or more and chromas of 6 or more; or values of 2 to 4 and a chroma of 4;
- if hue is 1OR, then values of 3 or more and chromas of 4 or more; or values of 3 to 4 and a chroma of 3; or 1OR 3/2;
- if hue is 7.5R, then values of 3 or more and chromas of 6 or more; or values of 2 to 4 and a chroma of 4; or 7.5R 3/2;
- a colour which is 'yellow' (see below) in the dry state and which qualifies in the moist state as 'red' as defined here, is diagnostic 'red' and not diagnostic 'yellow'.
- although colour must be substantially uniform, a slight variability is permitted, for example red mottles in a red matrix.

(ii) has structure that is weaker than moderate blocky or prismatic in the moist state; if structure is borderline, CEC per kg soil (NH4OAc, pH7) is less than 11 $cmol_c$;
(iii) is non-calcareous in any part of the horizon

which occurs within 1500 mm of the soil surface but may contain infrequent discrete, relict lime nodules in a non-calcareous soil matrix;
(iv) directly underlies a diagnostic topsoil horizon or a yellow-brown apedal B horizon;
(v) does not qualify as diagnostic stratified alluvium or as a diagnostic podzol B horizon.

Yellow-brown apedal B horizon

(i) does not have the 'grey' colours in the dry state as defined for the E horizon;
(ii) has one or more of the following colours in the moist state unless otherwise specified;
- if hue is 2.5Y, then values of 5 or more with chromas of 6 or more; or 2.5Y4/4; or 2.5Y5/4;
- if hue is 10YR, then a value of 3 and 4 and chroma of 3 or more; or a value of 5 to 6 with a chroma of 6 or more; 10YR5/4, 7/6, 7/8, 8/6 and 8/8 are permitted;
- if hue is 7.5Y, then a value of 4 with a chroma of 2 or more; or a value of 5 or more with a chroma of 6 or more; 7.5YR3/4, 5/4 and 8/6 are permitted;
- if hue is 5YR, then a value and chroma of 6 or more;
- a colour which is 'yellow' in the dry state as defined above and which qualifies as diagnostic red in the moist state, is diagnostic 'red' and not diagnostic 'yellow';
- although colour must be substantially uniform, some variability is permitted, for example mottles or concretions which are insufficient to qualify the horizon as a diagnostic plinthic B; faunal reworking may also result in acceptable colour variegations;
(iii) is non-calcareous within any part of the horizon which occurs within 1500 mm of the surface but may contain infrequent discrete, relict lime nodules in a non-calcareous soil matrix;
(iv) has structure as defined for the red apedal B;
(v) directly underlies a diagnostic topsoil horizon or an E horizon;
(vi) does not qualify as diagnostic stratified alluvium or as a diagnostic podzol B horizon.

Red structured B horizon

(i) has red colours as defined for the red apedal B; colour must be substantially uniform; the red colour is not directly inherited from the rock, but is the result of the relative accumulation of iron oxides following mineral weathering;
(ii) has structure more strongly developed than defined for the red apedal B; if structure is borderline, CEC (NH$_4$OAc, pH7) per kg soil (measured below any organic matter accumulation that forms part of an AB or BA horizon) is equal to or greater than 11 cmol$_c$.
(iii) directly underlies an orthic A, often with reddish hue;
(iv) has a transition with the overlying orthic A horizon which may be gradual but not clear with respect to texture and/or structure.

Soft plinthic B horizon

(i) has undergone localisation and accumulation of iron and manganese oxides under conditions of a fluctuating water table to give many (more than 10% by volume of the horizon) distinct reddish brown, yellowish brown and/or black mottles, with or without hardening to form sesquioxide concretions;
(ii) has grey colours caused by gleying, either in the horizon itself or immediately beneath it;
(iii) has, in the non-concretionary parts of the horizon, a loose, friable or slightly firm consistence; the horizon is non-indurated and can be cut with a spade when wet, even though individual mottles may have hardened irreversibly to form concretions;
(iv) occurs as the second or third of a vertical sequence of diagnostic horizons, provided that when it is the third, the second horizon is an E horizon, a red apedal B or a yellow-brown apedal B;
(v) does not qualify as a diagnostic soft carbonate horizon.

Hard plinthic B horizon

(i) consists of an indurated zone of accumulation of iron and manganese oxides which cannot be cut with a spade, even when wet;
(ii) occurs beneath an orthic A horizon, an E horizon or a yellow-brown apedal B.

Prismacutanic B horizon

(i) has an abrupt transition with an overlying orthic A or E horizon with respect to at least two of the following three properties:

- **Texture**
  If the clay content of the material above the abrupt transition is less than 20%, then the clay content below it must be at least twice as high (for example, 15% increasing abruptly to 30%); if the material above the transition has more than 20% clay, then the material below must show an absolute increase of at least 20% clay (for example, 25% increasing abruptly to at least 45%).
- **Structure**
  At least one grade stronger than that of the overlying horizon.
- **Consistence**
  At least two grades harder or firmer than that of the overlying horizon.

(ii) has prismatic or columnar structure (usually coarse); occasionally primary blocky structure is more pronounced than secondary prismatic or columnar structure;

(iii) lacks evidence of wetness in the form of low chromas or, if it has signs of wetness, then the vertical faces of prisms or columns have continuous coatings of uniform dark colour;

(iv) exhibits colour contrast between clayskins and ped interiors.

Pedocutanic B horizon

(i) underlies a diagnostic topsoil horizon, or an E horizon either directly or via a stoneline;

(ii) has moderately to strongly developed subangular or angular blocky structure in the moist state; if development of structure is borderline, CEC (NH_4OAc, pH7) per kg soil (measured below any organic matter accumulation that forms part of an AB or BA horizon) is equal to or greater than 11 $cmol_c$; a prismatic tendency is permitted if the upper boundary is not abrupt with respect to at least two of the following three properties: texture, structure, consistence;

(iii) has clearly expressed cutanic character resulting from illuviation of fine material and manifested as prominent cutans on most ped surfaces;

(iv) does not qualify as a diagnostic G horizon, a prismacutanic B, a plinthic B or a red structured B horizon, all of which have cutanic character to a greater or lesser degree.

Lithocutanic B horizon

(i) underlies a diagnostic topsoil horizon, either directly or via a stoneline, or an E horizon;

(ii) merges into underlying weathering rock;

(iii) has, at least in part, a general organisation in respect of colour, structure or consistence which has distinct affinities with the underlying parent rock;

(iv) has cutanic character expressed usually as tongues or prominent colour variegations caused by residual soil formation and illuviation resulting in the localisation of one or more of clay, iron and manganese oxides, and organic matter in a non-homogenised matrix of geological material (saprolite) in a variable but generally youthful stage of weathering;

(v) lacks a laterally continuous horizon which would qualify as either a diagnostic pedocutanic B or prismacutanic B;

(vi) does not qualify as a diagnostic podzol B, a neocarbonate B, a soft or hardpan carbonate horizon, or diagnostic dorbank;

(vii) if the horizon shows signs of wetness, then more than 25% by volume has saprolite character.

Neocutanic B horizon

(i) directly underlies a diagnostic topsoil or E horizon;

(ii) does not contain sufficient calcium or calcium-magnesium carbonate to effervesce visibly when treated with cold 10% hydrochloric acid but may contain infrequent discrete, relict lime nodules in a non-calcareous soil matrix;

(iii) occurs in unconsolidated material, usually transported, which has undergone pedogenesis to an extent which excludes the horizon from diagnostic stratified alluvium, regic sand and man-made soil deposit, but pedogenesis has been insufficient to produce any other diagnostic horizon;

Examples are:
- non-uniform colour by virtue of the presence of cutans and channel infillings sufficient to prevent the horizon qualifying as a diagnostic red apedal or yellow-brown apedal B horizon;
- disappearance of fine stratifications in a deposit which was initially stratified (contrast with an underlying stratified C);
- aggregation of soil particles to the extent that it is no longer loose, but insufficient to qualify as a diagnostic pedocutanic or prismacutanic B.

Neocarbonate B horizon

(i) directly underlies a diagnostic topsoil or E horizon;
(ii) contains, within 1 500 mm of the surface, sufficient calcium or calcium-magnesium carbonate to effervesce visibly when treated with cold 10% hydrochloric acid;
(iii) does not have the morphology required to qualify as a diagnostic soft or hardpan carbonate horizon;
(iv) occurs in unconsolidated material, usually transported, which has undergone pedogenesis to an extent which excludes the horizon from diagnostic stratified alluvium, regic sand and man-made soil deposit, and which has caused the presence of carbonates, but which has been insufficient to produce any other diagnostic horizon;

Examples are:
- horizons which, but for the presence of carbonates, would have qualified as diagnostic red apedal or yellow-brown apedal B horizons;
- disappearance of fine stratifications and the presence of carbonates in a deposit which was initially stratified (contrast with an underlying stratified C);
- aggregation of soil particles in the presence of carbonates to the extent that it is no longer loose, but insufficient to qualify as a diagnostic pedocutanic or prismacutanic B.

Podzol B horizon

(i) underlies an orthic A or E horizon;
(ii) does not qualify as a diagnostic placic pan;
(iii) occurs continuously or as mottles or as tongues (cutans on ped faces are excluded);
(iv) has, in its uppermost part or at the points of accumulation in the horizon, more pyrophosphate extractable Fe + Al than the overlying horizon and at least three of:
  - more than 0.5% pyrophosphate extractable C
  - more acid oxalate extractable fulvic acid than the overlying horizon
  - more than 0.3% pyrophosphate extractable Fe + Al
  - a value of 0.3 or more for the ratio: $(Fe + Al)_p / (Fe + Al)_d$ where the subscripts p and d refer, respectively, to pyrophosphate- and dithionite-extractable Fe + Al.
(v) has, at the points of accumulation, a darker colour than an overlying E horizon or the horizon which directly underlies the podzol B;
(vi) when moist, has either loose to slightly firm or hard and brittle (cemented ortstein) consistence; it is non-plastic.

Regic sand

(i) is a recent deposit, usually aeolian, which, except for a possible darkening of the topsoil by organic matter, shows little or no further evidence of pedogenesis;
(ii) is coarse textured and has little or no macroscopically visible structure; it may be massive or single grained; fine aeolian stratification may be present;
(iii) has a colour that does not qualify it for inclusion in a diagnostic red apedal or yellow-brown apedal B; colour is commonly 'grey' as defined for the E horizon;
(iv) has mineralogical composition little, if any, different from that of the parent material;
(v) has consistence that is loose, friable or soft;
(vi) directly underlies an orthic A horizon or, if this is absent, occurs at the surface;
(vii) does not qualify as a neocutanic B, a neocarbonate B or as an E horizon.

Stratified alluvium

(i) is unconsolidated and contains stratifications

caused by alluvial or colluvial deposition;
(ii) directly underlies a diagnostic orthic or melanic A horizon, or occurs at the surface.

Placic pan
(i) occurs in or within 100 mm of the lower limit of a podzol B horizon;
(ii) is a thin (usually 2–10 mm thick), wavy, continuous or discontinuous, black to dark reddish brown sheet, roughly parallel to the soil surface, cemented by iron, by iron and manganese, or by an iron-organic matter complex, and has hard or brittle consistence when moist;
(iii) has at least two of:
- more than 0.5% pyrophosphate extractable C
- more than 0.3% pyrophosphate extractable Fe + Al
- a value of 0.3 or more for the ratio: $(Fe + Al)_p / (Fe + Al)_d$
  where the subscripts p and d refer, respectively, to pyrophosphate- and dithionite-extractable Fe + Al.

Dorbank
(i) underlies an orthic A (unless exposed by erosion), a red apedal B, a neocutanic B or a neocarbonate B horizon;
(ii) is hard to extremely hard when moist and usually reddish brown in colour;
(iii) has a massive or laminated structure, the latter orientated parallel to the soil surface;
(iv) is cemented by silica, for the removal of which treatment with hot, concentrated alkali is needed; it does not slake if soaked only in either water or acid; when other cementing agents (mainly $CaCO_3$) are present, alternate treatments with alkali and acid are needed to cause slaking.

Saprolite
(i) is an horizon of weathering rock with a general organisation in respect of colour, structure or consistence, which still has distinct affinities with the parent rock;
(ii) grades into relatively unweathered and, eventually, fresh rock;

(iii) does not qualify as a diagnostic soft or hardpan carbonate horizon, dorbank or hard rock;
(iv) underlies a diagnostic podzol B or pedocutanic B horizon.

Soft carbonate horizon
(i) has morphology which is largely that of the calcium and/or calcium-magnesium carbonates present, whether in powder (here the colour of the carbonates dominates the colour of any non-carbonates present), nodular, honeycomb, or boulder form;
(ii) unless exposed by erosion, occurs beneath a melanic or orthic A, a red apedal B, a yellow-brown apedal B, a neocutanic B or a neocarbonate B horizon;
(iii) does not qualify as diagnostic dorbank or as a diagnostic hardpan carbonate horizon.

Hardpan carbonate horizon
(i) is continuous throughout the pedon;
(ii) is cemented by carbonates such as to be a barrier to roots and slowly permeable to water;
(iii) is massive, vesicular or platy and extremely hard when dry and hard or very firm when moist;
(iv) unless exposed by erosion, occurs beneath a melanic or orthic A, or yellow-brown apedal B, red apedal B, neocutanic B or neocarbonate B;
(v) does not qualify as diagnostic dorbank.

Unconsolidated material without signs of wetness
(i) underlies a diagnostic podzol B or pedocutanic B horizon;
(ii) may be any combination of organic matter, clay, silt sand and coarse fragments;
(iii) does not qualify as diagnostic hard rock or saprolite (weathering of coarse fragments in an unconsolidated matrix can give a false impression of saprolite);
(iv) lacks grey, low chroma colours, with or without sesquioxide mottles that are evidence of wetness as defined for G and E horizons.

Unconsolidated material with signs of wetness
(i) underlies a diagnostic podzol or pedocutanic

B horizon;
(ii) may be any combination of organic matter, clay, silt, sand and coarse fragments;
(iii) does not qualify as diagnostic hard rock or saprolite (weathering of coarse fragments in an unconsolidated matrix can give a false impression of saprolite);
(iv) has grey, low chroma colours, with or without sesquioxide mottles that are evidence of wetness as defined for G and E horizons.

Unspecified material with signs of wetness

(i) underlies red apedal, yellow-brown apedal, neocutanic or neocarbonate B horizons;
(ii) can vary from unconsolidated soil material to partly weathered rock;
(iii) does not qualify as a diagnostic lithocutanic or soft plinthic B or as a diagnostic soft or hardpan carbonate horizon or as a diagnostic E or G horizon;
(iv) has grey, low chroma matrix colours, due to reduction and iron loss, that have been caused by wetness; if present, sesquioxide mottling may be yellowish brown, olive brown, red or black.

Hard rock

(i) is a continuously hard layer of rock that cannot be cut with a spade, even when wet; the most important members are igneous, metamorphic and indurated sedimentary rocks and silcrete;
(ii) does not qualify as a diagnostic lithocutanic or hard plinthic B horizon, or as a hardpan carbonate horizon, or as diagnostic dorbank;
(iii) occurs beneath a diagnostic melanic or orthic topsoil horizon.

Man-made soil deposit

(i) is a man-made deposit of soil material, with or without rock fragments;
(ii) occurs beneath an orthic A or, if this is absent, at the surface;
(iii) does not qualify as any other diagnostic subsoil horizon or material.

PROPERTIES DIAGNOSTIC FOR THE SOIL FAMILIES

Eighteen sets of properties are used to distinguish soil families.

Fibrous and humified organic material

Fibrous organic material differs from humified organic material in that it is composed of well preserved plant remains that are readily identifiable as to botanical origin. To be diagnostic, the layer of fibrous material must be at least 200 mm thick.

Thin and thick humic A horizons

A thick humic A horizon is more than 450 mm thick and contains more than 1.8% organic carbon measured in a sample taken between the 250 mm and 450 mm depths. A thin humic A horizon does not meet the requirements of a thick humic A horizon, but contains more than 1.8% organic carbon in a sample taken over its entire thickness.

Dark and light coloured A horizons overlying the E horizon in Fernwood form

The formation of the E horizon is the result of weathering and the removal of iron. In some cases this has taken place in mildly reducing or podzolising conditions and the A horizon is light coloured. In other instances, conditions (usually marked wetness) have favoured the accumulation of organic matter and, in turn, a marked darkening in the colour of the A horizon. The latter, darker coloured topsoil horizons (with moist colour values of 4 or less and chromas of 1 or less) are distinguished at family level from those with lighter colours.

Bleached orthic A horizon

Many A horizons have a bleached 'grey' colour in the dry state, as defined for the diagnostic E horizon. Some of these A horizons are underlain by diagnostic subsoil horizons (e.g. E horizon, G horizon) which themselves have undergone reduction and removal of iron. In these soils, the reduced nature of the A horizon is covariant with that of the accompanying subsoil horizon. However, in many cases, these bleached A horizons overlie diagnostic subsoil horizons (e.g. pedocutanic, lithocutanic) which have not suffered marked reducing conditions. In these soils, a distinction is made at family level between members which have and which do not have, bleached A horizons. Bleached A horizons often have moist state colours that are darker (very dark grey 10YR 3/1 is a common moist colour) than the 'grey' dry state colours of the diagnostic E horizon. In some cases this diagnostic A horizon has an overall bleached appearance while parts retain the original unbleached colour, presenting a mottled appearance on close inspection. Important to note is that the bleached A is an A (i.e. a topsoil horizon) and not a subsoil horizon as are E and G horizons.

Dark, red and other colours found in vertic A horizons and in pedocutanic B horizons which occur beneath melanic A horizons

Dark colours (black, dark grey, dark brown and very dark brown), have in the most state, values of 4 or less and chromas of 1 or less and if hue is 10YR or 7.5YR, values of 4 or less with a chroma of 2. Red means red colours of hues 5YR, 2.5YR, 10R and 7.5R.

Grey and yellow E horizons

Some E horizons have, in the moist state only, a 'yellow' (occasionally 'red' with hue 5YR) colour as defined for the diagnostic yellow-brown apedal B horizon. In the dry state, however, they have a 'grey' E horizon colour. An incomplete covering of the mineral soil particles by ferric oxides is probably the reason for this difference in colour between the dry and moist state.

Presence and absence of lamellae in the E horizon of Fernwood form

Lamellae are wavy, horizontally orientated layers, in vertical section often branched, which, relative to the surrounding soil, are enriched in one or more of aluminosilicate clays, sesquioxides and organic matter. When present, the first lamellae in a profile usually occur within 400 mm to 1 000 mm of the soil surface. In the upper part of the profile, lamellae are thin (a few mm thick) becoming thicker (sometimes up to 120 mm) with depth and eventually, at greater depth, thinner. They are not the boundaries between depositional layers.

Dystrophic, mesotrophic, eutrophic

Dystrophic refers to soil that has suffered marked leaching such that the sum of the exchangeable (as opposed to soluble) Ca, Mg, K and Na, expressed in $cmol_c$ per kg clay, is less than 5. In mesotrophic soils, this figure is in the range 5–15, and in eutrophic soils the figure is greater than 15. It is calculated using the S-value and clay percentage as follows: exchangeable cations ($cmol_c$) kg^{-1} soil) x 100 ÷ % clay. Dystrophic, mesotrophic and eutrophic refer to soils with low, medium and high base status respectively. Diagnostic eutrophic horizons are not calcareous. Once sufficient data are available, it is possible that these criteria will be replaced by percentage base saturation.

Non-red and red colours in B horizons and stratified alluvium

Where iron oxides have imparted a red colour (hues of 5YR, 2.5YR, 10R, 7.5R) to the greater part of an horizon, the resultant soil structure is usually more water stable than similar soil which is not red. In many soil forms the non-red and red distinction is made at family level.

Luvic B horizon

A soil has a luvic B in the following circumstances:
- when any part of the A or E horizon has 15% or less, the B1 horizon must contain at least 5% more clay than the A, O or E;
- when any part of the A or E horizon has more than 15% clay, the ratio of clay percentage in the B1 to that in the A or E must be 1.3 or greater.

The presence of more clay in the B than in the A or E horizon is implicit in the definitions of many soil forms (e.g. Sterkspruit, Estcourt). However, there are several forms (with red apedal, yellow-brown apedal, neocutanic and neocarbonate B horizons) where this is not so, and it is considered necessary to distinguish between members which have markedly more clay in the B than in the A or E horizon from those which do not. The luvic concept is used for this purpose.

Subangular/fine angular and medium/coarse angular structure in pedocutanic B and red structured B horizons.

By definition such structure must be at least moderately developed in the moist state. A distinction is made at family level between medium and coarse angular structures on the one hand, and subangular and fine angular structures on the other. The former generally are more common and, in terms of root and water penetration, less suitable for crop production than the latter.

Continuous black cutans in prismacutanic B horizons

Prismacutanic B horizons in Estcourt form which have continuous black cutans on vertical ped faces are distinguished from those which do not have such black cutans. The presence of black (as opposed to other dark colours) cutans is usually an indication of a wet soil climate. Free lime is then usually absent from the C horizon which normally shows signs of gleying.

Ortstein hardening of podzol B horizons

Ortstein is a podzol B horizon or sub-horizon that, under conditions of periodic saturation with water, has been cemented by a combination of organic matter with iron and/or aluminium. For such hardening to be diagnostic, the horizon must be at least weakly cemented when moist. The latter is brittle and can normally be broken easily by hand.

Hard and not hard lithocutanic B horizons and saprolite

More than 70% by volume of a hard lithocutanic B or saprolite horizon is bedrock, fresh or partly weathered, with at least a hard consistence in the dry, moist and wet states. Horizons which do not meet these requirements are not hard. The latter often occur in higher rainfall areas where weathering has often taken place to considerable depth.

Signs of wetness

These signs consist of grey, low chroma colours, sometimes with blue or green tints, with or without sesquioxide mottles. The latter, if present, may be yellowish brown, olive brown, red or black. The signs of wetness must occur within 1500 mm of the surface and must not be of such a nature or in such a profile position as to qualify as a diagnostic E, G or soft plinthic B horizon or as undifferentiated material with signs of wetness.

Calcareous horizons and layers

An horizon or layer is calcareous if, in some part, it contains sufficient calcium carbonate or calcium-magnesium carbonate to effervesce visibly when treated with cold 10% hydrochloric acid. It is not considered calcareous if it contains discrete, relict lime nodules in a non-calcareous matrix. It does not qualify as a diagnostic neocarbonate B, or as a soft or hardpan carbonate horizon.

Podzolic character beneath a diagnostic yellow-brown apedal B horizon

Constantia form has an horizon sequence of orthic A – E horizon – yellow-brown apedal B horizon. Some members of the form have an horizon beneath the yellow-brown apedal B which, if it had been directly beneath an A or E horizon, would have qualified as a diagnostic podzol B horizon. Such members are distinguished at family level from those which do not have such podzolic character below the yellow-brown apedal B.

Friable and firm C horizons

Permeability to water is an important soil property. In some soils, the classification does not allow any deduction about the permeability of certain subsoil horizons and materials to be made. For these, it has been considered necessary to introduce a criterion indicative of permeability. Amount of clay, kind of clay mineral and the amount of ferric oxides and their contribution to structure are important determinants of permeability. Consistence in the moist state is a property which, in the context of the horizons for which this criterion is used, reflects, to a considerable extent, the combined effect of these factors on permeability. The term 'friable' is used here for soil that has a loose, friable or slightly firm consistence, while 'firm' refers to soil with a firmer consistence. Those coarse textured materials that are hard setting are included in the concept of 'friable'.

No	Group	Form	Family criteria	Definition
1	Organic			Organic O horizon
			fibric/sapric	Fibrous organic material differs from humified organic material (sapric) in that it is composed of well preserved plant remains that are readily identifiable as to botanical origin. To be diagnostic (fibric), the layer of fibrous material must be at least 200 mm thick.
			lithic/cumulic	A distinction is made between solid rock and saprolite that must occur within 500 mm below the lower boundary of the organic O horizon and within 1500 mm of the soil surface (lithic) on the one hand and unconsolidated material (cumulic) on the other.
2	Humic			Humic A horizon
		Xanthirhodic		Humic A horizon, yellow-brown apedal B over a red apedal B horizon
			thick/thin	A thick humic A horizon (thick) is more that 450 mm thick and contains more than 1.8% organic carbon measured in a sample taken between the 250 mm and 450 mm depths. A thin humic A horizon (thin) does not meet the requirements of a thick humic A horizon, but contains more than 1.8% organic carbon in a sample taken over its entire thickness.
			luvic/haplic	A soil has a luvic B in the following circumstances: when any part of the A or E horizon has 15% clay or less, the B1 horizon must contain at least 5% more clay than the A or E; when any part of the A or E horizon has more than 15% clay, the ratio of the clay percentage in the B1 to that in the A or E must be 1.3 or greater. If a soil is not luvic, it is haplic.
		Xanthic		Humic A horizon over a yellow-brown apedal B horizon
			thick/thin	See above
			luvic/haplic	See above
		Rhodic		Humic A horizon with a red apedal B
			thick/thin	See above
			luvic/haplic	See above
		Pedocutanic		Humic A horizon over a pedocutanic B
			thick/thin	See above
			rhodic/arhodic	A red coloured (rhodic) or a non-red (arhodic) B horizon (red as defined for red apedal B).
			micropedal/ macropedal	By definition such structure must be at least moderately developed in the moist state. A distinction is made at family level between medium and coarse angular peds on the one hand (macropedal), and subangular and fine angular peds on the other (micropedal).
		Neocutanic		Humic A horizon over a neocutanic B
			thick/thin	See above
			rhodic/arhodic	See above
			luvic/haplic	See above
		Lithocutanic		Humic A horizon over a lithocutanic B
			thick/thin	See above

No	Group	Form	Family criteria	Definition
			orthosaprolitic/ hypersaprolitic	For hypersaprolitic families, more than 70% by volume of the lithocutanic B or saprolite horizon is bedrock (fresh or partly weathered) with at least a hard consistence in the dry and a very firm consistence in the moist state. Horizons which do not meet these requirements are not hard (orthosaprolitic families).
3	Vertic			Vertic A horizon
		Hydromorphic		A vertic A horizon over a G horizon
			calcic/acalcic	An horizon or layer is calcareous if, in some part, it contains sufficient calcium carbonate or calcium-magnesium carbonate to effervesce visibly when treated with cold 10% hydrochloric acid. It is not considered calcareous if it contains discrete, relict lime nodules in a non-calcareous matrix. It does not qualify as a diagnostic neocarbonate B, or as a soft or hardpan carbonate horizon.
		Aeromorphic		Vertic A horizon
			melanic/rhodic/ alterchromic	Melanic colours in the surface horizon (black, dark grey, dark brown and very dark brown), have in the most state, values of 4 or less and chromas of 1 or less and if hue is 10YR or 7.5YR, values of 4 or less with a chroma of 2. Rhodic means red colours of hues 5YR, 2.5YR, 10R and 7.5R. Other colours fall under alterchromic.
			calcic/acalcic	See above
4	Melanic			Melanic A horizon
		Hydromorphic		Melanic A over a G horizon
			calcic/acalcic	See above
		Pedocutanic		Melanic A over a pedocutanic B
			sombric/rhodic/ alterchromic	Sombric colours (black, dark grey, dark brown and very dark brown), have in the most state, values of 4 or less and chromas of 1 or less and if hue is 10YR or 7.5YR, values of 4 or less with a chroma of 2. Rhodic means red colours of hues 5YR, 2.5YR, 10R and 7.5R. Other colours fall under alterchromic.
			micropedal/ macropedal	See above
			calcic/acalcic	See above
		Soft calcic		Melanic A over a soft carbonate horizon
			calcic/acalcic	The surface horizon is calcic or acalcic as defined above.
		Hard calcic		Melanic A over a hardpan carbonate horizon
			calcic/acalcic	The surface horizon is calcic or acalcic as defined above.
		Lithocutanic		Melanic A over a lithocutanic B horizon
			orthosprolitic/ hypersaprolitic	See above
			calcic/acalcic	See above
		Lithic		Melanic A over hard rock
			calcic/acalcic	The surface horizon is calcic or acalcic as defined above.

No	Group	Form	Family criteria	Definition
		Cumulic		The melanic A horizon is usually underlain either by material with alluvial stratifications or by unconsolidated sediments in which soil formation has not progressed sufficiently far to produce one of the following diagnostic horizons: G horizon, pedocutanic B horizon or a soft or hardpan carbonate horizon.
			aeromorphic/ hydromorphic	Signs of wetness (hydromorphic) consist of grey, low chroma colours, sometimes with blue or green tints, with or without sesquioxide mottles. The latter, if present, may be yellowish brown, olive brown, red or black. The signs of wetness must occur within 1500 mm of the surface and must not be of such a nature or in such a profile position as to qualify as a diagnostic E, G or soft plinthic B horizon or as undifferentiated material with signs of wetness. If there are no signs of wetness in or immediately below the A horizon it is aeromorphic.
			calcic/acalcic	See above
5	Silicic			Dorbank
		Rhodic		Orthic A, red apedal B over dorbank
			luvic/haplic	See above
		Neocutanic		Orthic A, neocutanic B over dorbank
			chromic/ achromic	A horizons with a bleached 'grey' colour in the dry state, as defined for the diagnostic E horizon, are achromic. Bleached A horizons often have moist state colours that are darker (very dark grey 10YR 3/1 is a common moist colour) than the 'grey' dry state colours of the diagnostic E horizon. In some cases this diagnostic A horizon has an overall bleached appearance while parts retain the original unbleached colour, presenting a mottled appearance on close inspection. Unbleached A horizons are chromic.
			rhodic/arhodic	See above
			luvic/haplic	See above
		Neocalcic		Orthic A, neocarbonate B over dorbank
			chromic/achromic	See above
			rhodic/arhodic	See above
			luvic/haplic	See above
		Orthic		Orthic A over dorbank
			calcic/acalcic	The surface horizon is calcic or acalcic as defined above.
6	Calcic			Carbonate or gypsic horizon
		Xanthic-soft		Orthic A, yellow-brown apedal B over soft carbonate horizon
			luvic/haplic	See above
			aeromorphic/ hydromorphic	See above
		Xanthic-hard		Orthic A, yellow-brown apedal B over hardpan carbonate horizon
			luvic/haplic	See above
		Rhodic-soft		Orthic A, red apedal B over soft carbonate horizon
			luvic/haplic	See above
			aeromorphic/ hydromorphic	See above

No	Group	Form	Family criteria	Definition
		Rhodic-hard		Orthic A, red apedal B over hardpan carbonate horizon
			luvic/haplic	See above
		Neocutanic-soft		Orthic A, neocutanic B over soft carbonate horizon
			chromic/achromic	See above
			rhodic/arhodic	A red coloured (rhodic) or a non-red (arhodic) B horizon (red as defined for red apedal B)
			luvic/haplic	See above
			aeromorphic/ hydromorphic	See above
		Neocutanic-hard		Orthic A, neocutanic B over hardpan carbonate horizon
			chromic/achromic	See above
			rhodic/arhodic	See above
			luvic/haplic	See above
		Neocalcic-soft		Orthic A, neocarbonate B over soft carbone horizon
			chromic/achromic	See above
			rhodic/arhodic	See above
			luvic/haplic	See above
			aeromorphic/ hydromorphic	See above
		Neocalcic-hard		Orthic A, neocarbonate B over hardpan carbonate horizon
			chromic/achromic	See above
			rhodic/arhodic	See above
			luvic/haplic	See above
		Soft		Orthic A over soft neocarbonate horizon
			aeromorphic/ hydromorphic	See above
		Hard		Orthic A over hardpan carbonate horizon
			calcic/acalcic	The surface horizon is calcic or acalcic as defined above.
7	Duplex			Prismacutanic or pedocutanic B
		Eluvic-prismacutanic		Orthic A, E horizon over prismacutanic B horizon
			albic/hypoxanthic	Some E horizons have, in the moist state only, a 'yellow' (occasionally 'red' with hue 5YR) colour as defined for the diagnostic yellow-brown apedal B horizon (hypoxanthic). In the dry state, however, they have a 'grey' E horizon colour. If they are not 'yellow' when moist then they are albic.
			sombric/asombric	Prismacutanic B horizons in Estcourt form which have continuous black cutans on vertical ped faces (sombric) are distinguished from those which do not have such black cutans (asombric). The presence of black (as opposed to other dark colours) cutans is usually an indication of a wet soil climate. Free lime is then usually absent from the C horizon which normally shows signs of gleying.
		Eluvic-pedocutanic		Orthic A, E horizon over pedocutanic B horizon
			albic/hypoxanthic	See above
			rhodic/arhodic	See above

No	Group	Form	Family criteria	Definition
			micropedal/macropedal	See above
		Prismacutanic		Orthic A over prismacutanic B horizon
			chromic/achromic	See above
			rhodic/arhodic	See above
		Pedocutanic-cumulic-hydromorphic		Orthic A, pedocutanic B over unconsolidated material with signs of wetness.
			chromic/achromic	See above
			micropedal/macropedal	See above
			calcic/acalcic	See above
		Pedocutanic-cumulic-aeromorphic		Orthic A, pedocutanic B over unconsolidated material without signs of wetness
			chromic/achromic	See above
			rhodic/arhodic	See above
			micropedal/macropedal	See above
			calcic/acalcic	See above
		Pedocutanic-lithic		Orthic A, pedocutanic B over saprolite
			chromic/achromic	See above
			rhodic/arhodic	See above
			micropedal/macropedal	See above
			calcic/acalcic	See above
		Eluvic-placic		Orthic A, E horizon over podzol B with placic pan
			aeromorphic/hydromorphic	See above
8	Podzolic			Podzol B
		Eluvic-cumulic-hydromorphic		Orthic A, E horizon, podzol B over unconsolidated material with signs of wetness
			densic/adensic	Ortstein is a podzol B horizon or subhorizon that, under conditions of periodic saturation with water, has been cemented by a combination of organic matter with iron and/or aluminium. For such hardening to be diagnostic (densic), the horizon must be at least weakly cemented when moist. The latter is brittle and can normally be broken easily by hand. In the absence of such ortstein hardening the soil is adensic.
			friable/firm	The term 'friable' is used here for soil that has a loose, friable or slightly firm consistence, while 'firm' refers to soil with a firm consistence. Those coarse textured materials that are hard setting are included in the concept of 'friable'.
		Eluvic-cumulic-aeromorphic		Orthic A, E horizon, podzol B over unconsolidated material without signs of wetness
			friable/firm	See above
		Eluvic-lithic		Orthic A, E horizon, podzol B over saprolite

No	Group	Form	Family criteria	Definition
			orthosaprolitic/ hypersaprolitic	See above
			aeromorphic/ hydromorphic	See above
		Placic		Orthic A over podzol B with placic pan
			aeromorphic/ hydromorphic	See above
		Cumulic-hydromorphic		Orthic A, podzol B over unconsolidated material with signs of wetness
			densic/adensic	See above
			friable/firm	See above
		Cumulic-aeromorphic		Orthic A, podzol B over unconsolidated material without signs of wetness
			friable/firm	See above
		Lithic		Orthic A, podzol B over saprolite
			orthosaprolitic/ hypersaprolitic	See above
			aeromorphic/ hydromorphic	See above
9	Plinthic			Plinthic B horizon
		Eluvic-soft		Orthic A, E horizon over soft plinthic B
			albic/hypoxanthic	See above
		Eluvic-hard		Orthic A, E horizon over hard plinthic B
			albic/hypoxanthic	See above
		Soft		Orthic A over soft plinthic B horizon
			luvic/haplic	See above
		Hard		Orthic A over hard plinthic B horizon
			chromic/achromic	See above
		Xanthic-soft		Orthic A, yellow-brown apedal B over soft plinthic B horizon
			dystrophic/ mesotrophic/ eutrophic	Dystrophic refers to soil that has suffered marked leaching such that the sum of the exchangeable (as opposed to soluble) Ca, Mg, K and Na, expressed in $cmol_c$ per kg clay, is less than 5. In mesotrophic soils, this figure is in the range 5–15, and in eutrophic soils the figure is greater than 15. It is calculated using the S-value and clay percentage as follows: exchangeable cations ($cmol_c$ kg^{-1} soil) x 100 ÷ % clay. Dystrophic, mesotrophic and eutrophic refer to soils with low, medium and high base status respectively. Diagnostic eutrophic horizons are not calcareous. Once sufficient data are available, it is possible that these criteria will be replaced by percentage base saturation.
			luvic/haplic	See above
		Xanthic-hard		Orthic A, yellow-brown apedal B over hard plinthic
			dystrophic/ mesotrophic/ eutrophic	See above
			luvic/haplic	See above
		Rhodic-soft		Orthic A, red apedal B over soft plinthic B horizon

No	Group	Form	Family criteria	Definition
			dystrophic/ mesotrophic/ eutrophic	See above
			luvic/haplic	See above
		Rhodic-hard		Orthic A, red apedal B over hard plinthic B horizon
			dystrophic/ mesotrophic/ eutrophic	See above
			luvic/haplic	See above
10	Oxidic			Red or yellow-brown apedal, or red structured B
		Xanthic-hydromorphic		Orthic A, yellow-brown apedal B over unspecified material with signs of wetness
			dystrophic/ mesotrophic/ eutrophic	See above
			luvic/haplic	See above
		Xanthirhodic		Orthic A, yellow-brown apedal B over red apedal B horizon
			dystrophic/ mesostrophic	Dystrophic refers to soil that has suffered marked leaching such that the sum of the exchangeable (as opposed to soluble) Ca, Mg, K and Na, expressed in $cmol_c$ per kg clay, is less than 5. In mesotrophic soils, this figure is in the range 5–15. It is calculated using the S-value and clay percentage as follows: exchangeable cations ($cmol_c$ kg^{-1} soil) x 100 ÷ % clay. Dystrophic and mesotrophic refer to soils with a low and medium base status.
			luvic/haplic	See above
		Xanthic		Orthic A, yellow-brown apedal B over unspecified material
			dystrophic/ mesotrophic/ eutrophic	See above
			luvic/haplic	See above
		Rhodic-hydromorphic		Orthic A, red apedal B over unspecified material with signs of wetness
			dystrophic/ mesotrophic/ eutrophic	See above
			luvic/haplic	See above
		Rhodic		Orthic A over red apedal B horizon
			dystrophic/ mesotrophic/ eutrophic	See above
			luvic/haplic	See above
		Pedorhodic		Orthic A over red structured B horizon

No	Group	Form	Family criteria	Definition
			mesostrophic/ eutrophic/calcic	Mesostrophic refers to soil that has suffered marked leaching such that the sum of the exchangeable (as opposed to soluble) Ca, Mg, K and Na, expressed in $cmol_c$ per kg clay, in the range 5–15. In eutrophic soils the figure is greater than 15. It is calculated using the S-value and clay percentage as follows: exchangeable cations ($cmol_c$ kg^{-1} soil) x 100 ÷ % clay. Mesotrophic and eutrophic refer to soils with Medium and high base status. Diagnostic eutrophic horizons are not calcareous. Calcic refers to an horizon or layer if, in some part, it contains sufficient calcium carbonate or calcium-magnesium carbonate to effervesce visibly when treated with cold 10% hydrochloric acid. It is not considered calcareous if it contains discrete, relict lime nodules in a non-calcareous matrix.
			luvic/haplic	See above
			micropedal/ macropedal	See above
11	Gleyic			G Horizon
		Eluvic		Orthic A, E horizon over G horizon
			albic/hypoxanthic	See above
		Orthic		Orthic A over G horizon.
			calcic/acalcic	See above
12	Cumulic			Neocutanic/neocarbonate B/ regic sand/ alluvium
		Neocutanic		Orthic A over neocutanic B horizon
			chromic/achromic	See above
			rhodic/arhodic	See above
			luvic/haplic	See above
		Eluvic-neocutanic		Orthic A, E horizon over neocutanic B horizon
			albic/hypoxanthic	See above
			rhodic/arhodic	See above
			luvic/haplic	See above
		Neocutanic-hydromorphic		Orthic A, neocutanic B over unspecified material with signs of wetness
			chromic/achromic	See above
			rhodic/arhodic	See above
			luvic/haplic	See above
		Neocalcic		Orthic A over neocarbonate B horizon
			chromic/achromic	See above
			rhodic/arhodic	See above
			luvic/haplic	See above
		Eluvic-neocalcic		Orthic A, E horizon over neocarbonate B horizon
			albic/hypoxanthic	See above
			rhodic/arhodic	See above
			luvic/haplic	See above
		Neocalcic-hydromorphic		Orthic A, neocarbonate B over material with signs of wetness
			chromic/achromic	See above
			rhodic/arhodic	See above

No	Group	Form	Family criteria	Definition
			luvic/haplic	See above
		Eluvic		Orthic A over an E horizon
			achromic/umbric	Dark coloured A horizons (umbric) have moist Munsell values of 4 or less and chromas of 1 or less, otherwise they are achromic.
			albic/hypoxanthic	See above
			lamellic/haplic	Lamellae are wavy, horizontally orientated layers, in vertical section often branched, which, relative to the surrounding soil, are enriched in one or more of aluminosilicate clays, sesquioxides and organic matter. When present, the first lamellae in a profile usually occur within 400 mm to 1 000 mm of the soil surface. In the upper part of the profile, lamellae are thin (a few mm thick) becoming thicker (sometimes up to 120 mm) with depth and eventually, at greater depth, thinner. They are not the boundaries between depositional layers. When lamellae are not present the soil is haplic.
		Fluvic		Orthic A over stratified alluvium
			rhodic/arhodic	See above
			aeromorphic/hydromorphic	See above
			calcic/acalcic	See above
		Arenic		Othic A over regic sand
			rhodic/arhodic	See above
			calcic/acalcic	See above
13	Lithic			Lithocutanic B or hard rock
		Orthic		Orthic A over hard rock
			chromic/achromic	See above
			calcic/acalcic	The surface horizon is calcic or acalcic as defined above.
		Glossic		Orthic A over lithocutanic B horizon
			chromic/achromic	See above
			orthosaprolitic/hypersaprolitic	See above
			aeromorphic/hydromorphic	See above
			calcic/acalcic	See above
		Eluvic-glossic		Orthic A, E horizon over lithocutanic B horizon
			albic/hypoxanthic	See above
			orthosaprolitic/hypersaprolitic	See above
14	Anthropic			Orthic A over man-made soil deposit
			calcic/acalcic	See above

APPENDIX: CRITERIA FOR GENERATING DISTRIBUTION MAPS FOR SOIL GROUPS

Digitised maps and data from the land type memoirs managed by the ARC Institute for Soil Climate and Water (ISCW) were interrogated using Arcview GIS software. The objective was to generate maps showing the distribution and abundance of each soil group. The soils information on the land type maps is based on the earlier binomial system of MacVicar *et al.* (1977) and it was necessary to make certain assumptions concerning the most appropriate group to which some soil series belong. This was especially the case for the humic soil group as well as oxidic soils with free carbonates which key out in the calcic group.

Selecting criteria for classifying soil series into groups and the mapping of abundance classes was done by M Van der Walt, T E Dohse and M V Fey. The following keywords (in quotation marks) formed the basis for defining each group. In most cases the keyword is a soil form abbreviation or a soil series code. Other criteria included pans 'P' in the gleyic group and dorbank as a depth limiting material in the silicic group.

ORGANIC SOILS
'Ch'

HUMIC SOILS
'Ia' Or 'Ma' Or 'Kp' Or 'Lu' Or (to approximate thin humic families) 'Hu16' Or 'Hu17' Or 'Hu18' Or 'Cv16' Or 'Cv17' Or 'Cv18' Or 'Cv26' Or 'Cv27' Or 'Cv28' Or 'Gf11' Or 'Gf12' Or 'Gf13' Or 'Gf21' Or 'Gf22' Or 'Gf23'

VERTIC SOILS
'Rg' Or 'Ar'

MELANIC SOILS
'Mw' Or 'My' Or 'Bo' Or 'Ik' Or 'Tk' Or 'Wo'

SILICIC SOILS
'Ms14' Or 'Ms24' Or using dorbank from Depth Limiting Material only, '*db*'

CALCIC SOILS
'Ms12' Or 'Ms22' Or 'Hu40' Or 'Hu41' Or 'Hu42' Or 'Hu43' Or 'Hu44' Or 'Hu45' Or 'Hu46' Or 'Hu47' Or 'Hu48' Or 'Sd30' Or 'Sd31' Or 'Sd32' Or 'Gs2'

DUPLEX SOILS
'Sw' Or 'Va' Or 'Ss' Or 'Es'

PODZOLIC SOILS
'Hh' Or 'Lt'

PLINTHIC SOILS
'Wa' Or 'Lo' Or 'We' Or 'Av' Or 'Gc' Or 'Bv'

OXIDIC SOILS
'Sd' Or 'Hu' Or 'Cv' Or 'Gf'
Excluding
'Hu16' Or 'Hu17' Or 'Hu18' Or 'Hu40' Or 'Hu41' Or 'Hu42' Or 'Hu43' Or 'Hu44' Or 'Hu45' Or 'Hu46' Or 'Hu47' Or 'Hu48' Or 'Cv16' Or 'Cv17' Or 'Cv18' Or 'Cv26' Or 'Cv27' Or 'Cv28' Or 'Sd30' Or 'Sd31' Or 'Sd32' Or 'Gf11' Or 'Gf12' Or 'Gf13' Or 'Gf21' Or 'Gf22' Or 'Gf23'

GLEYIC SOILS AND PANS
'Ka' Or 'Kd' Or 'P'

CUMULIC SOILS
'Oa' Or 'Du' Or 'Fw1' Or 'Fw2'

LITHIC SOILS
'Ms1' Or 'Gs1' Or 'R'

Index

A

aardvark 37, *149*, 158, *159*, 160, 161, 171, 262, 263
abruptic 73, 77, 119
acalcic 41, 48, 59, 69, *75*, 76, 77, 119, 127, 129, *136*, 139, 179, 188, 189, 190, 191, 192, 193, 250, 251, 252, 253, 274, 275, 276, 277, 280, 281
achromic 15, 57, 59, 68, 69, 74, *75*, 76, 77, 79, 81, 82, 98, 114, 116, 124, 125, 127, 128, 129, *138*, 139, 142, 179, 204, 205, 244, 245, 264, 275, 276, 277, 278, 280, 281
acric 28, 29, 99, 110, 119
Acrisols 28, 29, 99, 110, 112, 130, 139
Addo 11, 13, 68, 69, 70, 172, 179, 206, 207, 246
adensic 88, 179, 220, 221, 277, 278
aeolian 11, 13, 14, 18, *56*, 58, 60, 66, 67, *78*, 87, 90, *107*, 123, 124, *125*, 218, 220, 268
aeromorphic 41, 48, 68, 69, 76, 77, *85*, 86, 88, 89, 114, 127, 129, *136*, *137*, 139, *179*, 188, 189, 190, 191, 192, 193, 206, 207, 218, 219, 224, 225, 252, 253, 264, 274, 275, 276, 277, 278, 281
A horizon 12, 14, 15, 18, 20, 26, 29, *31*, 35, 36, 45, *46*, 53, 55, 58, 60, 64, 68, 73, 74, 75, 76, 83, 94, 98, *101*, 105, 106, 111, 116, 127, 136, *138*, 182–275
albic *75*, 76, 77, 79, *85*, 86, 89, 97, 98, 99, 103, *117*, 119, 120, 124, *125*, 127, 128, 129, 130, 136, *138*, 139, 179, 216, 217, 220, 221, 228, 229, 230, 240, 241, 244, 245, 276, 278, 280, 281
Alfisols 29, 49, 59, 60, 74, 77, 100, 111, 112, 119, 130
Alisols 110
allophane 45, 87, 261
alluvium 11, 14, 18, *40*, 47, 52, 54, 124, 266, 267, 268, 271, 280, 281
alterchromic 41, 48, 274
alumic 28, 29
aluminium, Al 12, 13, 26, 29, 31, 32, 42, 78, 83, 87, 89, 90, 94, 100–103, 111, 180–277
aluminosilicate 222, 271, 281
Andisols 45, 60
anion 74, 78, 106, 182, 186, 226, 234, 261
anthropic 7, 10, 11, 14, 18, 123, 143, *144*, *145*, *146*, 281
Anthrosols 143
apedal 11, 13, 18, 25, 26, 30, 31, 43, 58, 60, 64, 94, 98, 101, 102, 103, 105, 106, *107*, *108*, 111, 112, *113*, 114, 124, 127, 182, 184, 186, 188, 200, 202, 206, 218, 220, 222, 226, 230, 232, 234, 236, 238, 248, 264–279
Aqualfs 111
Aquepts 130
aquic 11, 18, 41, 70, 77, 119
Aquicambids 70
aquods 87, 89
Aquox 111
Aquults 111
Arcadia 11, 12, 39, 41, 179, 188–193
arenic 59, 123, 124, 127, 128, 129, 130, 244, 281
Arenosols 99, 110, 130
Argidurids 59
argillan 79
argillic 73, 74
Argiustolls 49
Argixerolls 49
arhodic 28, 59, 68, 69, *75*, 76, 83, *127*, 128, 129, 179, 248, 249, 273, 275, 276, 277, 280, 281
aridic 41, 60
Askham 11, 13, 68, 69, 70
asombric *75*, 76, *78*, 179, 216, 217, 276
Atterberg 36, 45, 181–253
Augrabies 11, 14, 124, *127*, 129, 130, 179, 246, 247
Avalon 11, 13, *95*, 98, 99, 100, *103*, 179, 226, 227

B

Bainsvlei 11, 13, *95*, 98, 99, 100, 103
basalt 27, 42, *46*, 190, 194, 196, 262
bat-eared fox 158, 160, *161*
bauxite 32, *33*, 100
beidellite 42
B horizon 13, 26, 28, 29, *32*, 46, 51, 52, 55, 58, 60, 64, 71–98, 102–108, 111–120, 124, 127, 135, 136, 141, 182–281
biomantle 5, 140, 150, 162, 174–5, 252, 261
biomass 12, 89
biotite *137*
bleaching 55, *57*, 60, 64, *75*, 79, 84, 124, 127, 136, *138*, 218
blocky 13, 36, 38, 39, *47*, 64, 73, 74, 79, 106, 112, 113, 182–267
Bloemdal 11, 14, 109, 110, 111
Bonheim 11, 12, 48, 49, 50, 51, 52, 179, 194–197
boron, B 56, 112, 202, 257
Brandvlei 11, 13, *64*, 69, 70
bushpig 171, *172*

C

Ca 33, 42, 63, 65, 70, 78, 79, 155, 181–280

calcaric 41, 48, 49, 110, 119, 130, 139
Calciaquolls 49
Calciargids 70
calcic 11, 12, 15, 18, *37*, 41, 48, 49–51, 57–59, 63–80, 105, 108, 109, 119, 124, 127, 129, 130, *137*, 139, 155, 179, 194–199, 206–215, 242, 243, 246, 274–277, 280, 281
Calcisols 49, 69
calcite 12, 58, 63, 65, 66, 70, *155*, 188, 189, 199, 204, 205
Calcixerolls 49
Cambisols 29, 110, 130, 139
carbon 13, 20, 21, 25, 26, 32, 34, 37, 45, 47, 50, 56, 66, 83, 101, 106, 155, 180, 190, 198, 204, 206, 216, 218, 256, 257, 264, 270, 273
Cartref 11, 14, 135, 136, 138, 139, 142
catena *102*, 111, 112
cation 20, 26, 37, 78, 79, 87, 111, 180, 186, 190, 224, 255
CBD 180–253
CEC 29, 37, 47, 74, 77, 97, 106, 111, 112, 116, 118, 181–253, 255, 256, 265, 266, 267
cementation 11, 18, 53, 80, 84, 91, 93, 94, 97, *127*, 200, 202, 204, 224, 230
Champagne 11, 12, 19, 21, 23, 179, 180, 181
Chernozems 49
chroma 97, 105, 115, 163, 264, 265, 266, 269, 270, 271, 272, 274, 275
chromic 29, 41, 49, 57, 59, 68, 69, *75*, 76, 77, 98, 99, 110, 127–130, 136, *137*, 139, 179, 206–215, 246–253, 264, 275–281
chronosequence 260
climate 5, 6, 12, 13, 21, 31, 32, 35, *37*, 41, 42, 43, 49, 58, 60, 66, 70, 71, 89, 90, 92, 100, 111, 114, 115, 140, *142*, 150, 154, 158, 163, 165, 178, 180–276
Clovelly 11, 14, *107*, 109, 110, 111, 179, 236, 237, 260
cobalt, Co 87, 173, 181–261
Coega 11, 13, *64*, 69, 70
colluvial 11, 13, 18, 20, 30, 31, 32, 52, 58, 67, 78, 90, *106*, 111, 118, 123, *138*, 180, 238, 269
colluvium 14, 32, *37*, 44, 47, 52, 54, 55, *75*, *81*, 84, *85*, *107*, *120*, *127*, *140*, 214, 222, 226, 228, 232
colour 11, 13, 14, 18, 26, 31, 32, *33*, 36, 37, 39, *47*, 51, *56*, 64, 73, 79, 80, 83, 84, 97, 101, 102, 105, 109, 111, 116, 119, *124*, 127, 135, 136, 162, 163, 188–276
Concordia 11, 13, 88, 89, 114
concretions 13, 55, 66, 67, *81*, 94, 97,

101, 119, *120*, 230, 242, 252, 266
consistence 34, 36, 56, 84, 116, 135, 136, 252, 265–269, 272, 274, 277
Constantia 11, 14, 114, 264, 272
copper, Cu 82, 173, 181–257
cracking 11, 12, 18, 31, 38, *39*, 45, *167*, 256
crusting 13, 38, 39, *57*, 64, 65, 74, 81, 82, 120, 150, 170, 216, 252, 260
cumulic 10, 11, 14, 18, *20*, 21, 47, 48, 49, 50, 52, 74, 76, 77, 81, 82, 85–9, 109, 114–6, 123–45, 179, 214–21, 242–49, 273–80
cutanic 13, 14, 29, 49, 69, 73, 77, 78, 130, 136, 139, 141, 216, 248, 267
cutans 14, 51, 52, 73, 74, 92, 105, 113, *118*, *127*, *141*, 186, 188, 198, 200, 204, 212, 214, 216, 234, 242, 246, 248, 252, 267, 268, 272, 276

D

denitrification 82, 118
densic 84, 88, 277, 278
dioctahedral 42
dithionite 258, 268, 269
dolerite 26, *31*, 32, *33*, 37, 42, 43, *46*, *81*, *106*, *107*, *108*, *140*, *141*, 173, 188, 198, 260
dolomite 32, 234
dorbank 11, 12, 18, 53, *54*, *55*, 56, 58, 60, *61*, 70, 105, 145, 155, 200, 202, 204, 267, 269, 270, 275
drawbar pull 188
Dresden 11, 13, *95*, 98, 99, 100
dryland 34, 103, 114, 142, 182, 226, 234
Dundee 11, 14, 123, *124*, 129, 130
dung beetles 160, 168, *170*
duplex 11, 13, 15, 18, 26, 32, 35, 50, 51, 57, 73–84, 91, 109, 112, 116, 119, 120, 124, 135, 136, *138*, 141, 179, 212–217, 240, 242, 276
duric 53, 54
durinodes 55, *56*, 196
duripan 12, 53, 55, 60
Durisols 58, 59
Durixeralfs 59
Dystric 21, 28, 29, 99, 110, 119, 130, 139
dystrophic 88, 97, 98, 103, 106, 109, 110, 112, 179, 226, 227, 236–239, 271, 278–9

E

earthworm 162, *163*, 165, 166–263
ECEC 226
edaphic 97, 180
Eh 101
E horizon 13, 57, 74, 75, 79, 83, 84, 87, *90*, 94, 102, 103, 114, 115, 116, *120*, 124, 135, *136*, 138, 216, 220, 240, 244, 264–281

elephant *170*, 172, *174*, *175*
eluviation 11, 13, 57, 58, 78, 84, 90, 116, 150
eluvic 74, *75*, 76, 77, 79, 80, 81, 82, 88, 89, *95*, 97, 98, 99, 100, 103, 114, 116, *117*, 118, 119, 120, 124, *125*, *127*, 128, 129, 130, *133*, 135, 136, *138*, 139, 142, 163, 179, 216, 217, 228, 229, 230, 240, 241, 244, 245, 276, 277, 278, 280, 281
Endoaqualfs 49, 77
Endoaquolls 41, 49
endogleyic 48, 69, 99, 110, 119, 130, 139
endoleptic 41
entic 59
Entisols 100, 119, 123, 130, 139
epipedon 45
erodibility 12, 13, 80, 113, 142, 150, 212, 214, 216, 232, 263
erosional 26
Estcourt 11, 13, *75*, 76, 77, 81, 179, 216, 217, 271, 272, 276
Etosha 11, 13, *65*, 68, 69, 70, 140, *174*, *175*
eutric 21, *39*, 41, 48, 49, 99, 110, 119, 130, 139
eutrophic 60, 98, 106, 109, 112, 114, 179, 234, 235, 271, 278, 279, 280
evaporation 12, 13, 22, 58, 63, 119

F

Fe, see iron
Fernwood 11, 14, 87, 124, *125*, 128, 129, 130, *133*, 179, 244, 245, 270, 271
ferralic 29, 110
Ferralsols 28, 29, 99, 110, 112
ferric 97, 99, 111, 116, 119, 271, 272
ferricrete 13, 93, 259
ferrihydrite 87, 119
ferrolysis 13, 31, 57, 60, 79, 80, 101, 116, 118, 216, 240, 260
fertigation 92
fertility 43, 128, 190, 260
fertiliser 82, 180, 220, 226, 230, 232
fibric 20, 21, 22, 273
firm 19, 50, 84, *86*, 88, 113, 184, 190, 192, 194, 214, 220, 222, 224, 232, 236, 238, 240, 264, 266, 268, 269, 272, 274, 277, 278
fluvic 49, 123, *124*, 127, 129, 130, *133*, 281
Fluvisols 49, 130
footslope 40, *44*, *55*, *144*, 188, 230
fractipetric 59
fragipan 265
friable 19, 26, 88, 112, 113, 116, 179, 182, 186, 218, 219, 220, 221, 222, 224, 226, 228, 238, 248, 264, 265, 266, 268, 272, 277, 278
fulvic 268

G

gabbro 37, 40, 44, *81*, *136*
Gamoep 11, 13, 68, 69, 70
Garies 11, 12, *54*, *55*, 58, 59, 60, *61*, 71, *156*, 179, 200, 201, 202, 203, 204
geology 150, 161, 260, 262
geomorphology 260
geophagy 172, 173
geric 28, 29
G horizon 14, 48, 50, 51, 97, 115, 116, 118, *119*, 120, 240, 242, 244, 264, 265, 267, 270, 274, 275, 280
gibbsite 26, 29, 32, *33*, 66, 87, *96*, *141*, 182, 186, 187, 226, 227, 237, 238, 239
gilgai 12, 36, 39, *40*, 162, 263
glaebules *108*
Glencoe 11, 13, *95*, 98, 99, 100, 179, 230, 231
Glenrosa 11, 14, 135, *136*, *137*, *138*, 139, *141*, 142, 179, 252, 253
gley (gleyed, gleying) 10, 14, 19, 20, 31, 39, 73, 80, *86*, 94, *108*, 115–119, *141*, 216, 190, 240, 264, 266, 272, 276
gleyic 11, 14, 18, 21, 41, 48, 49, 50, 51, 77, 89, 94, 99, 110, 115–121, *117*, *118*, *121*, 130, 139, *141*, 179, 240–3, 280
Gleysols 21, 41, 48, 119
glossic 135–142, *136*, *137*, *138*, *140*, *141*, 179, 252–3, 281
gneiss 142, 204, 206, 214
goethite 13, 26, 31, 32, *33*, 87, 94, *96*, 97, 101, 102, 106, 111, 119, 182, 226, 236
granite *61*, *137*, *142*, 186, 212, 248
Griffin 11, 14, *107*, 109–111, 179, 182, 238, 239
Groenkop 11, 13, 88, 89, 179, 224, 225
grumic 41
gypsic 11, 18, 63, 66, 70, 72, 262, 275
Gypsids 70
Gypsisol 69

H

Haplargids 77
haplic 21, 28, 29, 41, 48, 49, 59, 68, 69, 77, 89, 98, 99, 109, 110, 119, 124, *125*, 127, 128, 129, 130, 139, 179, 182, 183, 186, 187, 200–205, 208, 209, 210, 211, 226, 227, 230–237, 244, 245, 273–281
Haplocalcids 70
Haplocambids 70
Haplodurids 59
haploidisation 12
Haploxeralfs 49, 77
Haploxererts 41
Hapludalfs 111

Hapluderts 41
Hapludox 29
Hapludults 111
Haplustalfs 29, 49, 77, 111
Haplusterts 41
Haplustox 29
Haplustults 29, 111
hectorite 42
hemic 21
heuweltjies 58, 60, 70, 154–158, *155, 156, 157*, 162, 204
hillslope 260, 262
histic 21
Histosols 21
honey badger 168, *169*, 170, 260
horizonation 52, 78, 83, *145*
hortic 143
Houwhoek 11, 13, 88, 89
humate 10, 11, 18, 37, 83, 224
humic 11, 12, 13, 18, 19, 21, 25–34, *26, 27, 30, 31, 32, 33, 34*, 37, 45, 46, 79, 83, 89, 90, *91*, 100, *101*, 105, 106, 108, 111, 114, 116, 135, *141, 170*, 179–187, 218, 236, 238, 261, 264, 270, 273
humified 21, 270, 273
Hutton 11, 14, *105*, 107, 109, 110, 111, 179, 200, 202, 204, 206, 208, 210, 232, 233, 248
hydric 143
hydromorphic 21, 31, 35, 41, 48, 49, 68, 69, 76, *77*, 80, 82, 84, *85*, 86, 87, 88, 89, 90, 100, 106, 109, 110, 111, 114, 124, *127*, 128, 129, 130, 139, 140, *141*, 179, 190, 214, 215, 220, 221, 260, 262, 264, 274–281
hydrophobic 244
hyperalbic 130
hypercalcic 49, 69
hyperdystric 28, 29
hyperochric 59
hypersaprolitic 28, 48, 88, 136, *137*, 139, 142, 274, 278, 281
hyperskeletic 139
hyperthermic *142*
hypocalcic 49, 69
hypoluvic 110
hypoxanthic *75*, 76, 79, 97, 98, 116, *117*, 119, 120, 124, *125*, 127, 128, 129, 136, 139, 230, 276, 278, 280, 281

I

illite 42, 106, 260
illuvial 73, 74, 84, 87, 127, 216, 230
illuviation 13, 14, 26, 51, 55, 78, 124, 127, *137*, 218, 267
Immerpan 11, 12, 48, 49, 51, 52
imogolite 87, 261
Inanda 11, 12, *26*, 28, 29, *34*, 179, 186, 187

inceptic 18, 32, 128
inceptisols 29, 49, 59, 60, 70, 71, 74, 123, 130, 139
indurated 13, *30*, 54, 55, 56, 94, 196, 206, 228, 232, 266, 270
infiltration 37, 47, 55, 65, 74, 81, 103, 146, 154, 158, 162, 165, 170, 244, 260, 262
Inhoek 11, 12, 47, 48, 49, 52
interlayer 38, 42, 47, 218, 219, 221, 223
interstratified 37, 38, 201, 203, 219, 221, 225, 235, 247, 249, 260
Iron, Fe 10, 11, 13, 14, 18, 20, 26, 29, 31, 32, 42, 52, 55, 56, 57, 66, 71, 78, 79, *81*, 83, 84, 87, 91, 93, 94, 97, 100, 101, *102*, 103, 105, 106, 111, 113, 115, 116, *118*, 119, 120, 127, *137*, 141, *155*, 181–269

J

jarosite 119
Jonkersberg 11, 13, 88, 89, 179, 222, 223

K

kandic 106
kaolinite 26, 31, 47, 56, 87, *96*, 97, 106, 112, 113, 119, *137*, 183–253
Kastanozems 49
Katspruit 11, 14, 116, *117*, 118, 119, 120, 179, 242, 243
Kimberley 11, 13, *65*, 68, 69, 70, 150, 257
Kinkelbos 11, 13, 14, 124, 127, 129, 130
Klapmuts 11, 13, 76, 77
Knersvlakte 11, 12, 55, 58, 59, 60, *152*
kommetjies 162–166, *163, 164*
Kranskop 11, 12, 26–32, *27, 31*, 111, 179, 182, 183, 260
Kroonstad 11, 14, 116–120, *117*, 179, 240, 241, 257

L

lamellae 14, 91, 124, 200, 262, 263, 271, 281
lamellic 124, *125*, 127, 129, 130, 281
Lamotte 11, 13, 88, 89, *92*, 179, 220, 221
land type 6, 45, 49, 128, 178, 180–252, 257
laterite 13, *30*, 91, 93, 94, 100, *101*, 262
lepidocrocite 119
leptic 21, 29, 41, 49, 77, 89, 139
Leptosols 49, 139
lessivage 78, 80, 83, 111, 136, 141
lime 12, 22, 32, 34, 52, 56, 63, 82, 92, 103, 112, 180–250, 256, 257, 266, 267, 272, 274, 276, 280
limestone 33, *51*, 63

liquid limit 36, 52, *120*, 264
lithic 10, 11, 14, 15, 18, *20*, 21, *27*, 29, 32, 48, 49, 50, 51, 52, 74, *75*, 76, 77, 80, *81*, 82, *85*, 88, 89, 100, 116, 123, 135–142, *136, 137, 138, 139, 141, 142*, 163, 179, 180, 181, 212, 213, 224, 225, 250–253, 264, 273, 274, 277, 278, 281
lithocutanic 11, 14, 18, 26, 28, 29, 32, 47, 48, 49, 52, 64, 73, 116, 135, 136, *138*, 141, 179, 198, 199, 252, 264, 265, 267, 270, 272, 273, 274, 281
lithology 165, 180
litter 80, 160, 162, 166, 261
Lixisols 29, 69, 77, 99, 110, 130, 139
Longlands 11, 13, *95*, 98, 99, 100, 103, 179, 228, 229
Lusiki 11, 12, 26, *27*, 28, 29
luvic 28, 29, 49, 55, 59, 68, 69, 73, 84, 97, 98, 109, 110, 112, 114, 119, 124, *127*, 128, 129, 179, 184, 185, 206, 207, 226, 238, 239, 246, 247, 248, 249, 271, 273, 275, 276, 278, 279, 280, 281
Luvisols 29, 49, 69, 77, 110, 130

M

macropedal 28, 48, 74, 76, *77*, 79, 109, 179, 214, 215, 273, 274, 277, 280
macropores 78, 162, 170
maghemite 26, 97
magnesium, Mg 20, 42, 63, 64, 65, 70, 74, 78, 79, 155, 181–253, 256, 258, 259, 264, 271, 278, 279, 280
Magwa 11, 12, 26, *27*, 28, 29, 179, 184, 185
Manganese, Mn 10, 13, *32*, 84, 94, 97, 100, 101, 102, 105, *108*, 111, 119, 155, 173, 181–253, 258, 261, 266, 267, 269
matric 38
Mayo 11, 12, 48, 49, 52, 179, 198, 199
mazic 41
melanic 11, 12, 14, 18, 25, 35, 41–52, *46, 47*, *50, 51*, 112, 115, 118, 119, 135, 179, 190–199, 263–275
mesic 154
mesofauna 79, 80, 150
mesotrophic 98, 109, 112, 179, 230–233, 271, 278–280
mica 38, 42, 56, 65, 106, 182–253
micromorphological 261
micronutrients 181–258
micropedal 15, 28, 48, *75*, 76, 77, 109, 179, 194, 195, 196, 197, 212, 213, 234, 235, 273, 274, 277, 280
Milkwood 11, 12, 48, 49, 52, 140
mineralisation 34, 182
mineralogy 5, 12, 25, 32, 36, 38, 42, 47, 56, 65, 66, *75*, 78, 87, 94, *102*, 106, 112, 118, 178, 182, 188, 190, 194, 196, 198, 208, 220, 228, 232,

234, 260, 261, 262, 268
Mispah 11, 14, 135, 136, *138*, 139, 141, 142, 179, 250, 251
Mohr 45, 47, 262
mole 87, 150, 162, 165–8, *166*, *167*, 171, 188, 261
mollic 45, 48, 49
Mollisols 41, 49, 60, 71
Molopo 11, 13, 68, 69, 70
Montagu 11, 14, 124, *127*, 129, 130
montmorillonite 42, 44, 260
mor 84, 92
morphology 13, 38, 55, 64, 73, 80, 81, 91, 94, 96, 97, 105, 116, *118*, 140, *155*, 174, 212, 216, 228, 263, 268, 269
mottles 13, 64, 84, 93, 94, 97, 99, 100, *118*, 119, 188, 194, 196, 214, 216, 218, 220, 222, 226, 228, 230, 236, 240, 242, 252, 265, 266, 268, 269, 270, 272, 275
Munsell 254, 265, 281

N

Na, see sodium
Namib 11, 14, 123, 124, 128, 129, 130, 244
Natrargids 77
natric 60, 74
Natrixeralfs 77
Natrustalfs 77
neocalcic 59, 68, 69, 70, 124, *127*, 128, 129, 130, 179, 204–211, 246, 247, 275, 276, 280
neocarbonate 11, 14, 18, 58, 60, 64, 124, *127*, 204, 206, 208, 210, 246, 267–276, 280
neocutanic 11, 14, 18, 26, 27, 28, 29, 47, 52, 58, 59, 60, 64, 65, 68, 69, 70, 73, 114, 124, 127–133, *127*, *133*, 141, 145, 179, 242, 246, 248, 249, 267–276, 280
neoformation 42, 50, 119
nitic 110
Nitisols 110
nitrogen 34, 43, 97, 101, 158, 166, 182, 228
nodules 13, 26, 29, 32, 33, 36, 40, 55, 66, *67*, 93, 101, 106, 190, 194, 196, 198, 206, 208, 210, 214, 218, 228, 232, 242, 266, 267, 272, 274, 280
Nomanci 11, 12, 26, *27*, 28, 29, 31, 32
nontronite 42

O

Oakleaf 11, 14, 124, *127*, 128, 130, 179, 246, 248, 249
O horizon 12, 19, 21, 47, 264, 273
opal 58
Orthents 139
orthic 10, 11, 14, 18, 47, *50*, 53, 57, 58, 59, 60, 64, 68, 73–82, *75*, *78*, 84, 87, 94, 98, 103–119, *117*, 127, 135–139, 141, 142, 179, 200–281
Orthods 89
orthosaprolitic 28, 48, 88, *136*, *138*, 139, 142, 179, 198, 199, 224, 225, 252, 253, 264, 274, 278, 281
ortstein 84, 87, 90, 91, 268, 272, 277
ortsteinic 89
Oudtshoorn 11, 12, 55, 58, 59, 60
oxalate 87, 218, 220, 222, 268
oxidic 11, 12, 13, 18, 25, 32, 43, 50, 74, 83, 94, 103–115, *106*, *107*, *108*, *113*, *114*,127, 128, 179, 182, 232–239, 248, 279
Oxisols 29, 100, 105, 111, 112, 261

P

Paleudults 103, 111
Paleustults 111
Palexerolls 49
palygorskite 56, 208–210
ped 36–39, 50, 73, 78, 79, 80, 102, 113, 116, *118*, 136, 188, 194, 196, 240, 265–272, 276
pedocutanic 11, 15, 18, 26, *27*, 28, 29, *46*, 47, 48, 49, 51, 64, 73, 74, 75, 76, 77, *78*, 79, 80, 82, 116, 136, 141, 179, 194–197, 212–215, 242, 267–277
pedogenesis 60, 123, 260, 261, 267, 268
pedogenic 10, 12, 18, 41, 51, 63, 87, 93, 94, 111, 154, 155, 218, 260, 261
pedological 10, 15, 82, 146, 171
pedology 83, 102, 115, 143, 154
pedon 15, 102, 106, 194, 232, 236, 238, 242, 269
pedorhodic 106, *108*, 110, 114, 179, 234, 235, 279
pedoturbation 12, 27, 31, 55, 60, 111, 140, 150, 174
pellic 41
petric 49, 59, 69, 70, 99
Petroargids 70
petrocalcic 49, 70
Petrocalcids 70
Petrocambids 59, 70
petroduric 54
petrogypsic 70, 71
petroplinthic 28, 29, 94, 99
pH 13, 20, *23*, 26, 37, 42, 47, 56, 58, 60, 65, 66, 67, 71, 72, 74, 78, 79, 82, 87, 89, 90, 91, 101, 106, 111, 113, 116, 118, 128, 162, 180–261
Phaeozems 41, 48, 49
Phosphate, P 22, 26, 32, 34, 43, 65, 74, 78, 92, 112, 184, 208, 226, 232, 257
phyllosilicate 79, 106, 112, 260
Pinedene 11, 14, *108*, 109, 110, 111
Pinegrove 11, 13, 88, 89, 179, 218, 219

placic 83, 84, *86*, 87, 88, 89, 90, 91, 97, 179, 222, 223, 261, 264, 268, 269, 277, 278
Planosols 39, 77, 119
plasmic 79
plasticity index 36, 46, 47, 51, 74, 188, 194, 264
Plinthaqualfs 100
Plinthaquults 100
plinthic 11, 13, 18, 27, 47, 50, 52, 55, 80, 83, 90–103, *95*, *96*, *101*, *102*, *103*, 105, 109, 111, 114, 116, 119, 120, 128, 141, 163, 179, 226–30, 236, 240, 262–267, 270, 272, 275, 278, 279
plinthite, see plinthic
Plinthosols 99
Plooysburg 11, 13, *65*, 66, 68, 69, 70
podzol 6, 11, 13, 18, 64, 83–97, *90*, *91*, *92*, 218, 220, 222, 224, 244, 260, 261, 264, 266, 267, 268, 269, 272, 277, 278
podzolic 11, 13, 18, 83–93, *84*, *85*, *86*, 114, 116, 128, 135, 179, 218–225, 264, 272
podzolisation 13, 31, 83, 92, 261, 263
porcupine 160, 167, *168*, 171
posic 28, 29, 110
potassium 22, 42
precipitation 13, 66, 67, 89, 155
Prieska 11, 13, 69, 70, 179, 208, 209, 210, 211, 257
prismacutanic 11, 13, 18, 73–80, *75*, *78*, 112, 116, 136, 141, 179, 216, 217, 240, 264, 267, 268, 272, 276, 277
pyrophosphate 87, 218, 220, 222, 268, 269
pyrophyllite 206–7

Q

quartz 57, 58, 78, *79*, 80, 87, 94, 180–253
quartzite *138*, 180, 184

R

rainfall 12, 13, 22, 25, 31, 32, 34, *37*, 43, 49, 54, 55, 58, 66, 81, 83, *84*, 89, *90*, 93, 100, 103, 106, 111, 112, 119, *120*, *125*, 128, 142, 145, 154, 158, 160, 163, 180, 182, 188, 254, 272
redox 10, 13, 31, 79, 83, 90, 91, 97, 100, 101, *102*, 106, 111, 119, 140
regic 11, 14, 18, 19, 124, 244, 265, 267, 268, 280, 281
regolith 82, 83, 90, 100, *106*, 155, 244
rendzic 49
Rendzina 51
Rensburg 11, 12, *37*, 39, 41, 51, 190
retentivity 180–255
rhodic 15, *26*, 28, 29, *37*, 41, 48, 49,

54, 59, 60, *65*, 68–71, 74, 76, 77, 79, 80, 95–103, *95*, *96*, 106–111, *106*, *107*, 127–130, 167, 179–248, 273–281
Rhodoxeralfs 49, 77
Rhodudalfs 111
Rhodudults 111
Rhodustalfs 49, 77, 111
Rhodustults 29, 111
rock 11, 14, 18, 22, 27, 32, 42, 47, 49, 52, 63, 70, 78, 80, 100, 111, 123, 128, 135, *136*, *137*, 138, 140, 142, 163, 180, 250, 254, 260, 266, 267, 269, 270, 273, 274, 281
rodent(s) 55, 154, 166, 167, 168, 169, 171
root(s) 13, 14, 23, 29, 32, *34*, 43, 50, 55, 64, 65, 66, 73, 80, 81, *82*, 94, 97, *102*, *103*, *107*, *113*, *118*, 119, 120, 137, 140, 167, 174, *175*, 180, 184, 186, 188, 190, 200, 218, 220, 222, 224, 234, 246, 248, 269, 272
Rubic 110, 130

S

sandstone *26*, *27*, *30*, 31, 32, 84, *85*, 88, 89, 100, *101*, *102*, *107*, 120, *141*, 216, 222, 226, 228, 230, 236, 238, 240, 263
sapric 20, 21, 179, 180, 181, 273
saprolite 10, 14, 21, *27*, *30*, 31, 32, 52, *81*, *85*, 87, *102*, *108*, 111, 123, 135, *136*, 140, *141*, 182, 188, 212, 265, 267, 269, 270, 272, 273, 274, 277, 278
sauconite 42
selenium, Se 76, 77, 173, 179, 214, 215
Sepane 11, 13, 76, 77, 80, 179, 214, 215
sepiocrete 18, 155, 208
sepiolite 12, 13, 18, 53, 56, 58, 63, 65, *155*, 200, 210, 208, 209
sesquioxide(s) 10, 26, 31, 84, 180, 186, 218, 230, 242, 246, 252, 264, 265, 266, 269, 270, 271, 272, 275, 281
shale(s) 32, 75, *81*, 100, *107*, 120, *138*, *140*, *141*, 182, 226, 228, 250
Shortlands 11, 14, 106, 108, 109, 110, 111, 112, 114, 179, 234, 235
Si 42, 89, 111, 155, 181–253
silicic 11, 12, 13, 18, 53–65, *54*, 70, 80, 105, 108, 155, 179, 200, 201, 202, 203, 204, 205, 208, 275
Silicon, see Si
silty 65, 190, 198, 250
skeletic 29, 139
slickensides 12, 35, 36, *37*, *39*, 40, 41, 45, 51, 74, 79, 119, 188, 190, 192, 194, 264
smectite 37, 38, 41, 42, 44, 47, 56, 63, 65, 87, 112, 119, 188–260
snakes 168
sodic 41, 56, 60, 74, 77, 144, 190, 192, 194, 196, 202, 210, 216, 242
sodium, Na 12, 37, 42, 56, 58, 60, 74, 78, 79, 87, 154, 173, 181–280
solonetz 77
solum, sola 12, 13, 14, 25, 26, 27, 30, 31, 53, 60, 63, 65, 66, 78, 81, 87, 103, 105, 106, 114, 118, 119, 120, 127, 135, 136, 182, 226
sombric 48, 76, *78*, 79, 83, 92, 274, 276
spiders 170, *171*
spodic 87, 261
Spodosols 60, 87, 89
stagnic 69, 77, 99, 130
Stagnosols 99, 119
Steendal 11, 12, 48, 49, *51*, 52
Sterkspruit 11, 13, 76, 77, 271
stoneline(s) *46*, 78, 80, *81*, 97, 136, *141*, *156*, 162, 170, *175*, 252, 267
structure 10, 12, 13, 18, 21, 26, 30, 35, 36, 38, 39, 42, 43, 45, *47*, 50, 51, 52, 54, *55*, 58, 64, 70, 73, 74, 75, 79, 80, 94, 103, 106, 112, *113*, *118*, 119, 124, 135, 136, 146, 154, 165, 184–273
subangular 38, 182–273
Swartland 11, 13, 15, 74–77, *74*, *75*, 80–82, 154, 162, 179, 212, 213
Sweetwater 11, 12, 26, 28, 29, 32

T

talc 192, 193, 195, 231, 243
technic 143, *144*, 146
Technosols 143
temperature 111, 162, 257
termitaria 153, 154
termite 58, 60, 70, 71, *137*, 150, *152*, 154, *155*, 158, 158, 159, 160, 171, 174, *175*, 204, 262
terrain 5, 32, 44, *82*, 91, 101, 135, 140, *164*, 165, 178, 180–252
thick 12, 22, 26–31, *26*, *27*, 34, 54, 57, 84, 87, *106*, 116, *117*, 124, 179, 180, 184–187, 216, 222, 244, 248, 256, 262, 269–273, 281
thin 26, 27, 28, *31*, 32, 57, 64, 67, 74, 79, 83, 84, 92, 179, 182, 183, 262, 269–273, 281
thionic 21, 119
tillage 34, 82, 103, 112, 120, 146, 162, 166, 174, 188, 196, 234, 254
topography 12, 21, 40, 58, 70, 81, 136, *145*, 150, 163, 165
toposequence 38, 42, 102, 261
toxic 56, 71, 167, 182
trace element(s), see also micronutrients 26, 65, 71, 82, 87, 92, 112, 184, 186, 192, 200, 212, 218, 220, 244
Trawal 11, 12, 55, 58, 59, 60, 179, 204, 205
trioctahedral 42
Tsitsikamma 11, 13, 88, 89
Tukulu 11, 14, 124, *127*, 128, 129, 130

U

Udalfs 100, 111
udic 60
Udorthents 100
Udox 100, 111
Udults 100, 111
Ultisols 29, 100, 111, 112, 119, 130
umbric 28, 29, 124, *125*, 127, 129, 179, 196, 197, 244, 281
Umbrisols 29, 48
urbic 143
Ustalfs 100, 111
ustic 41
Ustorthents 100
Ustox 100, 111
Ustults 100, 111

V

Valsrivier 11, 13, 76, 77, 214
vegetation 13, 21, 22, 29, 30, 31, 32, 56, 58, 65, 81, 82, *84*, 87, 88, 89, 90, 92, 112, *133*, 137, 140, *144*, *145*, 150, 151, *153*, 154, 155, 171, 174, 180–262
vertic 11, 12, 14, 18, 35–52, *37*, *39*, *43*, *44*, 63, 74, 75, 77, 82, 106, 108, 112, 115, 118, 119, 135, 162, *167*, 173, 179, 188, 190, 191, 192, 193, 194, 196, 264, 265, 271, 274
Vertisols 35, *37*, 38, 41, 42, 60, 71, 188, 260, 263
vetic 28, 29
Vilafontes 11, 14, 114, 124, *127*, 128, 130

W

warthog 161, 171, 172
Wasbank 11, 13, *95*, 98, 99, 100, 103
Westleigh 11, 13, *95*, 98, 99, 100
Willowbrook 11, 12, 48, 49, 50, 52
Witbank 11, 14, 143
Witfontein 11, 13, 88, 89

X

xanthic 27, 28, 29, 68, 69, 70, *95*, 97, 98, 99, 100, 102, *103*, *107*, *108*, 110, 111, 179, 184, 185, 226, 227, 230, 231, 236, 237, 273, 275, 278, 279
xanthirhodic 27, 28, 29, *31*, 106, *107*, 109, 110, 111, 179, 182, 183, 238, 239, 273, 279
xeric 35, 41, 60, 155, 161, 188
XRD 258, 259

Z

Zinc, Zn 26, 112, 180–257